ON THE
ORIGIN
OF SPECIES
BY MEANS OF NATURAL SELECTION

ON THE
ORIGIN
OF SPECIES
BY MEANS OF NATURAL SELECTION

Charles Darwin

With a Preface by George Davidson

ARCTURUS

About George Davidson

George Davidson studied languages and linguistics at the universities of Glasgow, Edinburgh and Strasbourg, and is a graduate of Edinburgh University. A former senior editor with Chambers Harrap, he is now a freelance compiler and editor of dictionaries and other reference books. He is an elder of the Church of Scotland, and lives in Edinburgh.

ARCTURUS

This edition published in 2008 by Arcturus Publishing Limited
26/27 Bickels Yard, 151–153 Bermondsey Street,
London SE1 3HA

Copyright © 2008 Arcturus Publishing Limited

ISBN: 978-1-84193-989-6

Printed in China

Typeset by "talkingdesign", Sussex

After the Harvard Classics edition, published by P. F. Collier & Son Corporation, 1909

PREFACE

When asked if there was anything that could be inferred about God from the study of Creation, the 20th-century geneticist J.B.S. Haldane replied that He must have 'an inordinate fondness for beetles'. There are about 350,000 known species of beetle in the world, with estimates of millions more still to be discovered. But was each species separately created by God? Until the middle of the 19th century, most people believed so. Darwin, in *On the Origin of Species*, provided another explanation.

Charles Darwin was born in Shrewsbury, England, in 1809, the second son of Robert Darwin, a doctor, and his wife Susannah. He married his cousin Emma Wedgwood in 1839; they had ten children. His family being sufficiently well off to free him from the necessity of following a profession, he was able to devote himself to his studies and writing, despite bouts of severe ill-health throughout much of his adult life. He died in 1882, and was buried in Westminster Abbey, in London.

Darwin was educated at Shrewsbury School, after which he studied medicine for two years (1825–27) at Edinburgh University. He then attended Cambridge University to study divinity, intending to become a Church of England clergyman. He graduated in 1831, but realized he had no calling into the Church.

Both in Edinburgh and in Cambridge, Darwin's interest in and knowledge of natural history and geology developed, as did his acquaintance with eminent men in these fields, through one of whom he was in 1831 appointed as a naturalist on the surveying voyage of H.M.S. *Beagle* to South America and the Pacific Ocean under Captain Robert FitzRoy. This voyage was a key event in Darwin's life, affording him an unparalleled opportunity to collect zoological and geological specimens and make observations: according to Darwin himself, it determined his whole subsequent career.

One key part of the voyage, from the point of view of Darwin's later theories, was the Beagle's five-week visit to the Galapagos Islands, off the coast of South America, where he noted that many islands had each their own species of tortoises and mockingbirds. (It was also true of the islands' finches, now famous as 'Darwin's finches', though Darwin did not notice this at the time.) The relationship between the various island species, and between island species and those of the mainland, was the beginning of Darwin's thinking on transmutation of species. He had also noted that as one travelled southwards through South America, related species of animal replaced one another.

Once home, Darwin passed his specimens to a number of experts for examination. Their technical descriptions of the specimens were published as *The Zoology of the Voyage of H.M.S. Beagle* (1838–43). Darwin turned his diary of the voyage into a book (first published in 1837 as *Journal and Remarks, 1832–1836*, the third part of Captain FitzRoy's account of the voyage, and then separately in 1839 as the *Journal of Researches into the Geology and Natural History of the Various Countries Visited by H.M.S. Beagle*), and also wrote scientific papers based on his observations. Significantly, in July 1837 he also started his first 'note-book on Transmutation of Species', and began to collect facts about the formation of breeds of domesticated animals and plants. He saw that 'selection was the keystone of man's success' in breeding animals and plants, but how selection could apply to 'organisms living in a state of nature' was still a mystery to him.

Darwin found the answer to this problem in the writings of the economist Thomas Malthus (1766–1834), according to whom human population growth would always outstrip food production, making life a constant struggle for survival. From this, Darwin hypothesized that in the struggle for existence, 'favourable variations [in animals and plants] would tend to be preserved, and unfavourable ones to be destroyed', resulting eventually in the formation of new species. Here, then, he had the theory he needed – the principle of evolution of species through natural selection, according to which there is not only competition for survival between members of different species, but also between members of the same species, and any randomly occurring characteristic that gives an individual a better chance of survival in particular circumstances will thus be passed on to later generations. (It was not Darwin, however, but the philosopher Herbert Spencer (1820–1903) who coined the phrase 'the survival of the fittest'; Darwin first adopted it in the fifth (1869) edition of On the Origin of Species.)

By the summer of 1842 Darwin had prepared a short outline of his theory, and by early 1844 had written an expanded version, but he was not yet ready to risk bringing the theory to public attention, well aware of the hostility it would probably arouse in some quarters. It was not until 1856 that he turned his attention seriously to it again. However, in June 1858 Darwin was startled to receive from the naturalist Alfred Russel Wallace (1823–1913) an essay detailing Wallace's theory of evolution by natural selection that was essentially identical to his own. (Wallace too had derived his key idea from Malthus.) In order that neither scientist would have priority over the other, Wallace's essay and extracts from Darwin's writing were read to the Linnaean Society in London on 1 July 1858, the first public statement of their theory. Neither Wallace nor Darwin was actually present at this historic occasion.

Darwin's book On the Origin of Species by Means of Natural Selection, or, The Preservation of Favoured Races in the Struggle for Life was first published in November 1859. The first edition sold out immediately. As expected, Darwin's theory had a mixed reception, acclaimed by some but rejected and denounced by others, especially those who realized that what he was proposing was incompatible with the description of creation set out in the Bible. Darwin himself did not take part in the public controversy over the issue, his case being ably defended by the biologist and philosopher Thomas Huxley (1825–95) and others, but he did add to later editions of the book a detailed defence of his theory against criticisms that had been made of it (see this volume, page 156).

Darwin wrote many other books, including The Variation of Plants and Animals under Domestication (1868), The Descent of Man, and Selection in Relation to Sex (1871) in which he describes the ancestors of the human race as arboreal, hairy, and having tails and prehensile feet, and The Expression of the Emotions in Man and Animals (1872).

Darwin and Wallace were not the first to propose the evolution of species – Darwin's grandfather Erasmus Darwin had been a proponent of an earlier version of evolutionary theory – but it is Darwin and Wallace who put the modern theory of evolution on a firm foundation, and in particular Darwin through his seminal work On the Origin of Species. A century and a half after its first publication, it remains both a classic and a key text, one of the best-known and most influential books of all time.

The text of On the Origin of Species published here is that of the sixth (1872) edition, the last edition to be published in Darwin's lifetime and therefore considered to be the definitive edition. It is in this edition of the book that the word 'evolution' is first used by Darwin in place of the earlier term 'descent'.

George Davidson, 2008

Birds Pl 45.

Finches from *The Zoology of the Voyage of H M S Beagle*

'But with regard to the material world, we can at least
go so far as this – we can perceive that events are brought about not by
insulated interpositions of Divine power, exerted in each particular case, but
by the establishment of general laws.'

Whewell: *Bridgewater Treatise.*

'The only distinct meaning of the word 'natural' is *stated*, *fixed* or *settled*;
since what is natural as much requires and presupposes an intelligent agent
to render it so, *i.e.*, to effect it continually or at stated times, as what is
supernatural or miraculous does to effect it for once.'

Butler: *Analogy of Revealed Religion.*

'To conclude, therefore, let no man out of a weak conceit of sobriety, or an
ill-applied moderation, think or maintain, that a man can search too far or
be too well studied in the book of God's word, or in the book of God's
works; divinity or philosophy; but rather let men endeavour an endless
progress or proficience in both.'

Bacon: *Advancement of Learning.*

Down, Beckenham, Kent,
First Edition, November 24th, 1859.
Sixth Edition, January, 1872.

ON THE ORIGIN OF SPECIES BY MEANS OF NATURAL SELECTION

or

The Preservation of Favoured
Races in the Struggle for Life

By CHARLES DARWIN, M.A., F.R.S.

Sixth Edition, with all Additions and Corrections.

CONTENTS

small area and to naturalisation–Action of Natural Selection, through Divergence of Character and Extinction, on the descendants from a common parent–Explains the Grouping of all organic beings–Advance in organisation–Low forms preserved–Convergence of character–Indefinite multiplication of species–Summary.

Effects of changed conditions–Use and disuse, combined with natural selection; organs of flight and of vision–Acclimatisation–Correlated variation–Compensation and economy of growth–False correlations–Multiple, rudimentary, and lowly organised structures variable–Parts developed in an unusual manner are highly variable; specific characters more variable than generic: secondary sexual characters variable–Species of the same genus vary in an analogous manner–Reversions to long-lost characters–Summary.

Difficulties of the theory of descent with modification–Absence or rarity of transitional varieties –Transitions in habits of life–Diversified habits in the same species–Species with habits widely different from those of their allies–Organs of extreme perfection–Modes of transition–Cases of difficulty–Natura non facit saltum–Organs of small importance–Organs not in all cases absolutely perfect–The law of Unity of Type and of the Conditions of Existence embraced by the theory of Natural Selection.

Longevity–Modifications not necessarily simultaneous–Modifications apparently of no direct service–Progressive development–Characters of small functional importance, the most constant–Supposed incompetence of natural selection to account for the incipient stages of useful structures–Causes which interfere with the acquisition through natural selection of useful structures–Gradations of structure with changed functions–Widely different organs in members of the same class, developed from one and the same source–Reasons for disbelieving in great and abrupt modifications.

INTRODUCTION

When on board H.M.S. 'Beagle', as naturalist, I was much struck with certain facts in the distribution of the organic beings inhabiting South America, and in the geological relations of the present to the past inhabitants of that continent. These facts, as will be seen in the latter chapters of this volume, seemed to throw some light on the origin of species – that mystery of mysteries, as it has been called by one of our greatest philosophers. On my return home, it occurred to me, in 1837, that something might perhaps be made out on this question by patiently accumulating and reflecting on all sorts of facts which could possibly have any bearing on it. After five years' work I allowed myself to speculate on the subject, and drew up some short notes; these I enlarged in 1844 into a sketch of the conclusions, which then seemed to me probable; from that period to the present day I have steadily pursued the same object. I hope that I may be excused for entering on these personal details, as I give them to show that I have not been hasty in coming to a decision.

My work is now (1859) nearly finished; but as it will take me many more years to complete it, and as my health is far from strong, I have been urged to publish this Abstract. I have more especially been induced to do this, as Mr. Wallace, who is now studying the natural history of the Malay Archipelago, has arrived at almost exactly the same general conclusions that I have on the origin of species. In 1858 he sent me a memoir on this subject, with a request that I would forward it to Sir Charles Lyell, who sent it to the Linnean Society, and it is published in the third volume of the Journal of that Society. Sir C. Lyell and Dr. Hooker, who both knew of my work – the latter having read my sketch of 1844 – honoured me by thinking it advisable to publish, with Mr. Wallace's excellent memoir, some brief extracts from my manuscripts.

This Abstract, which I now publish, must necessarily be imperfect. I cannot here give references and authorities for my several statements; and I must trust to the reader reposing some confidence in my accuracy. No doubt errors will have crept in, though I hope I have always been cautious in trusting to good authorities alone. I can here give only the general conclusions at which I have arrived, with a few facts in illustration, but which, I hope, in most cases will suffice. No one can feel more sensible than I do of the necessity of hereafter publishing in detail all the facts, with references, on which my conclusions have been grounded; and I hope in a future work to do this. For I am well aware that scarcely a single point is discussed in this volume on which facts cannot be adduced, often apparently leading to conclusions directly opposite to those at which I have arrived. A fair result can be obtained only by fully stating and balancing the facts and arguments on both sides of each question; and this is here impossible.

I much regret that want of space prevents my having the satisfaction of acknowledging the generous assistance which I have received from very many naturalists, some of them personally unknown to me. I cannot, however, let this opportunity pass without expressing my deep obligations to Dr. Hooker, who, for the last fifteen years, has aided me in every possible way by his large stores of knowledge and his excellent judgment.

In considering the origin of species, it is quite conceivable that a naturalist, reflecting on the mutual affinities of organic beings, on their embryological relations, their geographical distribution, geological succession, and other such facts, might come to the conclusion that species had not been independently created, but had descended, like varieties, from other species. Nevertheless, such a conclusion, even if well

Charles Darwin, 1875

The H.M.S. *Beagle* in the Straits of Magellan in South America

founded, would be unsatisfactory, until it could be shown how the innumerable species inhabiting this world have been modified, so as to acquire that perfection of structure and coadaptation which justly excites our admiration. Naturalists continually refer to external conditions, such as climate, food, etc., as the only possible cause of variation. In one limited sense, as we shall hereafter see, this may be true; but it is preposterous to attribute to mere external conditions, the structure, for instance, of the woodpecker, with its feet, tail, beak, and tongue, so admirably adapted to catch insects under the bark of trees. In the case of the mistletoe, which draws its nourishment from certain trees, which has seeds that must be transported by certain birds, and which has flowers with separate sexes absolutely requiring the agency of certain insects to bring pollen from one flower to the other, it is equally preposterous to account for the structure

of this parasite, with its relations to several distinct organic beings, by the effects of external conditions, or of habit, or of the volition of the plant itself.

It is, therefore, of the highest importance to gain a clear insight into the means of modification and coadaptation. At the commencement of my observations it seemed to me probable that a careful study of domesticated animals and of cultivated plants would offer the best chance of making out this obscure problem. Nor have I been disappointed; in this and in all other perplexing cases I have invariably found that our knowledge, imperfect though it be, of variation under domestication, afforded the best and safest clue. I may venture to express my conviction of the high value of such studies, although they have been very commonly neglected by naturalists.

From these considerations, I shall devote the first chapter of this Abstract to Variation under

Domestication. We shall thus see that a large amount of hereditary modification is at least possible; and, what is equally or more important, we shall see how great is the power of man in accumulating by his Selection successive slight variations. I will then pass on to the variability of species in a state of nature; but I shall, unfortunately, be compelled to treat this subject far too briefly, as it can be treated properly only by giving long catalogues of facts. We shall, however, be enabled to discuss what circumstances are most favourable to variation. In the next chapter the Struggle for Existence amongst all organic beings throughout the world, which inevitably follows from the high geometrical ratio of their increase, will be considered. This is the doctrine of Malthus, applied to the whole animal and vegetable kingdoms. As many more individuals of each species are born than can possibly survive; and as, consequently, there is a frequently recurring struggle for existence, it follows that any being, if it vary however slightly in any manner profitable to itself, under the complex and sometimes varying conditions of life, will have a better chance of surviving, and thus be *naturally selected*. From the strong principle of inheritance, any selected variety will tend to propagate its new and modified form.

This fundamental subject of Natural Selection will be treated at some length in the fourth chapter; and we shall then see how Natural Selection almost inevitably causes much Extinction of the less improved forms of life, and leads to what I have called Divergence of Character. In the next chapter I shall discuss the complex and little known laws of variation. In the five succeeding chapters, the most apparent and gravest difficulties in accepting the theory will be given: namely, first, the difficulties of transitions, or how a simple being or a simple organ can be changed and perfected into a highly developed being or into an elaborately constructed organ; secondly, the subject of Instinct, or the mental

powers of animals; thirdly, Hybridism, or the infertility of species and the fertility of varieties when intercrossed; and fourthly, the imperfection of the Geological Record. In the next chapter I shall consider the geological succession of organic beings throughout time; in the twelfth and thirteenth, their geographical distribution throughout space; in the fourteenth, their classification or mutual affinities, both when mature and in an embryonic condition. In the last chapter I shall give a brief recapitulation of the whole work, and a few concluding remarks.

No one ought to feel surprise at much remaining as yet unexplained in regard to the origin of species and varieties, if he make due allowance for our profound ignorance in regard to the mutual relations of the many beings which live around us. Who can explain why one species ranges widely and is very numerous, and why another allied species has a narrow range and is rare? Yet these relations are of the highest importance, for they determine the present welfare, and, as I believe, the future success and modification of every inhabitant of this world. Still less do we know of the mutual relations of the innumerable inhabitants of the world during the many past geological epochs in its history. Although much remains obscure, and will long remain obscure, I can entertain no doubt, after the most deliberate study and dispassionate judgment of which I am capable, that the view which most naturalists until recently entertained, and which I formerly entertained – namely, that each species has been independently created – is erroneous. I am fully convinced that species are not immutable; but that those belonging to what are called the same genera are lineal descendants of some other and generally extinct species, in the same manner as the acknowledged varieties of any one species are the descendants of that species. Furthermore, I am convinced that Natural Selection has been the most important, but not the exclusive, means of modification.

Chapter I

VARIATION UNDER DOMESTICATION

Causes of variability–Effects of habit and the use or disuse of parts–Correlated variation–Inheritance–Character of domestic varieties–Difficulty of distinguishing between varieties and species–Origin of domestic varieties from one or more species–Domestic pigeons, their differences and origin–Principles of selection, anciently followed, their effects–Methodical and unconscious selection–Unknown origin of our domestic productions–Circumstances favourable to man's power of selection.

CAUSES OF VARIABILITY

When we compare the individuals of the same variety or sub-variety of our older cultivated plants and animals, one of the first points which strikes us is, that they generally differ more from each other than do the individuals of any one species or variety in a state of nature. And if we reflect on the vast diversity of the plants and animals which have been cultivated, and which have varied during all ages under the most different climates and treatment, we are driven to conclude that this great variability is due to our domestic productions having been raised under conditions of life not so uniform as, and somewhat different from, those to which the parent species had been exposed under nature. There is, also, some probability in the view propounded by Andrew Knight, that this variability may be partly connected with excess of food. It seems clear that organic beings must be exposed during several generations to new conditions to cause any great amount of variation; and that, when the organisation has once begun to vary, it generally continues varying for many generations. No case is on record of a variable organism ceasing to vary under cultivation. Our oldest cultivated plants, such as wheat, still yield new varieties: our oldest domesticated animals are still capable of rapid improvement or modification.

As far as I am able to judge, after long attending to the subject, the conditions of life appear to act in two ways, – directly on the whole organisation or on certain parts alone, and indirectly by affecting the reproductive system. With respect to the direct action, we must bear in mind that in every case, as Professor Weismann has lately insisted, and as I have incidentally shown in my work on 'Variation under Domestication,' there are two factors: namely, the nature of the organism, and the nature of the conditions. The former seems to be much the more important; for nearly similar variations sometimes arise under, as far as we can judge, dissimilar conditions; and, on the other hand, dissimilar variations arise under conditions which appear to be nearly uniform. The effects on the offspring are either definite or indefinite. They may be considered as definite when all or nearly all the offspring of individuals exposed to certain conditions during several generations are modified in the same manner. It is extremely difficult to come to any conclusion in regard to the extent of the changes which have been thus definitely induced. There can, however, be little doubt about many slight changes, such as size from the amount of food, colour from the nature of the food, thickness of the skin and hair from climate, etc. Each of the endless variations which we see in the plumage of our fowls must have had some efficient cause; and if the same cause were to act uniformly during a long series of generations on many individuals, all probably would be modified in the same manner. Such facts as the complex and extraordinary outgrowths which variably follow from the insertion of a minute drop of poison by a gall-producing insect, shows us what singular modifications might result in the case of plants from a chemical change in the nature of the sap.

Indefinite variability is a much more common result of changed conditions than definite variability, and has probably played a more important part in the formation of our domestic races. We see indefinite variability in the endless slight peculiarities which distinguish the individuals of the same species, and which cannot be accounted for by inheritance from either parent or from some more remote ancestor. Even strongly marked differences occasionally appear in the young of the same litter, and in seedlings from the same seed capsule. At long intervals of time, out of millions of individuals reared in the same country and fed on nearly the same food, deviations of structure so strongly pronounced as to deserve to be called monstrosities arise; but monstrosities cannot be separated by any distinct line from slighter variations. All such changes of structure, whether extremely slight or strongly marked, which appear amongst many individuals living together, may be considered as the indefinite effects of the conditions of life on each individual organism, in nearly the same manner as the chill affects different men in an indefinite manner, according to their state of body or constitution, causing coughs or colds, rheumatism or inflammation of various organs.

With respect to what I have called the indirect action of changed conditions, namely, through the reproductive system of being affected, we may infer that variability is thus induced, partly from the fact of this system being extremely sensitive to any change in the conditions, and partly from the similarity, as Kölreuter and others have remarked, between the variability which follows from the crossing of distinct species, and that which may be observed with plants and animals when reared under new or unnatural conditions. Many facts clearly show how eminently susceptible the reproductive system is to very slight changes in the surrounding conditions. Nothing is more easy than to tame an animal, and few things more difficult than to get it to breed freely under confinement, even when the male and female unite. How many animals there are which will not breed, though kept in an almost free state in their native country! This is generally, but erroneously, attributed to vitiated instincts. Many cultivated plants display the utmost vigour, and yet rarely or never seed! In some few cases it has been discovered that a very trifling change, such as a little more or less water at some particular period of growth, will determine whether or not a plant will produce seeds. I cannot here give the details which I have collected and elsewhere published on this curious subject; but to show how singular the laws are which determine the reproduction of animals under confinement, I may mention that carnivorous animals, even from the tropics, breed in this country pretty freely under confinement, with the exception of the plantigrades or bear family, which seldom produce young; whereas carnivorous birds, with the rarest exceptions, hardly ever lay fertile eggs. Many exotic plants have pollen utterly worthless, in the same condition as in the most sterile hybrids. When, on the one hand, we see domesticated animals and plants, though often weak and sickly, breeding freely under confinement; and when, on the other hand, we see individuals, though taken young from a state of nature perfectly tamed, long-lived and healthy (of which I could give numerous instances), yet having their reproductive system so seriously affected by unperceived causes as to fail to act, we need not be surprised at this system, when it does act under confinement, acting irregularly, and producing offspring somewhat unlike their

Craxirex Galapagoensis

The Galapagos Hawk, a hunter and scavenger who, as Darwin observed,
fearlessly investigated human visitors to the islands

parents. I may add, that as some organisms breed freely under the most unnatural conditions (for instance, rabbits and ferrets kept in hutches), showing that their reproductive organs are not easily affected; so will some animals and plants withstand domestication or cultivation, and vary very slightly – perhaps hardly more than in a state of nature.

Some naturalists have maintained that all variations are connected with the act of sexual reproduction; but this is certainly an error; for I have given in another work a long list of 'sporting plants,' as they are called by gardeners; that is, of plants which have suddenly produced a single bud with a new and sometimes widely different character from that of the other buds on the same plant. These bud variations, as they may be named, can be propagated by grafts, offsets, etc., and sometimes by seed. They occur rarely under nature, but are far from rare under culture. As a single bud out of many thousands, produced year after year on the same tree under uniform conditions, has been known suddenly to assume a new character; and as buds on distinct trees, growing under different conditions, have sometimes yielded nearly the same variety – for instance, buds on peach-trees producing nectarines, and buds on common roses producing moss roses – we clearly see that the nature of the conditions is of subordinate importance in comparison with the nature of the organism in determining each particular form of variation; perhaps of not more importance than the nature of the spark, by which a mass of combustible matter is ignited, has in determining the nature of the flames.

EFFECTS OF HABIT AND OF THE USE OR DISUSE OF PARTS; CORRELATED VARIATION; INHERITANCE

Changed habits produce an inherited effect, as in the period of the flowering of plants when transported from one climate to another. With animals the increased use or disuse of parts has

had a more marked influence; thus I find in the domestic duck that the bones of the wing weigh less and the bones of the leg more, in proportion to the whole skeleton, than do the same bones in the wild duck; and this change may be safely attributed to the domestic duck flying much less, and walking more, than its wild parents. The great and inherited development of the udders in cows and goats in countries where they are habitually milked, in comparison with these organs in other countries, is probably another instance of the effects of use. Not one of our domestic animals can be named which has not in some country drooping ears; and the view which has been suggested that the drooping is due to disuse of the muscles of the ear, from the animals being seldom much alarmed, seems probable.

Many laws regulate variation, some few of which can be dimly seen, and will hereafter be briefly discussed. I will here only allude to what may be called correlated variation. Important changes in the embryo or larva will probably entail changes in the mature animal. In monstrosities, the correlations between quite distinct parts are very curious; and many instances are given in Isidore Geoffroy St. Hilaire's great work on this subject. Breeders believe that long limbs are almost always accompanied by an elongated head. Some instances of correlation are quite whimsical: thus cats which are entirely white and have blue eyes are generally deaf; but it has been lately stated by Mr. Tait that this is confined to the males. Colour and constitutional peculiarities go together, of which many remarkable cases could be given amongst animals and plants. From facts collected by Heusinger, it appears that white sheep and pigs are injured by certain plants, whilst dark-coloured individuals escape: Professor Wyman has recently communicated to me a good illustration of this fact; on asking some farmers in Virginia how it was that all their pigs were black, they informed him that the pigs ate the paint-root (Lachnanthes), which coloured their bones pink, and which caused the hoofs of all but the black varieties to drop off;

and one of the 'crackers' (*i.e.* Virginia squatters) added, 'We select the black members of a litter for raising, as they alone have a good chance of living.' Hairless dogs have imperfect teeth; long-haired and coarse-haired animals are apt to have, as is asserted, long or many horns; pigeons with feathered feet have skin between their outer toes; pigeons with short beaks have small feet, and those with long beaks large feet. Hence if man goes on selecting, and thus augmenting, any peculiarity, he will almost certainly modify unintentionally other parts of the structure, owing to the mysterious laws of correlation.

The results of the various, unknown, or but dimly understood laws of variation are infinitely complex and diversified. It is well worth while carefully to study the several treatises on some of our old cultivated plants, as on the hyacinth, potato, even the dahlia, etc.; and it is really surprising to note the endless points of structure and constitution in which the varieties and sub-varieties differ slightly from each other. The whole organisation seems to have become plastic, and departs in a slight degree from that of the parental type.

Any variation which is not inherited is unimportant for us. But the number and diversity of inheritable deviations of structure, both those of slight and those of considerable physiological importance, are endless. Dr. Prosper Lucas' treatise, in two large volumes, is the fullest and the best on this subject. No breeder doubts how strong is the tendency to inheritance; that like produces like, is his fundamental belief: doubts have been thrown on this principle only by theoretical writers. When any deviation of structure often appears, and we see it in the father and child, we cannot tell whether it may not be due to the same cause having acted on both; but when amongst individuals, apparently exposed to the same conditions, any very rare deviation, due to some extraordinary combination of circumstances, appears in the parent – say, once amongst several million individuals – and it

reappears in the child, the mere doctrine of chances almost compels us to attribute its reappearance to inheritance. Every one must have heard of cases of albinism, prickly skin, hairy bodies, etc., appearing in several members of the same family. If strange and rare deviations of structure are really inherited, less strange and commoner deviations may be freely admitted to be inheritable. Perhaps the correct way of viewing the whole subject would be, to look at the inheritance of every character whatever as the rule, and non-inheritance as the anomaly.

The laws governing inheritance are for the most part unknown. No one can say why the same peculiarity in different individuals of the same species, or in different species, is sometimes inherited and sometimes not so; why the child often reverts in certain characteristics to its grandfather or grandmother or more remote ancestor; why a peculiarity is often transmitted from one sex to both sexes, or to one sex alone, more commonly but not exclusively to the like sex. It is a fact of some importance to us, that peculiarities appearing in the males of our domestic breeds are often transmitted, either exclusively or in a much greater degree, to the males alone. A much more important rule, which I think may be trusted, is that, at whatever period of life a peculiarity first appears, it tends to reappear in the offspring at a corresponding age, though sometimes earlier. In many cases this could not be otherwise; thus the inherited peculiarities in the horns of cattle could appear only in the offspring when nearly mature; peculiarities in the silk-worm are known to appear at the corresponding caterpillar or cocoon stage. But hereditary diseases and some other facts make me believe that the rule has a wider extension, and that, when there is no apparent reason why a peculiarity should appear at any particular age, yet that it does tend to appear in the offspring at the same period at which it first appeared in the parent. I believe this rule to be of the highest importance in explaining the laws of embryology. These remarks are of course confined to the first *appearance* of the peculiarity, and not to the primary cause which may have acted on the ovules or on the male element; in nearly the same manner as the increased length of the horns in the offspring from a short-horned cow by a long-horned bull, though appearing late in life, is clearly due to the male element.

Having alluded to the subject of reversion, I may here refer to a statement often made by naturalists – namely, that our domestic varieties, when run wild, gradually but invariably revert in character to their aboriginal stocks. Hence it has been argued that no deductions can be drawn from domestic races to species in a state of nature. I have in vain endeavoured to discover on what decisive facts the above statement has so often and so boldly been made. There would be great difficulty in proving its truth: we may safely conclude that very many of the most strongly marked domestic varieties could not possibly live in a wild state. In many cases we do not know what the aboriginal stock was, and so could not tell whether or not nearly perfect reversion had ensued. It would be necessary, in order to prevent the effects of intercrossing, that only a single variety should be turned loose in its new home. Nevertheless, as our varieties certainly do occasionally revert in some of their characters to ancestral forms, it seems to me not improbable that if we could succeed in naturalising, or were to cultivate, during many generations, the several races, for instance, of the cabbage, in very poor soil (in which case, however, some effect would have to be attributed to the *definite* action of the poor soil), that they would, to a large extent, or even wholly, revert to the wild aboriginal stock. Whether or not the experiment would succeed, is not of great importance for our line of argument; for by the experiment itself the conditions of life are changed. If it could be shown that our domestic varieties manifested a strong tendency to reversion, – that is, to lose their acquired characters, whilst kept under the same conditions, and whilst kept in a considerable body, so that free

intercrossing might check, by blending together, any slight deviations in their structure, in such case, I grant that we could deduce nothing from domestic varieties in regard to species. But there is not a shadow of evidence in favour of this view: to assert that we could not breed our cart and race horses, long and short horned cattle, and poultry of various breeds, and esculent vegetables, for an unlimited number of generations, would be opposed to all experience.

CHARACTER OF DOMESTIC VARIETIES; DIFFICULTY OF DISTINGUISHING BETWEEN VARIETIES AND SPECIES; ORIGIN OF DOMESTIC VARIETIES FROM ONE OR MORE SPECIES

When we look to the hereditary varieties or races of our domestic animals and plants, and compare them with closely allied species, we generally perceive in each domestic race, as already remarked, less uniformity of character than in true species. Domestic races often have a somewhat monstrous character; by which I mean, that, although differing from each other, and from other species of the same genus, in several trifling respects, they often differ in an extreme degree in some one part, both when compared one with another, and more especially when compared with the species under nature to which they are nearest allied. With these exceptions (and with that of the perfect fertility of varieties when crossed, – a subject hereafter to be discussed), domestic races of the same species differ from each other in the same manner as do the closely allied species of the same genus in a state of nature, but the differences in most cases are less in degree. This must be admitted as true, for the domestic races of many animals and plants have been ranked by some competent judges as the descendants of aboriginally distinct species, and by other competent judges as mere varieties. If any well-marked distinction existed between a domestic race and a species, this source of doubt would not so perpetually recur. It has often been stated that domestic races do not differ from each other in characters of generic value. It can be shown that this statement is not correct; but naturalists differ much in determining what characters are of generic value; all such valuations being at present empirical. When it is explained how genera originate under nature, it will be seen that we have no right to expect often to find a generic amount of difference in our domesticated races.

In attempting to estimate the amount of structural difference between allied domestic races, we are soon involved in doubt, from not knowing whether they are descended from one or several parent species. This point, if it could be cleared up, would be interesting; if, for instance, it could be shown that the greyhound, bloodhound, terrier, spaniel and bulldog, which we all know propagate their kind truly, were the offspring of any single species, then such facts would have great weight in making us doubt about the immutability of the many closely allied natural species – for instance, of the many foxes – inhabiting the different quarters of the world. I do not believe, as we shall presently see, that the whole amount of difference between the several breeds of the dog has been produced under domestication; I believe that a small part of the difference is due to their being descended from distinct species. In the case of strongly marked races of some other domesticated species, there is presumptive or even strong evidence, that all are descended from a single wild stock.

It has often been assumed that man has chosen for domestication animals and plants having an extraordinary inherent tendency to vary, and likewise to withstand diverse climates. I do not dispute that these capacities have added largely to the value of most of our domesticated productions; but how could a savage possibly know, when he first tamed an animal, whether it would vary in succeeding generations, and whether it would endure other climates? Has the little variability of the ass and goose, or the small power of endurance of warmth by the reindeer, or

of cold by the common camel, prevented their domestication? I cannot doubt that if other animals and plants, equal in number to our domesticated productions, and belonging to equally diverse classes and countries, were taken from a state of nature, and could be made to breed for an equal number of generations under domestication, they would on an average vary as largely as the parent species of our existing domesticated productions have varied.

In the case of most of our anciently domesticated animals and plants, it is not possible to come to any definite conclusion, whether they are descended from one or several wild species. The argument mainly relied on by those who believe in the multiple origin of our domestic animals is, that we find in the most ancient times, on the monuments of Egypt, and in the lake habitations of Switzerland, much diversity in the breeds; and that some of these ancient breeds closely resemble, or are even identical with, those still existing. But this only throws far backward the history of civilisation, and shows that animals were domesticated at a much earlier period than has hitherto been supposed. The lake inhabitants of Switzerland cultivated several kinds of wheat and barley, the pea, the poppy for oil and flax; and they possessed several domesticated animals. They also carried on commerce with other nations. All this clearly shows, as Heer has remarked, that they had at this early age progressed considerably in civilisation; and this again implies a long continued previous period of less advanced civilisation, during which the domesticated animals, kept by different tribes in different districts, might have varied and given rise to distinct races. Since the discovery of flint tools in the superficial formations of many parts of the world, all geologists believe that barbarian man existed at an enormously remote period and we know that at the present day there is hardly a tribe so barbarous, as not to have domesticated at least the dog.

The origin of most of our domestic animals will probably forever remain vague. But I may

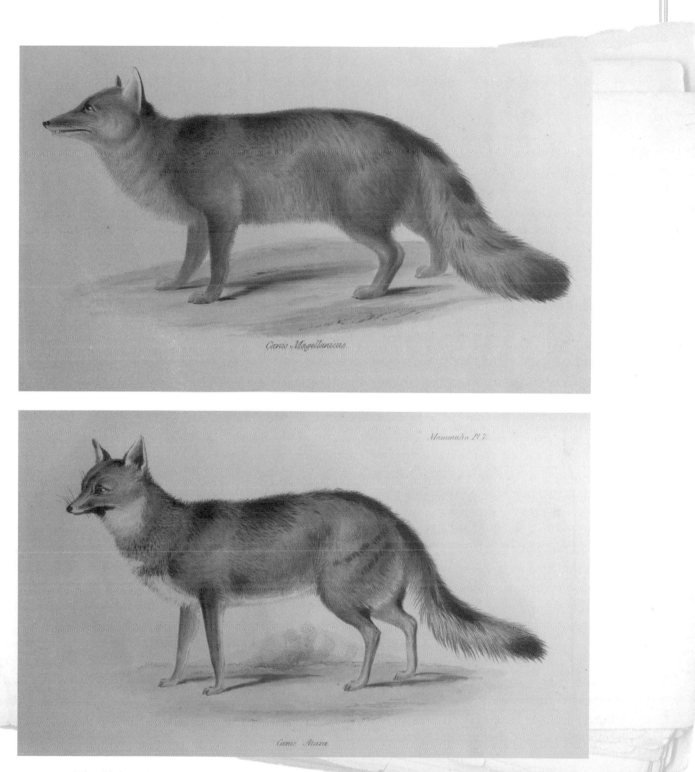

The Chilean Fox (top) and the Crab-eating or Brazilian Fox (bottom). Both appeared in
The Zoology of the Voyage of H.M.S. Beagle, edited by Darwin

here state, that, looking to the domestic dogs of the whole world, I have, after a laborious collection of all known facts, come to the conclusion that several wild species of Canidae have been tamed, and that their blood, in some cases mingled together, flows in the veins of our domestic breeds. In regard to sheep and goats I can form no decided opinion. From facts communicated to me by Mr. Blyth, on the habits, voice, constitution and structure of the humped Indian cattle, it is almost certain that they are descended from a different aboriginal stock from our European cattle and some competent judges believe that these latter have had two or three wild progenitors, – whether or not these deserve to be called species. This conclusion, as well as that of the specific distinction between the humped and common cattle, may, indeed, be looked upon as established by the admirable researches of Professor Rütimeyer. With respect to horses, from reasons which I cannot here give, I am doubtfully inclined to believe, in opposition to several authors, that all the races belong to the same species. Having kept nearly all the English breeds of the fowl alive, having bred and crossed them, and examined their skeletons, it appears to me almost certain that all are the descendants of the wild Indian fowl, Gallus bankiva; and this is the conclusion of Mr. Blyth, and of others who have studied this bird in India. In regard to ducks and rabbits, some breeds of which differ much from each other, the evidence is clear that they are all descended from the common duck and wild rabbit.

The doctrine of the origin of our several domestic races from several aboriginal stocks, has been carried to an absurd extreme by some authors. They believe that every race which breeds true, let the distinctive characters be ever so slight, has had its wild prototype. At this rate there must have existed at least a score of species of wild cattle, as many sheep, and several goats, in Europe alone, and several even within Great Britain. One author believes that there formerly

existed eleven wild species of sheep peculiar to Great Britain! When we bear in mind that Britain has now not one peculiar mammal, and France but few distinct from those of Germany, and so with Hungary, Spain, etc., but that each of these kingdoms possesses several peculiar breeds of cattle, sheep, etc., we must admit that many domestic breeds must have originated in Europe; from whence otherwise could they have been derived? So it is in India. Even in the case of the breeds of the domestic dog throughout the world, which I admit are descended from several wild species, it cannot be doubted that there has been an immense amount of inherited variation; for who will believe that animals closely resembling the Italian greyhound, the bloodhound, the bulldog, pugdog, or Blenheim spaniel, etc. – so unlike all wild Canidae – ever existed in a state of nature? It has often been loosely said that all our races of dogs have been produced by the crossing of a few aboriginal species; but by crossing we can only get forms in some degree intermediate between their parents; and if we account for our several domestic races by this process, we must admit the former existence of the most extreme forms, as the Italian greyhound, bloodhound, bulldog, etc., in the wild state. Moreover, the possibility of making distinct races by crossing has been greatly exaggerated. Many cases are on record showing that a race may be modified by occasional crosses, if aided by the careful selection of the individuals which present the desired character; but to obtain a race intermediate between two quite distinct races would be very difficult. Sir J. Sebright expressly experimented with this object and failed. The offspring from the first cross between two pure breeds is tolerably and sometimes (as I have found with pigeons) quite uniform in character, and everything seems simple enough; but when these mongrels are crossed one with another for several generations, hardly two of them are alike, and then the difficulty of the task becomes manifest.

BREEDS OF THE DOMESTIC PIGEON, THEIR DIFFERENCES AND ORIGIN

Believing that it is always best to study some special group, I have, after deliberation, taken up domestic pigeons. I have kept every breed which I could purchase or obtain, and have been most kindly favoured with skins from several quarters of the world, more especially by the Hon. W. Elliot, from India, and by the Hon. C. Murray, from Persia. Many treatises in different languages have been published on pigeons, and some of them are very important, as being of considerable antiquity. I have associated with several eminent fanciers, and have been permitted to join two of the London Pigeon Clubs. The diversity of the breeds is something astonishing. Compare the English carrier and the short-faced tumbler, and see the wonderful difference in their beaks, entailing corresponding differences in their skulls. The carrier, more especially the male bird, is also remarkable from the wonderful development of the carunculated skin about the head; and this is accompanied by greatly elongated eyelids, very large external orifices to the nostrils, and a wide gape of mouth. The short-faced tumbler has a beak in outline almost like that of a finch; and the common tumbler has the singular inherited habit of flying at a great height in a compact flock, and tumbling in the air head over heels. The runt is a bird of great size, with long, massive beak and large feet; some of the sub-breeds of runts have very long necks, others very long wings and tails, others singularly short tails. The barb is allied to the carrier, but, instead of a long beak, has a very short and broad one. The pouter has a much elongated body, wings, and legs; and its enormously developed crop, which it glories in inflating, may well excite astonishment and even laughter. The turbit has a short and conical beak, with a line of reversed feathers down the breast; and it has the habit of continually expanding, slightly, the upper part of the oesophagus. The jacobin has the feathers so much reversed along the back of the neck that they form a hood; and

it has, proportionally to its size, elongated wing and tail feathers. The trumpeter and laugher, as their names express, utter a very different coo from the other breeds. The fantail has thirty or even forty tail-feathers, instead of twelve or fourteen – the normal number in all the members of the great pigeon family: these feathers are kept expanded, and are carried so erect, that in good birds the head and tail touch: the oil-gland is quite aborted. Several other less distinct breeds might be specified.

In the skeletons of the several breeds, the development of the bones of the face, in length and breadth and curvature, differs enormously. The shape, as well as the breadth and length of the ramus of the lower jaw, varies in a highly remarkable manner. The caudal and sacral vertebrae vary in number; as does the number of the ribs, together with their relative breadth and the presence of processes. The size and shape of the apertures in the sternum are highly variable; so is the degree of divergence and relative size of the two arms of the furcula. The proportional width of the gape of mouth, the proportional length of the eyelids, of the orifice of the nostrils, of the tongue (not always in strict correlation with the length of beak), the size of the crop and of the upper part of the oesophagus; the development and abortion of the oil-gland; the number of the primary wing and caudal feathers; the relative length of the wing and tail to each other and to the body; the relative length of the leg and foot; the number of scutellae on the toes, the development of skin between the toes, are all points of structure which are variable. The period at which the perfect plumage is acquired varies, as does the state of the down with which the nestling birds are clothed when hatched. The shape and size of the eggs vary. The manner of flight, and in some breeds the voice and disposition, differs remarkably. Lastly, in certain breeds, the males and females have come to differ in a slight degree from each other.

Altogether at least a score of pigeons might be chosen, which, if shown to an ornithologist,

Six varieties of domestic pigeon:
1 Blue rock *2* Tumbler *3* Owl *4* Jacobin *5* Fantail *6* Pouter

and he were told that they were wild birds, would certainly be ranked by him as well-defined species. Moreover, I do not believe that any ornithologist would in this case place the English carrier, the short-faced tumbler, the runt, the barb, pouter and fantail in the same genus; more especially as in each of these breeds several truly inherited sub-breeds, or species, as he would call them, could be shown him.

Great as are the differences between the breeds of the pigeon, I am fully convinced that the common opinion of naturalists is correct, namely, that all are descended from the rock-pigeon (Columba livia), including under this term several geographical races or sub-species, which differ from each other in the most trifling respects. As several of the reasons which have led me to this belief are in some degree applicable in other cases, I will here briefly give them. If the several breeds are not varieties, and have not proceeded from the rock-pigeon, they must have descended from at least seven or eight aboriginal stocks; for it is impossible to make the present domestic breeds by the crossing of any lesser number: how, for instance, could a pouter be produced by crossing two breeds unless one of the parent-stocks possessed the characteristic enormous crop? The supposed aboriginal stocks must all have been rock-pigeons, that is, they did not breed or willingly perch on trees. But besides C. livia, with its geographical sub-species, only two or three other species of rock-pigeons are known; and these have not any of the characters of the domestic breeds. Hence the supposed aboriginal stocks must either still exist in the countries where they were originally domesticated, and yet be unknown to ornithologists; and this, considering their size, habits and remarkable characters, seems improbable; or they must have become extinct in the wild state. But birds breeding on precipices, and good fliers, are unlikely to be exterminated; and the common rock-pigeon, which has the same habits with the domestic breeds, has not been exterminated even on several of the smaller British

islets, or on the shores of the Mediterranean. Hence the supposed extermination of so many species having similar habits with the rock-pigeon seems a very rash assumption. Moreover, the several above-named domesticated breeds have been transported to all parts of the world, and, therefore, some of them must have been carried back again into their native country; but not one has become wild or feral, though the dovecot pigeon, which is the rock-pigeon in a very slightly altered state, has become feral in several places. Again, all recent experience shows that it is difficult to get wild animals to breed freely under domestication; yet on the hypothesis of the multiple origin of our pigeons, it must be assumed that at least seven or eight species were so thoroughly domesticated in ancient times by half-civilised man, as to be quite prolific under confinement.

An argument of great weight, and applicable in several other cases, is, that the above-specified breeds, though agreeing generally with the wild rock-pigeon in constitution, habits, voice, colouring, and in most parts of their structure, yet are certainly highly abnormal in other parts; we may look in vain through the whole great family of Columbidae for a beak like that of the English carrier, or that of the short-faced tumbler, or barb; for reversed feathers like those of the Jacobin; for a crop like that of the pouter; for tail-feathers like those of the fantail. Hence it must be assumed, not only that half-civilised man succeeded in thoroughly domesticating several species, but that he intentionally or by chance picked out extraordinarily abnormal species; and further, that these very species have since all become extinct or unknown. So many strange contingencies are improbable in the highest degree.

Some facts in regard to the colouring of pigeons well deserve consideration. The rock-pigeon is of a slaty-blue, with white loins; but the Indian sub-species, C. intermedia of Strickland, has this part bluish. The tail has a terminal dark bar, with the outer feathers externally edged at the base with white. The wings have two black bars. Some semi-domestic breeds, and some truly wild breeds, have, besides the two black bars, the wings chequered with black. These several marks do not occur together in any other species of the whole family. Now, in every one of the domestic breeds, taking thoroughly well-bred birds, all the above marks, even to the white edging of the outer tail-feathers, sometimes concur perfectly developed. Moreover, when birds belonging to two or more distinct breeds are crossed, none of which are blue or have any of the above-specified marks, the mongrel offspring are very apt suddenly to acquire these characters. To give one instance out of several which I have observed: I crossed some white fantails, which breed very true, with some black barbs – and it so happens that blue varieties of barbs are so rare that I never heard of an instance in England; and the mongrels were black, brown and mottled. I also crossed a barb with a spot, which is a white bird with a red tail and red spot on the forehead, and which notoriously breeds very true; the mongrels were dusky and mottled. I then crossed one of the mongrel barb-fantails with a mongrel barb-spot, and they produced a bird of as beautiful a blue colour, with the white loins, double black wing-bar, and barred and white-edged tail-feathers, as any wild rock-pigeon! We can understand these facts, on the well-known principle of reversion to ancestral characters, if all the domestic breeds are descended from the rock-pigeon. But if we deny this, we must make one of the two following highly improbable suppositions. Either, first, that all the several imagined aboriginal stocks were coloured and marked like the rock-pigeon, although no other existing species is thus coloured and marked, so that in each separate breed there might be a tendency to revert to the very same colours and markings. Or, secondly, that each breed, even the purest, has within a dozen, or at most within a score, of generations, been crossed by the rock-pigeon: I say within a dozen or twenty generations, for no instance is known of crossed

descendants reverting to an ancestor of foreign blood, removed by a greater number of generations. In a breed which has been crossed only once, the tendency to revert to any character derived from such a cross will naturally become less and less, as in each succeeding generation there will be less of the foreign blood; but when there has been no cross, and there is a tendency in the breed to revert to a character which was lost during some former generation, this tendency, for all that we can see to the contrary, may be transmitted undiminished for an indefinite number of generations. These two distinct cases of reversion are often confounded together by those who have written on inheritance.

Lastly, the hybrids or mongrels from between all the breeds of the pigeon are perfectly fertile, as I can state from my own observations, purposely made, on the most distinct breeds. Now, hardly any cases have been ascertained with certainty of hybrids from two quite distinct species of animals being perfectly fertile. Some authors believe that long-continued domestication eliminates this strong tendency to sterility in species. From the history of the dog, and of some other domestic animals, this conclusion is probably quite correct, if applied to species closely related to each other. But to extend it so far as to suppose that species, aboriginally as distinct as carriers, tumblers, pouters, and fantails now are, should yield offspring perfectly fertile, *inter se*, would be rash in the extreme.

From these several reasons, namely, – the improbability of man having formerly made seven or eight supposed species of pigeons to breed freely under domestication; – these supposed species being quite unknown in a wild state, and their not having become anywhere feral; – these species presenting certain very abnormal characters, as compared with all other Columbidae, though so like the rock-pigeon in most other respects; – the occasional re-appearance of the blue colour and various black marks in all the breeds, both when kept pure and when crossed; – and lastly, the mongrel offspring being perfectly fertile; – from these several reasons, taken together, we may safely conclude that all our domestic breeds are descended from the rock-pigeon or Columba livia with its geographical sub-species.

In favour of this view, I may add, firstly, that the wild C. livia has been found capable of domestication in Europe and in India; and that it agrees in habits and in a great number of points of structure with all the domestic breeds. Secondly, that although an English carrier or a short-faced tumbler differs immensely in certain characters from the rock-pigeon, yet that, by comparing the several sub-breeds of these two races, more especially those brought from distant countries, we can make, between them and the rock-pigeon, an almost perfect series; so we can in some other cases, but not with all the breeds. Thirdly, those characters which are mainly distinctive of each breed are in each eminently variable, for instance, the wattle and length of beak of the carrier, the shortness of that of the tumbler, and the number of tail-feathers in the fantail; and the explanation of this fact will be obvious when we treat of Selection. Fourthly, pigeons have been watched and tended with the utmost care, and loved by many people. They have been domesticated for thousands of years in several quarters of the world; the earliest known record of pigeons is in the fifth Egyptian dynasty, about 3000 B.C., as was pointed out to me by Professor Lepsius; but Mr. Birch informs me that pigeons are given in a bill of fare in the previous dynasty. In the time of the Romans, as we hear from Pliny, immense prices were given for pigeons; 'nay, they are come to this pass, that they can reckon up their pedigree and race.' Pigeons were much valued by Akber Khan in India, about the year 1600; never less than 20,000 pigeons were taken with the court. 'The monarchs of Iran and Turan sent him some very rare birds'; and, continues the courtly historian, 'His Majesty, by crossing the breeds, which method was never practised before, has improved

them astonishingly.' About this same period the Dutch were as eager about pigeons as were the old Romans. The paramount importance of these considerations in explaining the immense amount of variation which pigeons have undergone, will likewise be obvious when we treat of Selection. We shall then, also, see how it is that the several breeds so often have a somewhat monstrous character. It is also a most favourable circumstance for the production of distinct breeds, that male and female pigeons can be easily mated for life; and thus different breeds can be kept together in the same aviary.

I have discussed the probable origin of domestic pigeons at some, yet quite insufficient, length; because when I first kept pigeons and watched the several kinds, well knowing how truly they breed, I felt fully as much difficulty in believing that since they had been domesticated they had all proceeded from a common parent, as any naturalist could in coming to a similar conclusion in regard to the many species of finches, or other groups of birds, in nature. One circumstance has struck me much; namely, that nearly all the breeders of the various domestic animals and the cultivators of plants, with whom I have conversed, or whose treatises I have read, are firmly convinced that the several breeds to which each has attended, are descended from so many aboriginally distinct species. Ask, as I have asked, a celebrated raiser of Hereford cattle, whether his cattle might not have descended from long-horns, or both from a common parent-stock, and he will laugh you to scorn. I have never met a pigeon, or poultry, or duck, or rabbit fancier, who was not fully convinced that each main breed was descended from a distinct species. Van Mons, in his treatise on pears and apples, shows how utterly he disbelieves that the several sorts, for instance a Ribston-pippin or Codlin-apple, could ever have proceeded from the seeds of the same tree. Innumerable other examples could be given. The explanation, I think, is simple: from long-continued study they are strongly impressed with

The Medium Ground Finch, one of 13 species of 'Darwin's finches' found on the Galapagos Islands the differences between the several races; and though they well know that each race varies slightly, for they win their prizes by selecting such slight differences, yet they ignore all general arguments, and refuse to sum up in their minds slight differences accumulated during many successive generations. May not those naturalists who, knowing far less of the laws of inheritance than does the breeder, and knowing no more than he does of the intermediate links in the long lines of descent, yet admit that many of our domestic races are descended from the same parents – may they not learn a lesson of caution, when they deride the idea of species in a state of nature being lineal descendants of other species?

PRINCIPLES OF SELECTION ANCIENTLY FOLLOWED, AND THEIR EFFECTS

Let us now briefly consider the steps by which domestic races have been produced, either from one or from several allied species. Some effect may be attributed to the direct and definite action of the external conditions of life, and some to habit; but he would be a bold man who would account by such agencies for the differences between a dray and race horse, a greyhound and bloodhound, a carrier and tumbler pigeon. One of the most remarkable features in our domesticated races is that we see in them adaptation, not indeed to the animal's or plant's own good, but to man's use or fancy. Some variations useful to him have probably arisen suddenly, or by one step; many botanists, for instance, believe that the fuller's teasel, with its hooks, which can not be rivalled by any mechanical contrivance, is only a variety of the wild Dipsacus; and this amount of change may have suddenly arisen in a seedling. So it has probably been with the turnspit dog; and this is known to have been the case with the ancon sheep. But when we compare the dray horse and race horse, the dromedary and camel, the various breeds of sheep fitted either for cultivated land or mountain pasture, with the wool of one breed good for one purpose, and that of another breed

for another purpose; when we compare the many breeds of dogs, each good for man in different ways; when we compare the game cock, so pertinacious in battle, with other breeds so little quarrelsome, with 'everlasting layers' which never desire to sit, and with the bantam so small and elegant; when we compare the host of agricultural, culinary, orchard, and flower garden races of plants, most useful to man at different seasons and for different purposes, or so beautiful in his eyes, we must, I think, look further than to mere variability. We can not suppose that all the breeds were suddenly produced as perfect and as useful as we now see them; indeed, in many cases, we know that this has not been their history. The key is man's power of accumulative selection: nature gives successive variations; man adds them up in certain directions useful to him. In this sense he may be said to have made for himself useful breeds.

The great power of this principle of selection is not hypothetical. It is certain that several of our eminent breeders have, even within a single lifetime, modified to a large extent their breeds of cattle and sheep. In order fully to realise what they have done it is almost necessary to read several of the many treatises devoted to this subject, and to inspect the animals. Breeders habitually speak of an animal's organisation as something plastic, which they can model almost as they please. If I had space I could quote numerous passages to this effect from highly competent authorities. Youatt, who was probably better acquainted with the works of agriculturalists than almost any other individual, and who was himself a very good judge of animals, speaks of the principle of selection as 'that which enables the agriculturist, not only to modify the character of his flock, but to change it altogether. It is the magician's wand, by means of which he may summon into life whatever form and mould he pleases.' Lord Somerville, speaking of what breeders have done for sheep, says: – 'It would seem as if they had chalked out upon a wall a form perfect in itself, and then had given it existence.' In Saxony the importance of the principle of selection in regard to merino sheep is so fully recognised, that men follow it as a trade: the sheep are placed on a table and are studied, like a picture by a connoisseur; this is done three times at intervals of months, and the sheep are each time marked and classed, so that the very best may ultimately be selected for breeding.

What English breeders have actually effected is proved by the enormous prices given for animals with a good pedigree; and these have been exported to almost every quarter of the world. The improvement is by no means generally due to crossing different breeds; all the best breeders are strongly opposed to this practice, except sometimes amongst closely allied sub-breeds. And when a cross has been made, the closest selection is far more indispensable even than in ordinary cases. If selection consisted merely in separating some very distinct variety and breeding from it, the principle would be so obvious as hardly to be worth notice; but its importance consists in the great effect produced by the accumulation in one direction, during successive generations, of differences absolutely inappreciable by an uneducated eye – differences which I for one have vainly attempted to appreciate. Not one man in a thousand has accuracy of eye and judgment sufficient to become an eminent breeder. If gifted with these qualities, and he studies his subject for years, and devotes his lifetime to it with indomitable perseverance, he will succeed, and may make great improvements; if he wants any of these qualities, he will assuredly fail. Few would readily believe in the natural capacity and years of practice requisite to become even a skilful pigeon-fancier.

The same principles are followed by horticulturists; but the variations are here often more abrupt. No one supposes that our choicest productions have been produced by a single variation from the aboriginal stock. We have proofs that this is not so in several cases in which exact records have been kept; thus, to give a very trifling instance, the steadily increasing size of the common gooseberry may be quoted. We see an astonishing improvement in many florists' flowers, when the flowers of the present day are compared with drawings made only twenty or thirty years ago. When a race of plants is once pretty well established, the seed-raisers do not pick out the best plants, but merely go over their seed-beds, and pull up the 'rogues,' as they call the plants that deviate from the proper standard. With animals this kind of selection is, in fact, likewise followed; for hardly any one is so careless as to breed from his worst animals.

In regard to plants, there is another means of observing the accumulated effects of selection – namely, by comparing the diversity of flowers in the different varieties of the same species in the flower garden; the diversity of leaves, pods, or tubers, or whatever part is valued, in the kitchen garden, in comparison with the flowers of the same varieties; and the diversity of fruit of the same species in the orchard, in comparison with the leaves and flowers of the same set of varieties. See how different the leaves of the cabbage are, and how extremely alike the flowers; how unlike the flowers of the heartsease are, and how alike the leaves; how much the fruit of the different kinds of gooseberries differ in size, colour, shape, and hairiness, and yet the flowers present very slight differences. It is not that the varieties which differ largely in some one point do not differ at all in other points; this is hardly ever, – I speak after careful observation, – perhaps never, the case. The law of correlated variation, the importance of which should never be overlooked, will ensure some differences; but, as a general rule, it cannot be doubted that the continued selection of slight variations, either in the leaves, the flowers, or the fruit, will produce races differing from each other chiefly in these characters.

It may be objected that the principle of selection has been reduced to methodical practice for scarcely more than three-quarters of a century; it has certainly been more attended to of late years, and many treatises have been published on the subject; and the result has been, in a corresponding degree, rapid and important. But it is very far from true that the principle is a modern discovery. I could give several references to works of high antiquity, in which the full importance of the principle is acknowledged. In rude and barbarous periods of English history choice animals were often imported, and laws were passed to prevent their exportation: the destruction of horses under a certain size was ordered, and this may be compared to the 'roguing' of plants by nurserymen. The principle of selection I find distinctly given in an ancient Chinese encyclopaedia. Explicit rules are laid down by some of the Roman classical writers. From passages in Genesis, it is clear that the colour of domestic animals was at that early period attended to. Savages now sometimes cross their dogs with wild canine animals, to improve the breed, and they formerly did so, as is attested by passages in Pliny. The savages in South Africa match their draught cattle by colour, as do some of the Esquimaux their teams of dogs. Livingstone states that good domestic breeds are highly valued by the negroes in the interior of Africa who have not associated with Europeans. Some of these facts do not show actual selection, but they show that the breeding of domestic animals was carefully attended to in ancient times, and is now attended to by the lowest savages. It would, indeed, have been a strange fact, had attention not been paid to breeding, for the inheritance of good and bad qualities is so obvious.

UNCONSCIOUS SELECTION

At the present time, eminent breeders try by methodical selection, with a distinct object in view, to make a new strain or sub-breed, superior to anything of the kind in the country. But, for our purpose, a form of Selection, which may be called Unconscious, and which results from every one trying to possess and breed from the best individual animals, is more important. Thus, a man who intends keeping pointers naturally tries to get as good dogs as he can, and afterwards breeds from his own best dogs, but he has no wish

or expectation of permanently altering the breed. Nevertheless we may infer that this process, continued during centuries, would improve and modify any breed, in the same way as Bakewell, Collins, etc., by this very same process, only carried on more methodically, did greatly modify, even during their lifetimes, the forms and qualities of their cattle. Slow and insensible changes of this kind could never be recognised unless actual measurements or careful drawings of the breeds in question have been made long ago, which may serve for comparison. In some cases, however, unchanged, or but little changed, individuals of the same breed exist in less civilised districts, where the breed has been less improved. There is reason to believe that King Charles' spaniel has been unconsciously modified to a large extent since the time of that monarch. Some highly competent authorities are convinced that the setter is directly derived from the spaniel, and has probably been slowly altered from it. It is known that the English pointer has been greatly changed within the last century, and in this case the change has, it is believed, been chiefly effected by crosses with the foxhound; but what concerns us is, that the change has been effected unconsciously and gradually, and yet so effectually that, though the old Spanish pointer certainly came from Spain, Mr. Borrow has not seen, as I am informed by him, any native dog in Spain like our pointer. By a similar process of selection, and by careful training, English race horses have come to surpass in fleetness and size the parent Arabs, so that the latter, by the regulations for the Goodwood Races, are favoured in the weights which they carry. Lord Spencer and others have shown how the cattle of England have increased in weight and in early maturity, compared with the stock formerly kept in this country. By comparing the accounts given in various old treatises of the former and present state of carrier and tumbler pigeons in Britain, India, and Persia, we can trace the stages through which they have insensibly passed, and come to differ so greatly from the rock-pigeon.

Youatt gives an excellent illustration of the effects of a course of selection which may be considered as unconscious, in so far that the breeders could never have expected, or even wished, to produce the result which ensued – namely, the production of the distinct strains. The two flocks of Leicester sheep kept by Mr. Buckley and Mr. Burgess, as Mr. Youatt remarks, 'have been purely bred from the original stock of Mr. Bakewell for upwards of fifty years. There is not a suspicion existing in the mind of any one at all acquainted with the subject that the owner of either of them has deviated in any one instance from the pure blood of Mr. Bakewell's flock, and yet the difference between the sheep possessed by these two gentlemen is so great that they have the appearance of being quite different varieties.'

If there exist savages so barbarous as never to think of the inherited character of the offspring of their domestic animals, yet any one animal particularly useful to them, for any special purpose, would be carefully preserved during famines and other accidents, to which savages are so liable, and such choice animals would thus generally leave more offspring than the inferior ones; so that in this case there would be a kind of unconscious selection going on. We see the value set on animals even by the barbarians of Tierra del Fuego, by their killing and devouring their old women, in times of dearth, as of less value than their dogs.

In plants the same gradual process of improvement through the occasional preservation of the best individuals, whether or not sufficiently distinct to be ranked at their first appearance as distinct varieties, and whether or not two or more species or races have become blended together by crossing, may plainly be recognised in the increased size and beauty which we now see in the varieties of the heartsease, rose, pelargonium, dahlia, and other plants, when compared with the older varieties or with their parent-stocks. No one would ever expect to get a first-rate heartsease or dahlia from the seed of a wild plant. No one

would expect to raise a first-rate melting pear from the seed of a wild pear, though he might succeed from a poor seedling growing wild, if it had come from a garden-stock. The pear, though cultivated in classical times, appears, from Pliny's description, to have been a fruit of very inferior quality. I have seen great surprise expressed in horticultural works at the wonderful skill of gardeners in having produced such splendid results from such poor materials; but the art has been simple, and, as far as the final result is concerned, has been followed almost unconsciously. It has consisted in always cultivating the best known variety, sowing its seeds, and, when a slightly better variety chanced to appear, selecting it, and so onwards. But the gardeners of the classical period, who cultivated the best pears which they could procure, never thought what splendid fruit we should eat; though we owe our excellent fruit in some small degree to their having naturally chosen and preserved the best varieties they could anywhere find.

A large amount of change, thus slowly and unconsciously accumulated, explains, as I believe, the well-known fact, that in a number of cases we cannot recognise, and therefore do not know, the wild parent-stocks of the plants which have been longest cultivated in our flower and kitchen gardens. If it has taken centuries or thousands of years to improve or modify most of our plants up to their present standard of usefulness to man, we can understand how it is that neither Australia, the Cape of Good Hope, nor any other region inhabited by quite uncivilised man, has afforded us a single plant worth culture. It is not that these countries, so rich in species, do not by a strange chance possess the aboriginal stocks of any useful plants, but that the native plants have not been improved by continued selection up to a standard of perfection comparable with that acquired by the plants in countries anciently civilised.

In regard to the domestic animals kept by uncivilised man, it should not be overlooked that they almost always have to struggle for their own food, at least during certain seasons. And in two countries very differently circumstanced, individuals of the same species, having slightly different constitutions or structure, would often succeed better in the one country than in the other, and thus by a process of 'natural selection,' as will hereafter be more fully explained, two sub-breeds might be formed. This, perhaps, partly explains why the varieties kept by savages, as has been remarked by some authors, have more of the character of true species than the varieties kept in civilised countries.

On the view here given of the important part which selection by man has played, it becomes at once obvious, how it is that our domestic races show adaptation in their structure or in their habits to man's wants or fancies. We can, I think, further understand the frequently abnormal character of our domestic races, and likewise their differences being so great in external characters, and relatively so slight in internal parts or organs. Man can hardly select, or only with much difficulty, any deviation of structure excepting such as is externally visible; and indeed he rarely cares for what is internal. He can never act by selection, excepting on variations which are first given to him in some slight degree by nature. No man would ever try to make a fantail till he saw a pigeon with a tail developed in some slight degree in an unusual manner, or a pouter till he saw a pigeon with a crop of somewhat unusual size; and the more abnormal or unusual any character was when it first appeared, the more likely it would be to catch his attention. But to use such an expression as trying to make a fantail is, I have no doubt, in most cases, utterly incorrect. The man who first selected a pigeon with a slightly larger tail, never dreamed what the descendants of that pigeon would become through long-continued, partly unconscious and partly methodical, selection. Perhaps the parent bird of all fantails had only fourteen tail-feathers somewhat expanded, like the present Java fantail, or like individuals of other and distinct breeds, in which

as many as seventeen tail-feathers have been counted. Perhaps the first pouter pigeon did not inflate its crop much more than the turbit now does the upper part of its oesophagus, – a habit which is disregarded by all fanciers, as it is not one of the points of the breed.

Nor let it be thought that some great deviation of structure would be necessary to catch the fancier's eye: he perceives extremely small differences, and it is in human nature to value any novelty, however slight, in one's own possession. Nor must the value which would formerly have been set on any slight differences in the individuals of the same species, be judged of by the value which is now set on them, after several breeds have fairly been established. It is known that with pigeons many slight variations now occasionally appear, but these are rejected as faults or deviations from the standard of perfection in each breed. The common goose has not given rise to any marked varieties; hence the Toulouse and the common breed, which differ only in colour, that most fleeting of characters, have lately been exhibited as distinct at our poultry shows.

These views appear to explain what has sometimes been noticed, namely, that we know hardly anything about the origin or history of any of our domestic breeds. But, in fact, a breed, like a dialect of a language, can hardly be said to have a distinct origin. A man preserves and breeds from an individual with some slight deviation of structure, or takes more care than usual in matching his best animals, and thus improves them, and the improved animals slowly spread in the immediate neighbourhood. But they will as yet hardly have a distinct name, and from being only slightly valued, their history will have been disregarded. When further improved by the same slow and gradual process, they will spread more widely, and will be recognised as something distinct and valuable, and will then probably first receive a provincial name. In semi-civilised countries, with little free communication, the spreading of a new sub-breed will be a slow process. As soon as the points of value are once acknowledged, the principle, as I have called it, of unconscious selection will always tend, – perhaps more at one period than at another, as the breed rises or falls in fashion, – perhaps more in one district than in another, according to the state of civilisation of the inhabitants, – slowly to add to the characteristic features of the breed, whatever they may be. But the chance will be infinitely small of any record having been preserved of such slow, varying, and insensible changes.

CIRCUMSTANCES FAVOURABLE TO MAN'S POWER OF SELECTION

I will now say a few words on the circumstances, favourable or the reverse, to man's power of selection. A high degree of variability is obviously favourable, as freely giving the materials for selection to work on; not that mere individual differences are not amply sufficient, with extreme care, to allow of the accumulation of a large amount of modification in almost any desired direction. But as variations manifestly useful or pleasing to man appear only occasionally, the chance of their appearance will be much increased by a large number of individuals being kept. Hence number is of the highest importance for success. On this principle Marshall formerly remarked, with respect to the sheep of part of Yorkshire, 'as they generally belong to poor people, and are mostly *in small lots*, they never can be improved.' On the other hand, nurserymen, from keeping large stocks of the same plant, are generally far more successful than amateurs in raising new and valuable varieties. A large number of individuals of an animal or plant can be reared only where the conditions for its propagation are favourable. When the individuals are scanty all will be allowed to breed, whatever their quality may be, and this will effectually prevent selection. But probably the most important element is that the animal or plant should be so highly valued by man, that the closest attention is paid to even the slightest deviations in its qualities or structure.

Unless such attention be paid nothing can be effected. I have seen it gravely remarked, that it was most fortunate that the strawberry began to vary just when gardeners began to attend to this plant. No doubt the strawberry had always varied since it was cultivated, but the slight varieties had been neglected. As soon, however, as gardeners picked out individual plants with slightly larger, earlier, or better fruit, and raised seedlings from them, and again picked out the best seedlings and bred from them, then (with some aid by crossing distinct species) those many admirable varieties of the strawberry were raised which have appeared during the last half-century.

With animals, facility in preventing crosses is an important element in the formation of new races, – at least, in a country which is already stocked with other races. In this respect enclosure of the land plays a part. Wandering savages or the inhabitants of open plains rarely possess more than one breed of the same species. Pigeons can be mated for life, and this is a great convenience to the fancier, for thus many races may be improved and kept true, though mingled in the same aviary; and this circumstance must have largely favoured the formation of new breeds. Pigeons, I may add, can be propagated in great numbers and at a very quick rate, and inferior birds may be freely rejected, as when killed they serve for food. On the other hand, cats, from their nocturnal rambling habits, can not be easily matched, and, although so much valued by women and children, we rarely see a distinct breed long kept up; such breeds as we do sometimes see are almost always imported from some other country. Although I do not doubt that some domestic animals vary less than others, yet the rarity or absence of distinct breeds of the cat, the donkey, peacock, goose, etc., may be attributed in main part to selection not having been brought into play: in cats, from the difficulty in pairing them; in donkeys, from only a few being kept by poor people, and little attention paid to their breeding; for recently in certain parts of Spain and of the United States this animal has been surprisingly modified and improved by careful selection; in peacocks, from not being very easily reared and a large stock not kept; in geese, from being valuable only for two purposes, food and feathers, and more especially from no pleasure having been felt in the display of distinct breeds; but the goose, under the conditions to which it is exposed when domesticated, seems to have a singularly inflexible organisation, though it has varied to a slight extent, as I have elsewhere described.

Some authors have maintained that the amount of variation in our domestic productions is soon reached, and can never afterward be exceeded. It would be somewhat rash to assert that the limit has been attained in any one case; for almost all our animals and plants have been greatly improved in many ways within a recent period; and this implies variation. It would be equally rash to assert that characters now increased to their utmost limit, could not, after remaining fixed for many centuries, again vary under new conditions of life. No doubt, as Mr. Wallace has remarked with much truth, a limit will be at last reached. For instance, there must be a limit to the fleetness of any terrestrial animal, as this will be determined by the friction to be overcome, the weight of the body to be carried, and the power of contraction in the muscular fibres. But what concerns us is that the domestic varieties of the same species differ from each other in almost every character, which man has attended to and selected, more than do the distinct species of the same genera. Isidore Geoffroy St. Hilaire has proved this in regard to size, and so it is with colour, and probably with the length of hair. With respect to fleetness, which depends on many bodily characters, Eclipse was far fleeter, and a dray horse is comparably stronger, than any two natural species belonging to the same genus. So with plants, the seeds of the different varieties of the bean or maize probably differ more in size than do the seeds of the distinct species in any one genus in the same two families. The same remark holds good in regard to the fruit of the several varieties of the plum, and still more strongly with the melon, as well as in many other analogous cases.

To sum up on the origin of our domestic races of animals and plants. Changed conditions of life are of the highest importance in causing variability, both by acting directly on the organisation, and indirectly by affecting the reproductive system. It is not probable that variability is an inherent and necessary contingent, under all circumstances. The greater or less force of inheritance and reversion determine whether variations shall endure. Variability is governed by many unknown laws, of which correlated growth is probably the most important. Something, but how much we do not know, may be attributed to the definite action of the conditions of life. Some, perhaps a great, effect may be attributed to the increased use or disuse of parts. The final result is thus rendered infinitely complex. In some cases the intercrossing of aboriginally distinct species appears to have played an important part in the origin of our breeds. When several breeds have once been formed in any country, their occasional intercrossing, with the aid of selection, has, no doubt, largely aided in the formation of new sub-breeds; but the importance of crossing has been much exaggerated, both in regard to animals and to those plants which are propagated by seed. With plants which are temporarily propagated by cuttings, buds, etc., the importance of crossing is immense; for the cultivator may here disregard the extreme variability both of hybrids and of mongrels, and the sterility of hybrids; but plants not propagated by seed are of little importance to us, for their endurance is only temporary. Over all these causes of change, the accumulative action of Selection, whether applied methodically and quickly, or unconsciously and slowly, but more efficiently, seems to have been the predominant power.

Chapter II

VARIATION UNDER NATURE

Variability–Individual differences–Doubtful species–Wide ranging, much diffused, and common species, vary most–Species of the larger genera in each country vary more frequently than the species of the smaller genera–Many of the species of the larger genera resemble varieties in being very closely, but unequally, related to each other, and in having restricted ranges–Summary.

Before applying the principles arrived at in the last chapter to organic beings in a state of nature, we must briefly discuss whether these latter are subject to any variation. To treat this subject properly, a long catalogue of dry facts ought to be given; but these I shall reserve for a future work. Nor shall I here discuss the various definitions which have been given of the term species. No one definition has satisfied all naturalists; yet every naturalist knows vaguely what he means when he speaks of a species. Generally the term includes the unknown element of a distinct act of creation. The term 'variety' is almost equally difficult to define; but here community of descent is almost universally implied, though it can rarely be proved. We have also what are called monstrosities; but they graduate into varieties. By a monstrosity I presume is meant some considerable deviation of structure, generally injurious, or not useful to the species. Some authors use the term 'variation' in a technical sense, as implying a modification directly due to the physical conditions of life; and 'variations' in this sense are supposed not to be inherited; but who can say that the dwarfed condition of shells in the brackish waters of the Baltic, or dwarfed plants on Alpine summits, or the thicker fur of an animal from far northwards, would not in some cases be inherited for at least a few generations? And in this case I presume that the form would be called a variety.

It may be doubted whether sudden and considerable deviations of structure, such as we occasionally see in our domestic productions, more especially with plants, are ever permanently propagated in a state of nature. Almost every part of every organic being is so beautifully related to its complex conditions of life that it seems as improbable that any part should have been suddenly produced perfect, as that a complex machine should have been invented by man in a perfect state. Under domestication monstrosities sometimes occur which resemble normal structures in widely different animals. Thus pigs have occasionally been born with a sort of proboscis, and if any wild species of the same genus had naturally possessed a proboscis, it might have been argued that this had appeared as a monstrosity; but I have as yet failed to find, after diligent search, cases of monstrosities resembling normal structures in nearly allied forms, and these alone bear on the question. If monstrous forms of this kind ever do appear in a state of nature and are capable of reproduction (which is not always the case), as they occur rarely and singly, their preservation would depend on unusually favourable circumstances. They would, also, during the first and succeeding generations cross with the ordinary form, and thus their abnormal character would almost inevitably be lost. But I shall have to return in a future chapter to the preservation and perpetuation of single or occasional variations.

INDIVIDUAL DIFFERENCES

The many slight differences which appear in the offspring from the same parents, or which it may be presumed have thus arisen, from being observed in the individuals of the same species inhabiting the same confined locality, may be called individual differences. No one supposes that all the individuals of the same species are cast in the same actual mould. These individual differences are of the highest importance for us, for they are often inherited, as must be familiar to every one; and they thus afford materials for natural selection to act on and accumulate, in the same manner as man accumulates in any given direction individual differences in his domesticated productions. These individual differences generally affect what naturalists consider unimportant parts; but I could show, by a long catalogue of facts, that parts which must be called important, whether viewed under a physiological or classificatory point of view, sometimes vary in the individuals of the same species. I am convinced that the most experienced naturalist would be surprised at the number of the cases of variability, even in important parts of structure, which he could collect on good authority, as I have collected, during a course of years. It should be remembered that systematists are far from being pleased at finding variability in important characters, and that there are not many men who will laboriously examine internal and important organs, and compare them in many specimens of the same species. It would never have been expected that the branching of the main nerves close to the great central ganglion of an insect would have been variable in the same species; it might have been thought that changes of this nature could have been effected only by slow

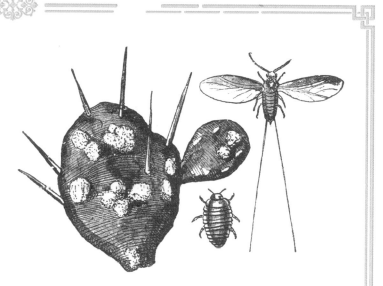

Cochineal insects (enlarged) beside the cactus pads on which they live and feed:
left wingless female *right* winged male

degrees; yet Sir J. Lubbock has shown a degree of variability in these main nerves in Coccus, which may almost be compared to the irregular branching of the stem of a tree. This philosophical naturalist, I may add, has also shown that the muscles in the larvae of certain insects are far from uniform. Authors sometimes argue in a circle when they state that important organs never vary; for these same authors practically rank those parts as important (as some few naturalists have honestly confessed) which do not vary; and, under this point of view, no instance will ever be found of an important part varying; but under any other point of view many instances assuredly can be given.

There is one point connected with individual differences which is extremely perplexing: I refer to those genera which have been called 'protean' or 'polymorphic,' in which species present an inordinate amount of variation. With respect to many of these forms, hardly two naturalists agree whether to rank them as species or as varieties. We may instance Rubus, Rosa, and Hieracium amongst plants, several genera of insects and of Brachiopod shells. In most polymorphic genera some of the species have fixed and definite characters. Genera which are polymorphic in one country seem to be, with a few exceptions, polymorphic in other countries, and likewise, judging from Brachiopod shells, at former periods of time. These facts are very perplexing, for they seem to show that this kind of variability is independent of the conditions of life. I am inclined to suspect that we see, at least in some of these polymorphic genera, variations which are of no service or disservice to the species, and which consequently have not been seized on and rendered definite by natural selection, as hereafter to be explained.

Individuals of the same species often present, as is known to every one, great differences of structure, independently of variation, as in the two sexes of various animals, in the two or three castes of sterile females or workers amongst insects, and in the immature and larval states of many of the lower animals.

There are, also, cases of dimorphism and trimorphism, both with animals and plants. Thus, Mr. Wallace, who has lately called attention to the subject, has shown that the females of certain species of butterflies, in the Malayan Archipelago, regularly appear under two or even three conspicuously distinct forms, not connected by intermediate varieties. Fritz Müller has described analogous but more extraordinary cases with the males of certain Brazilian Crustaceans: thus, the male of a Tanais regularly occurs under two distinct forms; one of these has strong and differently shaped pincers, and the other has antennae much more abundantly furnished with smelling-hairs. Although in most of these cases, the two or three forms, both with animals and plants, are not now connected by intermediate gradations, it is possible that they were once thus connected. Mr. Wallace, for instance, describes a certain butterfly which presents in the same island a great range of varieties connected by intermediate links, and the extreme links of the chain closely resemble the two forms of an allied dimorphic species inhabiting another part of the Malay Archipelago. Thus also with ants, the several worker-castes are generally quite distinct;

but in some cases, as we shall hereafter see, the castes are connected together by finely graduated varieties. So it is, as I have myself observed, with some dimorphic plants. It certainly at first appears a highly remarkable fact that the same female butterfly should have the power of producing at the same time three distinct female forms and a male; and that an hermaphrodite plant should produce from the same seed-capsule three distinct hermaphrodite forms, bearing three different kinds of females and three or even six different kinds of males. Nevertheless these cases are only exaggerations of the common fact that the female produces offspring of two sexes which sometimes differ from each other in a wonderful manner.

DOUBTFUL SPECIES
The forms which possess in some considerable degree the character of species, but which are so closely similar to other forms, or are so closely linked to them by intermediate gradations, that naturalists do not like to rank them as distinct species, are in several respects the most important for us. We have every reason to believe that many of these doubtful and closely allied forms have permanently retained their characters for a long time; for as long, as far as we know, as have good and true species. Practically, when a naturalist can unite by means of intermediate links any two forms, he treats the one as a variety of the other, ranking the most common, but sometimes the one first described, as the species, and the other as the variety. But cases of great difficulty, which I will not here enumerate, sometimes arise in deciding whether or not to rank one form as a variety of another, even when they are closely connected by intermediate links; nor will the commonly assumed hybrid nature of the intermediate forms always remove the difficulty. In very many cases, however, one form is ranked as a variety of another, not because the intermediate links have actually been found, but because analogy leads the observer to suppose either that they do now somewhere exist, or may formerly have existed;

and here a wide door for the entry of doubt and conjecture is opened.

Hence, in determining whether a form should be ranked as a species or a variety, the opinion of naturalists having sound judgment and wide experience seems the only guide to follow. We must, however, in many cases, decide by a majority of naturalists, for few well-marked and well-known varieties can be named which have not been ranked as species by at least some competent judges.

That varieties of this doubtful nature are far from uncommon cannot be disputed. Compare the several floras of Great Britain, of France, or of the United States, drawn up by different botanists, and see what a surprising number of forms have been ranked by one botanist as good species, and by another as mere varieties. Mr. H.C. Watson, to whom I lie under deep obligation for assistance of all kinds, has marked for me 182 British plants, which are generally considered as varieties, but which have all been ranked by botanists as species; and in making this list he has omitted many trifling varieties, but which nevertheless have been ranked by some botanists as species, and he has entirely omitted several highly polymorphic genera. Under genera, including the most polymorphic forms, Mr. Babington gives 251 species, whereas Mr. Bentham gives only 112 – a difference of 139 doubtful forms! Amongst animals which unite for each birth, and which are highly locomotive, doubtful forms, ranked by one zoologist as a species and by another as a variety, can rarely be found within the same country, but are common in separated areas. How many of the birds and insects in North America and Europe, which differ very slightly from each other, have been ranked by one eminent naturalist as undoubted species, and by another as varieties, or, as they are often called, geographical races! Mr. Wallace, in several valuable papers on the various animals, especially on the Lepidoptera, inhabiting the islands of the great Malayan Archipelago, shows that they may be classed under four heads, namely, as variable forms, as local forms, as geographical races or sub-

species, and as true representative species. The first or variable forms vary much within the limits of the same island. The local forms are moderately constant and distinct in each separate island; but when all from the several islands are compared together, the differences are seen to be so slight and graduated that it is impossible to define or describe them, though at the same time the extreme forms are sufficiently distinct. The geographical races or sub-species are local forms completely fixed and isolated; but as they do not differ from each other by strongly marked and important characters, 'There is no possible test but individual opinion to determine which of them shall be considered as species and which as varieties.' Lastly, representative species fill the same place in the natural economy of each island as do the local forms and sub-species; but as they are distinguished from each other by a greater amount of difference than that between the local forms and sub-species, they are almost universally ranked by naturalists as true species. Nevertheless, no certain criterion can possibly be given by which variable forms, local forms, sub-species and representative species can be recognised.

Many years ago, when comparing, and seeing others compare, the birds from the closely neighbouring islands of the Galapagos archipelago, one with another, and with those from the American mainland, I was much struck how entirely vague and arbitrary is the distinction between species and varieties. On the islets of the little Madeira group there are many insects which are characterized as varieties in Mr. Wollaston's admirable work, but which would certainly be ranked as distinct species by many entomologists. Even Ireland has a few animals, now generally regarded as varieties, but which have been ranked as species by some zoologists. Several experienced ornithologists consider our British red grouse as only a strongly marked race of a Norwegian species, whereas the greater number rank it as an undoubted species peculiar to Great Britain. A wide distance between the homes of two doubtful forms leads many naturalists to rank them as distinct species; but what distance, it has been well asked, will suffice; if that between America and Europe is ample, will that between Europe and the Azores, or Madeira, or the Canaries, or between the several islets of these small archipelagos, be sufficient?

Mr. B.D. Walsh, a distinguished entomologist of the United States, has described what he calls Phytophagic varieties and Phytophagic species. Most vegetable-feeding insects live on one kind of plant or on one group of plants; some feed indiscriminately on many kinds, but do not in consequence vary. In several cases, however, insects found living on different plants, have been observed by Mr. Walsh to present in their larval or mature state, or in both states, slight, though constant differences in colour, size, or in the nature of their secretions. In some instances the males alone, in other instances, both males and females, have been observed thus to differ in a slight degree. When the differences are rather more strongly marked, and when both sexes and all ages are affected, the forms are ranked by all entomologists as good species. But no observer can determine for another, even if he can do so for himself, which of these Phytophagic forms ought to be called species and which varieties. Mr. Walsh ranks the forms which it may be supposed would freely intercross, as varieties; and those which appear to have lost this power, as species. As the differences depend on the insects having long fed on distinct plants, it cannot be expected that intermediate links connecting the several forms should now be found. The naturalist thus loses his best guide in determining whether to rank doubtful forms as varieties or species. This likewise necessarily occurs with closely allied organisms, which inhabit distinct continents or islands. When, on the other hand, an animal or plant ranges over the same continent, or inhabits many islands in the same archipelago, and presents different forms in the different areas, there is always a good chance that intermediate forms will be discovered which will link together the extreme states; and these are

then degraded to the rank of varieties.

Some few naturalists maintain that animals never present varieties; but then these same naturalists rank the slightest difference as of specific value; and when the same identical form is met with in two distant countries, or in two geological formations, they believe that two distinct species are hidden under the same dress. The term species thus comes to be a mere useless abstraction, implying and assuming a separate act of creation. It is certain that many forms, considered by highly competent judges to be varieties, resemble species so completely in character that they have been thus ranked by other highly competent judges. But to discuss whether they ought to be called species or varieties, before any definition of these terms has been generally accepted, is vainly to beat the air.

Many of the cases of strongly marked varieties or doubtful species well deserve consideration; for several interesting lines of argument, from geographical distribution, analogical variation, hybridism, etc., have been brought to bear in the attempt to determine their rank; but space does not here permit me to discuss them. Close investigation, in many cases, will no doubt bring naturalists to agree how to rank doubtful forms. Yet it must be confessed that it is in the best known countries that we find the greatest number of them. I have been struck with the fact that if any animal or plant in a state of nature be highly useful to man, or from any cause closely attracts his attention, varieties of it will almost universally be found recorded. These varieties, moreover, will often be ranked by some authors as species. Look at the common oak, how closely it has been studied; yet a German author makes more than a dozen species out of forms, which are almost universally considered by other botanists to be varieties; and in this country the highest botanical authorities and practical men can be quoted to show that the sessile and pedunculated oaks are either good and distinct species or mere varieties.

The oak tree, or quercus, of which there are several hundred varieties

I may here allude to a remarkable memoir lately published by A. de Candolle, on the oaks of the whole world. No one ever had more ample materials for the discrimination of the species, or could have worked on them with more zeal and sagacity. He first gives in detail all the many points of structure which vary in the several species, and estimates numerically the relative frequency of the variations. He specifies above a dozen characters which may be found varying even on the same branch, sometimes according to age or development, sometimes without any assignable reason. Such characters are not of course of specific value, but they are, as Asa Gray has remarked in commenting on this memoir, such as generally enter into specific definitions. De Candolle then goes on to say that he gives the rank of species to the forms that differ by characters never varying on the same tree, and never found connected by intermediate states. After this discussion, the result of so much labour, he emphatically remarks: 'They are mistaken, who repeat that the greater part of our species are clearly limited, and that the doubtful species are in a feeble minority. This seemed to be true, so long as a genus was imperfectly known, and its species were founded upon a few specimens, that is to say, were provisional. Just as we come to know them better, intermediate forms flow in, and doubts as to specific limits augment.' He also adds that it is the best known species which present the greatest number of spontaneous varieties and sub-varieties. The Quercus robur has twenty-eight varieties, all of which, excepting six, are clustered round three sub-species, namely Q. pedunculata, sessiliflora and pubescens. The forms which connect these three sub-species are comparatively rare; and, as Asa Gray again remarks, if these connecting forms which are now rare were to become totally extinct the three sub-species would hold exactly the same relation to each other as do the four or five provisionally admitted species which closely surround the typical Quercus robur. Finally, De Candolle admits that out of the 300 species, which will be enumerated in his Prodromus as belonging

to the oak family, at least two-thirds are provisional species, that is, are not known strictly to fulfil the definition above given of a true species. It should be added that De Candolle no longer believes that species are immutable creations, but concludes that the derivative theory is the most natural one, 'and the most accordant with the known facts in palaeontology, geographical botany and zoology, of anatomical structure and classification.'

When a young naturalist commences the study of a group of organisms quite unknown to him he is at first much perplexed in determining what differences to consider as specific and what as varietal; for he knows nothing of the amount and kind of variation to which the group is subject; and this shows, at least, how very generally there is some variation. But if he confine his attention to one class within one country he will soon make up his mind how to rank most of the doubtful forms. His general tendency will be to make many species, for he will become impressed, just like the pigeon or poultry fancier before alluded to, with the amount of difference in the forms which he is continually studying; and he has little general knowledge of analogical variation in other groups and in other countries by which to correct his first impressions. As he extends the range of his observations he will meet with more cases of difficulty; for he will encounter a greater number of closely allied forms. But if his observations be widely extended, he will in the end generally be able to make up his own mind; but he will succeed in this at the expense of admitting much variation, and the truth of this admission will often be disputed by other naturalists. When he comes to study allied forms brought from countries not now continuous, in which case he cannot hope to find intermediate links, he will be compelled to trust almost entirely to analogy, and his difficulties will rise to a climax.

Certainly no clear line of demarcation has as yet been drawn between species and sub-species – that is, the forms which in the opinion of some naturalists come very near to, but do not quite

Plant varieties of the genus Rosa (top left), Hypericum (top right),
Hyssopus (bottom left) and Fuschia (bottom right)

arrive at, the rank of species; or, again, between sub-species and well-marked varieties, or between lesser varieties and individual differences. These differences blend into each other by an insensible series; and a series impresses the mind with the idea of an actual passage.

Hence I look at individual differences, though of small interest to the systematist, as of the highest importance for us, as being the first step towards such slight varieties as are barely thought worth recording in works on natural history. And I look at varieties which are in any degree more distinct and permanent, as steps toward more strongly marked and permanent varieties; and at the latter, as leading to sub-species, and then to species. The passage from one stage of difference to another may, in many cases, be the simple result of the nature of the organism and of the different physical conditions to which it has long been exposed; but with respect to the more important and adaptive characters, the passage from one stage of difference to another may be safely attributed to the cumulative action of natural selection, hereafter to be explained, and to the effects of the increased use or disuse of parts. A well-marked variety may therefore be called an incipient species; but whether this belief is justifiable must be judged by the weight of the various facts and considerations to be given throughout this work.

It need not be supposed that all varieties or incipient species attain the rank of species. They may become extinct, or they may endure as varieties for very long periods, as has been shown to be the case by Mr. Wollaston with the varieties of certain fossil land-shells in Madeira, and with plants by Gaston de Saporta. If a variety were to flourish so as to exceed in numbers the parent species, it would then rank as the species, and the species as the variety; or it might come to supplant and exterminate the parent species; or both might co-exist, and both rank as independent species. But we shall hereafter return to this subject.

From these remarks it will be seen that I look at the term species as one arbitrarily given, for the sake of convenience, to a set of individuals closely resembling each other, and that it does not essentially differ from the term variety, which is given to less distinct and more fluctuating forms. The term variety, again, in comparison with mere individual differences, is also applied arbitrarily, for convenience' sake.

WIDE-RANGING, MUCH DIFFUSED, AND COMMON SPECIES VARY MOST

Guided by theoretical considerations, I thought that some interesting results might be obtained in regard to the nature and relations of the species which vary most, by tabulating all the varieties in several well-worked floras. At first this seemed a simple task; but Mr. H.C. Watson, to whom I am much indebted for valuable advice and assistance on this subject, soon convinced me that there were many difficulties, as did subsequently Dr. Hooker, even in stronger terms. I shall reserve for a future work the discussion of these difficulties, and the tables of the proportional numbers of the varying species. Dr. Hooker permits me to add that after having carefully read my manuscript, and examined the tables, he thinks that the following statements are fairly well established. The whole subject, however, treated as it necessarily here is with much brevity, is rather perplexing, and allusions cannot be avoided to the 'struggle for existence,' 'divergence of character,' and other questions, hereafter to be discussed.

Alphonse de Candolle and others have shown that plants which have very wide ranges generally present varieties; and this might have been expected, as they are exposed to diverse physical conditions, and as they come into competition (which, as we shall hereafter see, is a far more important circumstance) with different sets of organic beings. But my tables further show that, in any limited country, the species which are the most common, that is abound most in individuals, and the species which are most widely diffused within their own country (and this is a different consideration from wide range, and to a certain extent from commonness), oftenest give rise to varieties sufficiently well-marked to have been recorded in botanical works. Hence it is the most flourishing, or, as they may be called, the dominant species, – those which range widely, are the most diffused in their own country, and are the most numerous in individuals, – which oftenest produce well-marked varieties, or, as I consider them, incipient species. And this, perhaps, might have been anticipated; for, as varieties, in order to become in any degree permanent, necessarily have to struggle with the other inhabitants of the country, the species which are already dominant will be the most likely to yield offspring, which, though in some slight degree modified, still inherit those advantages that enabled their parents to become dominant over their compatriots. In these remarks on predominance, it should be understood that reference is made only to the forms which come into competition with each other, and more especially to the members of the same genus or class having nearly similar habits of life. With respect to the number of individuals or commonness of species, the comparison of course relates only to the members of the same group. One of the higher plants may be said to be dominant if it be more numerous in individuals and more widely diffused than the other plants of the same country, which live under nearly the same conditions. A plant of this kind is not the less dominant because some conferva inhabiting the water or some parasitic fungus is infinitely more numerous in individuals, and more widely diffused. But if the conferva or parasitic fungus exceeds its allies in the above respects, it will then be dominant within its own class.

SPECIES OF THE LARGER GENERA IN EACH COUNTRY VARY MORE FREQUENTLY THAN THE SPECIES OF THE SMALLER GENERA

If the plants inhabiting a country, as described in any Flora, be divided into two equal masses, all those in the larger genera (i.e., those including many species) being placed on one side, and all those in the smaller genera on the other side, the former will be found to include a somewhat larger number of the very common and much diffused or dominant species. This might have been anticipated, for the mere fact of many species of the same genus inhabiting any country, shows that there is something in the organic or inorganic conditions of that country favourable to the genus; and, consequently, we might have expected to have found in the larger genera, or those including many species, a larger proportional number of dominant species. But so many causes tend to obscure this result, that I am surprised that my tables show even a small majority on the side of the larger genera. I will here allude to only two causes of obscurity. Fresh-water and salt-loving plants generally have very wide ranges and are much diffused, but this seems to be connected with the nature of the stations inhabited by them, and has little or no relation to the size of the genera to which the species belong. Again, plants low in the scale of organisation are generally much more widely diffused than plants higher in the scale; and here again there is no close relation to the size of the genera. The cause of lowly-organised plants ranging widely will be discussed in our chapter on Geographical Distribution.

From looking at species as only strongly marked and well-defined varieties, I was led to anticipate that the species of the larger genera in each country would oftener present varieties, than the species of the smaller genera; for wherever many closely related species (i.e., species of the same genus) have been formed, many varieties or incipient species ought, as a general rule, to be now forming. Where many large trees grow, we expect to find saplings. Where many species of a genus have been formed through variation, circumstances have been favourable for variation; and hence we might expect that the circumstances would generally still be favourable to variation. On the other hand, if we look at each species as a special act of creation, there is no apparent reason why more varieties should occur in a group having many species, than in one having few.

To test the truth of this anticipation I have arranged the plants of twelve countries, and the coleopterous insects of two districts, into two nearly equal masses, the species of the larger genera on one side, and those of the smaller genera on the other side, and it has invariably proved to be the case that a larger proportion of the species on the side of the larger genera presented varieties, than on the side of the smaller genera. Moreover, the species of the large genera which present any varieties, invariably present a larger average number of varieties than do the species of the small genera. Both these results follow when another division is made, and when all the least genera, with from only one to four species, are altogether excluded from the tables. These facts are of plain signification on the view that species are only strongly-marked and permanent varieties; for wherever many species of the same genus have been formed, or where, if we may use the expression, the manufactory of species has been active, we ought generally to find the manufactory still in action, more especially as we have every reason to believe the process of manufacturing new species to be a slow one. And this certainly holds true if varieties be looked at as incipient species; for my tables clearly show, as a general rule, that, wherever many species of a genus have been formed, the species of that genus present a number of varieties, that is, of incipient species, beyond the average. It is not that all large genera are now varying much, and are thus increasing in the number of their species, or that no small genera are now varying and increasing; for if this had been so, it would have been fatal to my theory; inasmuch as geology plainly tells us that small genera have in the lapse of time often increased greatly in size; and that large genera have often come to their maxima, declined, and disappeared. All that we want to show is, that where many species of a genus have been formed, on an average many are still forming; and this certainly holds good.

MANY OF THE SPECIES INCLUDED WITHIN THE LARGER GENERA RESEMBLE VARIETIES IN BEING VERY CLOSELY, BUT UNEQUALLY, RELATED TO EACH OTHER, AND IN HAVING RESTRICTED RANGES

There are other relations between the species of large genera and their recorded varieties which deserve notice. We have seen that there is no infallible criterion by which to distinguish species and well-marked varieties; and when intermediate links have not been found between doubtful forms, naturalists are compelled to come to a determination by the amount of difference between them, judging by analogy whether or not the amount suffices to raise one or both to the rank of species. Hence the amount of difference is one very important criterion in settling whether two forms should be ranked as species or varieties. Now Fries has remarked in regard to plants, and Westwood in regard to insects, that in large genera the amount of difference between the species is often exceedingly small. I have endeavoured to test this numerically by averages, and, as far as my imperfect results go, they confirm the view. I have also consulted some sagacious and experienced observers, and, after deliberation, they concur in this view. In this respect, therefore, the species of the larger genera resemble varieties, more than do the species of the smaller genera. Or the case may be put in another way, and it may be said, that in the larger genera, in which a number of varieties or incipient species greater than the average are now manufacturing, many of the species already manufactured still to a certain extent resemble varieties, for they differ from each other by a less than the usual amount of difference.

Moreover, the species of the larger genera are related to each other, in the same manner as the varieties of any one species are related to each other. No naturalist pretends that all the species of a genus are equally distinct from each other; they may generally be divided into sub-genera, or sections, or lesser groups. As Fries has well remarked, little groups of species are generally clustered like satellites around other species. And what are varieties but groups of forms, unequally related to each other, and clustered round certain forms – that is, round their parent-species. Undoubtedly there is one most important point of difference between varieties and species, namely, that the amount of difference between varieties, when compared with each other or with their parent-species, is much less than that between the species of the same genus. But when we come to discuss the principle, as I call it, of Divergence of Character, we shall see how this may be explained, and how the lesser differences between varieties tend to increase into the greater differences between species.

There is one other point which is worth notice. Varieties generally have much restricted ranges. This statement is indeed scarcely more than a truism, for if a variety were found to have a wider range than that of its supposed parent-species, their denominations would be reversed. But there is reason to believe that the species which are very closely allied to other species, and in so far resemble varieties, often have much restricted ranges. For instance, Mr. H.C. Watson has marked for me in the well-sifted London catalogue of Plants (4th edition) sixty three plants which are therein ranked as species, but which he considers as so closely allied to other species as to be of doubtful value: these sixty-three reputed species range on an average over 6.9 of the provinces into which Mr. Watson has divided Great Britain. Now, in this same catalogue, fifty-three acknowledged varieties are recorded, and these range over 7.7 provinces; whereas, the species to which these varieties belong range over 14.3 provinces. So that the acknowledged varieties have very nearly the same restricted average range, as have the closely allied forms, marked for me by Mr. Watson as doubtful species, but which are almost universally ranked by British botanists as good and true species.

SUMMARY

Finally, varieties cannot be distinguished from species, – except, first, by the discovery of intermediate linking forms; and, secondly, by a certain indefinite amount of difference between them; for two forms, if differing very little, are generally ranked as varieties, notwithstanding that they cannot be closely connected; but the amount of difference considered necessary to give to any two forms the rank of species cannot be defined. In genera having more than the average number of species in any country, the species of these genera have more than the average number of varieties. In large genera the species are apt to be closely but unequally, allied together, forming little clusters round other species. Species very closely allied to other species apparently have restricted ranges. In all these respects the species of large genera present a strong analogy with varieties. And we can clearly understand these analogies, if species once existed as varieties, and thus originated; whereas, these analogies are utterly inexplicable if species are independent creations.

We have, also, seen that it is the most flourishing or dominant species of the larger genera within each class which on an average yield the greatest number of varieties; and varieties, as we shall hereafter see, tend to become converted into new and distinct species. Thus the larger genera tend to become larger; and throughout nature the forms of life which are now dominant tend to become still more dominant by leaving many modified and dominant descendants. But, by steps hereafter to be explained, the larger genera also tend to break up into smaller genera. And thus, the forms of life throughout the universe become divided into groups subordinate to groups.

Chapter III

STRUGGLE FOR EXISTENCE

Its bearing on natural selection–The term used in a wide sense–Geometrical ratio of increase–Rapid increase of naturalised animals and plants–Nature of the checks to increase–Competition universal–Effects of climate–Protection from the number of individuals–Complex relations of all animals and plants throughout nature–Struggle for life most severe between individuals and varieties of the same species: often severe between species of the same genus–The relation of organism to organism the most important of all relations.

Before entering on the subject of this chapter, I must make a few preliminary remarks to show how the struggle for existence bears on Natural Selection. It has been seen in the last chapter that amongst organic beings in a state of nature there is some individual variability: indeed I am not aware that this has ever been disputed. It is immaterial for us whether a multitude of doubtful forms be called species or sub-species or varieties; what rank, for instance, the two or three hundred doubtful forms of British plants are entitled to hold, if the existence of any well-marked varieties be admitted. But the mere existence of individual variability and of some few well-marked varieties, though necessary as the foundation for the work, helps us but little in understanding how species arise in nature. How have all those exquisite adaptations of one part of the organisation to another part, and to the conditions of life, and of one organic being to another being, been perfected? We see these beautiful co-adaptations most plainly in the woodpecker and the mistletoe; and only a little less plainly in the humblest parasite which clings to the hairs of a quadruped or feathers of a bird; in the structure of the beetle which dives through the water; in the plumed seed which is wafted by the gentlest breeze; in short, we see beautiful adaptations everywhere and in every part of the organic world.

Again, it may be asked, how is it that varieties, which I have called incipient species, become ultimately converted into good and distinct species, which in most cases obviously differ from each other far more than do the varieties of the same species? How do those groups of species, which constitute what are called distinct genera, and which differ from each other more than do the species of the same genus, arise? All these results, as we shall more fully see in the next chapter, follow from the struggle for life. Owing to this struggle, variations, however slight and from whatever cause proceeding, if they be in any degree profitable to the individuals of a species, in their infinitely complex relations to other organic beings and to their physical conditions of life, will tend to the preservation of such individuals, and will generally be inherited by the offspring. The offspring, also, will thus have a better chance of surviving, for, of the many individuals of any species which are periodically born, but a small number can survive. I have called this principle, by which each slight variation, if useful, is preserved, by the term Natural Selection, in order to mark its relation to man's power of selection. But the expression often used by Mr. Herbert Spencer, of the Survival of the Fittest, is more accurate, and is sometimes equally convenient. We have seen that man by selection

Various coat-markings of the Cat family:
1 Tiger *2* Lioness *3* Panther *4* Jaguar *5* Ocelot *6* Pampas Cat

can certainly produce great results, and can adapt organic beings to his own uses, through the accumulation of slight but useful variations, given to him by the hand of Nature. But Natural Selection, we shall hereafter see, is a power incessantly ready for action, and is as immeasurably superior to man's feeble efforts, as the works of Nature are to those of Art.

We will now discuss in a little more detail the struggle for existence. In my future work this subject will be treated, as it well deserves, at greater length. The elder De Candolle and Lyell have largely and philosophically shown that all organic beings are exposed to severe competition. In regard to plants, no one has treated this subject with more spirit and ability than W. Herbert, Dean of Manchester, evidently the result of his great horticultural knowledge. Nothing is easier than to admit in words the truth of the universal struggle for life, or more difficult – at least, I have found it so – than constantly to bear this conclusion in mind. Yet unless it be thoroughly engrained in the mind, the whole economy of nature, with every fact on distribution, rarity, abundance, extinction, and variation, will be dimly seen or quite misunderstood. We behold the face of nature bright with gladness, we often see superabundance of food; we do not see or we forget that the birds which are idly singing round us mostly live on insects or seeds, and are thus constantly destroying life; or we forget how largely these songsters, or their eggs, or their nestlings, are destroyed by birds and beasts of prey; we do not always bear in mind, that, though food may be now superabundant, it is not so at all seasons of each recurring year.

THE TERM, STRUGGLE FOR EXISTENCE, USED IN A LARGE SENSE

I should premise that I use this term in a large and metaphorical sense, including dependence of one being on another, and including (which is more important) not only the life of the individual, but success in leaving progeny. Two canine animals, in

The semi-parastic plant Mistletoe

a time of dearth, may be truly said to struggle with each other which shall get food and live. But a plant on the edge of a desert is said to struggle for life against the drought, though more properly it should be said to be dependent on the moisture. A plant which annually produces a thousand seeds, of which only one of an average comes to maturity, may be more truly said to struggle with the plants of the same and other kinds which already clothe the ground. The mistletoe is dependent on the apple and a few other trees, but can only in a far-fetched sense be said to struggle with these trees, for, if too many of these parasites grow on the same tree, it languishes and dies. But several seedling mistletoes, growing close together on the same branch, may more truly be said to struggle with each other. As the mistletoe is disseminated by birds, its existence depends on them; and it may metaphorically be said to struggle with other fruit-bearing plants, in tempting the birds to devour and thus disseminate its seeds. In these several senses, which pass into each other, I use for convenience sake the general term of Struggle for Existence.

GEOMETRICAL RATIO OF INCREASE

A struggle for existence inevitably follows from the high rate at which all organic beings tend to increase. Every being, which during its natural lifetime produces several eggs or seeds, must suffer destruction during some period of its life, and during some season or occasional year, otherwise, on the principle of geometrical increase, its numbers would quickly become so inordinately great that no country could support the product. Hence, as more individuals are produced than can possibly survive, there must in every case be a struggle for existence, either one individual with another of the same species, or with the individuals of distinct species, or with the physical conditions of life. It is the doctrine of Malthus applied with manifold force to the whole animal and vegetable kingdoms; for in this case there can be no artificial increase of food, and no prudential restraint from marriage. Although some species may be now increasing, more or less rapidly, in numbers, all cannot do so, for the world would not hold them.

There is no exception to the rule that every organic being naturally increases at so high a rate, that, if not destroyed, the earth would soon be covered by the progeny of a single pair. Even slow-breeding man has doubled in twenty-five years, and at this rate, in less than a thousand years, there would literally not be standing room for his progeny. Linnaeus has calculated that if an annual plant produced only two seeds – and there is no plant so unproductive as this – and their seedlings next year produced two, and so on, then in twenty years there would be a million plants. The elephant is reckoned the slowest breeder of all known animals, and I have taken some pains to estimate its probable minimum rate of natural

increase; it will be safest to assume that it begins breeding when thirty years old, and goes on breeding till ninety years old, bringing forth six young in the interval, and surviving till one hundred years old; if this be so, after a period of from 740 to 750 years there would be nearly nineteen million elephants alive descended from the first pair.

But we have better evidence on this subject than mere theoretical calculations, namely, the numerous recorded cases of the astonishingly rapid increase of various animals in a state of nature, when circumstances have been favourable to them during two or three following seasons. Still more striking is the evidence from our domestic animals of many kinds which have run wild in several parts of the world; if the statements of the rate of increase of slow-breeding cattle and horses in South America, and latterly in Australia, had not been well authenticated, they would have been incredible. So it is with plants; cases could be given of introduced plants which have become common throughout whole islands in a period of less than ten years. Several of the plants, such as the cardoon and a tall thistle, which are now the commonest over the wide plains of La Plata, clothing square leagues of surface almost to the exclusion of every other plant, have been introduced from Europe; and there are plants which now range in India, as I hear from Dr. Falconer, from Cape Comorin to the Himalaya, which have been imported from America since its discovery. In such cases, and endless others could be given, no one supposes that the fertility of the animals or plants has been suddenly and temporarily increased in any sensible degree. The obvious explanation is that the conditions of life have been highly favourable, and that there has consequently been less destruction of the old and young and that nearly all the young have been enabled to breed. Their geometrical ratio of increase, the result of which never fails to be surprising, simply explains their extraordinarily rapid increase and wide diffusion in their new homes.

In a state of nature almost every full-grown plant annually produces seed, and amongst animals there are very few which do not annually pair. Hence we may confidently assert that all plants and animals are tending to increase at a geometrical ratio – that all would rapidly stock every station in which they could anyhow exist, and that this geometrical tendency to increase must be checked by destruction at some period of life. Our familiarity with the larger domestic animals tends, I think, to mislead us; we see no great destruction falling on them, and we do not keep in mind that thousands are annually slaughtered for food, and that in a state of nature an equal number would have somehow to be disposed of.

The only difference between organisms which annually produce eggs or seeds by the thousand, and those which produce extremely few, is, that the slow breeders would require a few more years to people, under favourable conditions, a whole district, let it be ever so large. The condor lays a couple of eggs and the ostrich a score, and yet in the same country the condor may be the more numerous of the two. The Fulmar petrel lays but one egg, yet it is believed to be the most numerous bird in the world. One fly deposits hundreds of eggs, and another, like the hippobosca, a single one. But this difference does not determine how many individuals of the two species can be supported in a district. A large number of eggs is of some importance to those species which depend on a fluctuating amount of food, for it allows them rapidly to increase in number. But the real importance of a large number of eggs or seeds is to make up for much destruction at some period of life; and this period in the great majority of cases is an early one. If an animal can in any way protect its own eggs or young, a small number may be produced, and yet the average stock be fully kept up; but if many eggs or young are destroyed, many must be produced or the species will become extinct. It would suffice to keep up the full number of a tree,

which lived on an average for a thousand years, if a single seed were produced once in a thousand years, supposing that this seed were never destroyed and could be ensured to germinate in a fitting place; so that, in all cases, the average number of any animal or plant depends only indirectly on the number of its eggs or seeds.

In looking at Nature, it is most necessary to keep the foregoing considerations always in mind – never to forget that every single organic being may be said to be striving to the utmost to increase in numbers; that each lives by a struggle at some period of its life; that heavy destruction inevitably falls either on the young or old during each generation or at recurrent intervals. Lighten any check, mitigate the destruction ever so little, and the number of the species will almost instantaneously increase to any amount.

NATURE OF THE CHECKS TO INCREASE

The causes which check the natural tendency of each species to increase are most obscure. Look at the most vigorous species; by as much as it swarms in numbers, by so much will it tend to increase still further. We know not exactly what the checks are even in a single instance. Nor will this surprise any one who reflects how ignorant we are on this head, even in regard to mankind, although so incomparably better known than any other animal. This subject of the checks to increase has been ably treated by several authors, and I hope in a future work to discuss it at considerable length, more especially in regard to the feral animals of South America. Here I will make only a few remarks, just to recall to the reader's mind some of the chief points. Eggs or very young animals seem generally to suffer most, but this is not invariably the case. With plants there is a vast destruction of seeds, but from some observations which I have made, it appears that the seedlings suffer most from germinating in ground already thickly stocked with other plants. Seedlings, also, are destroyed in vast numbers by various enemies; for instance, on a piece of ground three feet long

A pair of Goldcrests, or golden-crowned wrens, whose nest holds several eggs and two wide, dug and cleared, and where there could be no choking from other plants, I marked all the seedlings of our native weeds as they came up, and out of 357 no less than 295 were destroyed, chiefly by slugs and insects. If turf which has long been mown (and the case would be the same with turf closely browsed by quadrupeds) be let to grow, the more vigorous plants gradually kill the less vigorous, though fully grown plants; thus out of twenty species grown on a little plot of mown turf (three feet by four) nine species perished, from the other species being allowed to grow up freely.

The amount of food for each species, of course, gives the extreme limit to which each can increase; but very frequently it is not the obtaining food, but the serving as prey to other animals, which determines the average number of a species. Thus, there seems to be little doubt that the stock of partridges, grouse, and hares on any large estate depends chiefly on the destruction of vermin. If not one head of game were shot during the next twenty years in England, and, at the same time, if no vermin were destroyed, there would, in all probability, be less game than at present, although hundreds of thousands of game animals are now annually shot. On the other hand, in some cases, as with the elephant, none are destroyed by beasts of prey; for even the tiger in India most rarely dares to attack a young elephant protected by its dam.

Climate plays an important part in determining the average numbers of a species, and periodical seasons of extreme cold or drought seem to be the most effective of all checks. I estimated (chiefly from the greatly reduced numbers of nests in the spring) that the winter of 1854–5 destroyed four-fifths of the birds in my own grounds; and this is a tremendous destruction, when we remember that ten per cent. is an extraordinarily severe mortality from epidemics with man. The action of climate seems at first sight to be quite independent of the struggle for existence; but in so far as climate chiefly acts in reducing food, it brings on the most severe struggle between the individuals, whether of the same or of distinct species, which subsist on the same kind of food. Even when climate, for instance, extreme cold, acts directly, it will be the least vigorous individuals, or those which have got least food through the advancing winter, which will suffer the most. When we travel from south to north, or from a damp region to a dry, we invariably see some species gradually getting rarer and rarer, and finally disappearing; and the change of climate being conspicuous, we are tempted to attribute the whole effect to its direct action. But this is a false view; we forget that each species, even where it most abounds, is constantly suffering enormous destruction at some period of its life, from enemies or from competitors for the same place and food; and if these enemies or competitors be in the least degree favoured by any slight change of climate, they will increase in numbers; and as each area is already fully stocked with inhabitants, the other species must decrease. When we travel southward and see a species decreasing in numbers, we may feel sure that the cause lies quite as much in other species being favoured, as in this one being hurt. So it is when we travel northward, but in a somewhat lesser degree, for the number of species of all kinds, and therefore of competitors, decreases northward; hence in going northward, or in ascending a mountain, we far oftener meet with stunted forms, due to the *directly* injurious action of climate, than we do in proceeding southward or in descending a mountain. When we reach the arctic regions, or snow-capped summits, or absolute deserts, the struggle for life is almost exclusively with the elements.

That climate acts in main part indirectly by favouring other species we clearly see in the prodigious number of plants which in our gardens can perfectly well endure our climate, but which never become naturalised, for they cannot compete with our native plants nor resist destruction by our native animals.

When a species, owing to highly favourable circumstances, increases inordinately in numbers in a small tract, epidemics – at least, this seems generally to occur with our game animals – often ensue; and here we have a limiting check independent of the struggle for life. But even some of these so-called epidemics appear to be due to parasitic worms, which have from some cause, possibly in part through facility of diffusion amongst the crowded animals, been disproportionally favoured: and here comes in a sort of struggle between the parasite and its prey.

On the other hand, in many cases, a large stock of individuals of the same species, relatively to the numbers of its enemies, is absolutely necessary for its preservation. Thus we can easily raise plenty of corn and rape-seed, etc., in our fields, because the seeds are in great excess compared with the number of birds which feed on them; nor can the birds, though having a superabundance of food at this one season, increase in number proportionally to the supply of seed, as their numbers are checked during the winter; but any one who has tried knows how troublesome it is to get seed from a few wheat or other such plants in a garden; I have in this case lost every single seed. This view of the necessity of a large stock of the same species for its preservation, explains, I believe, some singular facts in nature such as that of very rare plants being sometimes extremely abundant, in the few spots where they do exist; and that of some social plants being social, that is abounding in individuals, even on the extreme verge of their range. For in such cases, we may believe, that a plant could exist only where the conditions of its life were so favourable that many could exist together, and thus save the species from utter destruction. I should add that the good effects of intercrossing, and the ill effects of close interbreeding, no doubt come into play in many of these cases; but I will not here enlarge on this subject.

COMPLEX RELATIONS OF ALL ANIMALS AND PLANTS TO EACH OTHER IN THE STRUGGLE FOR EXISTENCE

Many cases are on record showing how complex and unexpected are the checks and relations between organic beings, which have to struggle together in the same country. I will give only a single instance, which, though a simple one, interested me. In Staffordshire, on the estate of a relation, where I had ample means of investigation, there was a large and extremely barren heath, which had never been touched by the hand of man; but several hundred acres of exactly the same nature had been enclosed twenty-five years previously and planted with Scotch fir. The change in the native vegetation of the planted part of the heath was most remarkable, more than is generally seen in passing from one quite different soil to another: not only the proportional numbers of the heath-plants were wholly changed, but twelve species of plants (not counting grasses and carices) flourished in the plantations, which could not be found on the heath. The effect on the insects must have been still greater, for six insectivorous birds were very common in the plantations, which were not to be seen on the heath; and the heath was frequented by two or three distinct insectivorous birds. Here we see how potent has been the effect of the introduction of a single tree, nothing whatever else having been done, with the exception of the land having been enclosed, so that cattle could not enter. But how important an element enclosure is, I plainly saw near Farnham, in Surrey. Here there are extensive heaths, with a few clumps of old Scotch firs on the distant hill-tops: within the last ten years large spaces have been enclosed, and self-sown firs are now springing up in multitudes, so close together that all cannot live. When I ascertained that these young trees had not been sown or planted I was so much surprised at their numbers that I went to several points of view, whence I could examine hundreds of acres of the unenclosed heath, and literally I could not see a single Scotch fir, except the old planted clumps. But on looking closely between the stems of the heath, I found a multitude of seedlings and little trees, which had been perpetually browsed down by the cattle. In one square yard, at a point some hundred yards distant from one of the old clumps, I counted thirty-two little trees; and one of them, with twenty-six rings of growth, had, during many years tried to raise its head above the stems of the heath, and had failed. No wonder that, as soon as the land was enclosed, it became thickly clothed with vigorously growing young firs. Yet the heath was so extremely barren and so extensive that no one would ever have imagined that cattle would have so closely and effectually searched it for food.

Here we see that cattle absolutely determine the existence of the Scotch fir; but in several parts of the world insects determine the existence of cattle. Perhaps Paraguay offers the most curious instance of this; for here neither cattle nor horses nor dogs have ever run wild, though they swarm southward and northward in a feral state; and Azara and Rengger have shown that this is caused by the greater number in Paraguay of a certain fly, which lays its eggs in the navels of these animals when first born. The increase of these flies, numerous as they are, must be habitually checked by some means, probably by other parasitic insects. Hence, if certain insectivorous birds were to decrease in Paraguay, the parasitic insects would probably increase; and this would lessen the number of the navel-frequenting flies – then cattle and horses would become feral, and this would certainly greatly alter (as indeed I have observed in parts of South America) the vegetation: this again would largely affect the insects; and this, as we have just seen in Staffordshire, the insectivorous birds, and so onwards in ever-increasing circles of complexity. Not that under nature the relations will ever be as simple as this. Battle within battle must be continually recurring with varying success; and yet in the long-run the forces are so nicely balanced that the face of nature remains for long periods of time uniform, though assuredly the merest trifle

would give the victory to one organic being over another. Nevertheless, so profound is our ignorance, and so high our presumption, that we marvel when we hear of the extinction of an organic being; and as we do not see the cause, we invoke cataclysms to desolate the world, or invent laws on the duration of the forms of life!

I am tempted to give one more instance showing how plants and animals, remote in the scale of nature, are bound together by a web of complex relations. I shall hereafter have occasion to show that the exotic Lobelia fulgens is never visited in my garden by insects, and consequently, from its peculiar structure, never sets a seed. Nearly all our orchidaceous plants absolutely require the visits of insects to remove their pollen-masses and thus to fertilise them. I find from experiments that humble-bees are almost indispensable to the fertilisation of the heartsease (Viola tricolor), for other bees do not visit this flower. I have also found that the visits of bees are necessary for the fertilisation of some kinds of clover; for instance twenty heads of Dutch clover (Trifolium repens) yielded 2,290 seeds, but twenty other heads, protected from bees, produced not one. Again, 100 heads of red clover (T. pratense) produced 2,700 seeds, but the same number of protected heads produced not a single seed. Humble-bees alone visit red clover, as other bees cannot reach the nectar. It has been suggested that moths may fertilise the clovers; but I doubt whether they could do so in the case of the red clover, from their weight not being sufficient to depress the wing petals. Hence we may infer as highly probable that, if the whole genus of humble-bees became extinct or very rare in England, the heartsease and red clover would become very rare, or wholly disappear. The number of humble-bees in any district depends in a great measure upon the number of field mice, which destroy their combs and nests; and Colonel Newman, who has long attended to the habits of humble-bees, believes that 'more than two-thirds of them are thus destroyed all over England.'

Bees cut holes and extract nectar from the base of broad bean flowers: *left* wood bee *right* humble-bee

Now the number of mice is largely dependent, as every one knows, on the number of cats; and Colonel Newman says, 'Near villages and small towns I have found the nests of humble-bees more numerous than elsewhere, which I attribute to the number of cats that destroy the mice.' Hence it is quite credible that the presence of a feline animal in large numbers in a district might determine, through the intervention first of mice and then of bees, the frequency of certain flowers in that district!

In the case of every species, many different checks, acting at different periods of life, and during different seasons or years, probably come into play; some one check or some few being generally the most potent, but all will concur in determining the average number, or even the existence of the species. In some cases it can be shown that widely different checks act on the same species in different districts. When we look at the plants and bushes clothing an entangled bank, we are tempted to attribute their proportional numbers and kinds to what we call chance. But how false a view is this! Every one has heard that when an American forest is cut down, a very different vegetation springs up; but it has been observed that ancient Indian ruins in the southern United States, which must formerly have been cleared of trees, now display the same beautiful

diversity and proportion of kinds as in the surrounding virgin forests. What a struggle must have gone on during long centuries between the several kinds of trees, each annually scattering its seeds by the thousand; what war between insect and insect – between insects, snails, and other animals with birds and beasts of prey – all striving to increase, all feeding on each other, or on the trees, their seeds and seedlings, or on the other plants which first clothed the ground and thus checked the growth of the trees! Throw up a handful of feathers, and all fall to the ground according to definite laws; but how simple is the problem where each shall fall compared to that of the action and reaction of the innumerable plants and animals which have determined, in the course of centuries, the proportional numbers and kinds of trees now growing on the old Indian ruins!

The dependency of one organic being on another, as of a parasite on its prey, lies generally between beings remote in the scale of nature. This is likewise sometimes the case with those which may strictly be said to struggle with each other for existence, as in the case of locusts and grass-feeding quadrupeds. But the struggle will almost invariably be most severe between the individuals of the same species, for they frequent the same districts, require the same food, and are exposed to the same dangers. In the case of varieties of the same species, the struggle will generally be almost equally severe, and we sometimes see the contest soon decided: for instance, if several varieties of wheat be sown together, and the mixed seed be resown, some of the varieties which best suit the soil or climate, or are naturally the most fertile, will beat the others and so yield more seed, and will consequently in a few years supplant the other varieties. To keep up a mixed stock of even such extremely close varieties as the variously-coloured sweet peas, they must be each year harvested separately, and the seed then mixed in due proportion, otherwise the weaker kinds will steadily decrease in number and disappear. So again with the varieties of

The Swamp or Water Rat, published in *The Zoology of the Voyage of H.M.S. Beagle*

sheep; it has been asserted that certain mountain varieties will starve out other mountain varieties, so that they cannot be kept together. The same result has followed from keeping together different varieties of the medicinal leech. It may even be doubted whether the varieties of any of our domestic plants or animals have so exactly the same strength, habits, and constitution, that the original proportions of a mixed stock (crossing being prevented) could be kept up for half-a-dozen generations, if they were allowed to struggle together, in the same manner as beings in a state of nature, and if the seed or young were not annually preserved in due proportion.

STRUGGLE FOR LIFE MOST SEVERE BETWEEN INDIVIDUALS AND VARIETIES OF THE SAME SPECIES

As the species of the same genus usually have, though by no means invariably, much similarity in habits and constitution, and always in structure, the struggle will generally be more severe between them, if they come into competition with each other, than between the species of distinct genera. We see this in the recent extension over parts of the United States of one species of swallow having caused the decrease of another species. The recent increase of the missel thrush in parts of Scotland has caused the decrease of the song thrush. How frequently we hear of one species of rat taking the place of another species under the most different climates! In Russia the small Asiatic cockroach has everywhere driven before it its great congener. In Australia the imported hive bee is rapidly exterminating the small, stingless native bee. One species of charlock has been known to supplant another species; and so in other cases. We can dimly see why the competition should be most severe between allied forms, which fill nearly the same place in the economy of nature; but probably in no one case could we precisely say why one species has been victorious over another in the great battle of life.

A corollary of the highest importance may be deduced from the foregoing remarks, namely, that the structure of every organic being is related, in the most essential yet often hidden manner, to that of all other organic beings, with which it comes into competition for food or residence, or from which it has to escape, or on which it preys. This is obvious in the structure of the teeth and talons of the tiger; and in that of the legs and claws of the parasite which clings to the hair on the tiger's body. But in the beautifully plumed seed of the dandelion, and in the flattened and fringed legs of the water beetle, the relation seems at first confined to the elements of air and water. Yet the advantage of the plumed seeds no doubt stands in the closest relation to the land being already thickly clothed with other plants; so that the seeds may be widely distributed and fall on unoccupied ground. In the water beetle, the structure of its legs, so well adapted for diving, allows it to compete with other aquatic insects, to hunt for its own prey, and to escape serving as prey to other animals.

The store of nutriment laid up within the seeds of many plants seems at first sight to have no sort of relation to other plants. But from the strong growth of young plants produced from such seeds, as peas and beans, when sown in the midst of long grass, it may be suspected that the chief use of the nutriment in the seed is to favour the growth of the seedlings, whilst struggling with other plants growing vigorously all around.

Look at a plant in the midst of its range! Why does it not double or quadruple its numbers? We know that it can perfectly well withstand a little more heat or cold, dampness or dryness, for elsewhere it ranges into slightly hotter or colder, damper or drier districts. In this case we can clearly see that if we wish in imagination to give the plant the power of increasing in numbers, we should have to give it some advantage over its competitors, or over the animals which prey on it. On the confines of its geographical range, a change of constitution with respect to climate would clearly be an advantage to our plant; but we have reason to believe that only a few plants or animals range so far, that they are destroyed exclusively by the rigour of the climate. Not until we reach the extreme confines of life, in the Arctic regions or on the borders of an utter desert, will competition cease. The land may be extremely cold or dry, yet there will be competition between some few species, or between the individuals of the same species, for the warmest or dampest spots.

Hence we can see that when a plant or animal is placed in a new country, amongst new competitors, the conditions of its life will generally be changed in an essential manner, although the climate may be exactly the same as in its former home. If its average numbers are to increase in its new home, we should have to modify it in a different way to what we should have had to do in its native country; for we should have to give it some advantage over a different set of competitors or enemies.

It is good thus to try in imagination to give any one species an advantage over another. Probably in no single instance should we know what to do. This ought to convince us of our ignorance on the mutual relations of all organic beings; a conviction as necessary, as it is difficult to acquire. All that we can do is to keep steadily in mind that each organic being is striving to increase in a geometrical ratio; that each, at some period of its life, during some season of the year, during each generation, or at intervals, has to struggle for life and to suffer great destruction. When we reflect on this struggle we may console ourselves with the full belief that the war of nature is not incessant, that no fear is felt, that death is generally prompt, and that the vigorous, the healthy, and the happy survive and multiply.

Chapter IV

Natural Selection; or the Survival of the Fittest

Natural Selection–its power compared with man's selection–its power on characters of trifling importance–its power at all ages and on both sexes–Sexual selection–On the generality of intercrosses between individuals of the same species–Circumstances favourable and unfavourable to the results of Natural Selection, namely, intercrossing, isolation, number of individuals–Slow action–Extinction caused by Natural Selection–Divergence of Character, related to the diversity of inhabitants of any small area and to naturalisation–Action of Natural Selection, through Divergence of Character and Extinction, on the descendants from a common parent–Explains the Grouping of all organic beings–Advance in organisation–Low forms preserved–Convergence of Character–Indefinite multiplication of species–Summary.

How will the struggle for existence, briefly discussed in the last chapter, act in regard to variation? Can the principle of selection, which we have seen is so potent in the hands of man, apply under nature? I think we shall see that it can act most efficiently. Let the endless number of slight variations and individual differences occurring in our domestic productions, and, in a lesser degree, in those under nature, be borne in mind; as well as the strength of the hereditary tendency. Under domestication, it may be truly said that the whole organisation becomes in some degree plastic. But the variability, which we almost universally meet with in our domestic productions, is not directly produced, as Hooker and Asa Gray have well remarked, by man; he can neither originate varieties, nor prevent their occurrence; he can only preserve and accumulate such as do occur. Unintentionally he exposes organic beings to new and changing conditions of life, and variability ensues; but similar changes of conditions might and do occur under nature. Let it also be borne in mind how infinitely complex and close-fitting are the mutual relations of all organic beings to each other and to their physical conditions of life; and consequently what infinitely varied diversities of structure might be of use to each being under changing conditions of life. Can it, then, be thought improbable, seeing that

variations useful to man have undoubtedly occurred, that other variations useful in some way to each being in the great and complex battle of life, should occur in the course of many successive generations? If such do occur, can we doubt (remembering that many more individuals are born than can possibly survive) that individuals having any advantage, however slight, over others, would have the best chance of surviving and of procreating their kind? On the other hand, we may feel sure that any variation in the least degree injurious would be rigidly destroyed. This preservation of favourable individual differences and variations, and the destruction of those which are injurious, I have called Natural Selection, or the Survival of the Fittest. Variations neither useful nor injurious would not be affected by natural selection, and would be left either a fluctuating element, as perhaps we see in certain polymorphic species, or would ultimately become fixed, owing to the nature of the organism and the nature of the conditions.

Several writers have misapprehended or objected to the term Natural Selection. Some have even imagined that natural selection induces variability, whereas it implies only the preservation of such variations as arise and are beneficial to the being under its conditions of life. No one objects to agriculturists speaking of the potent effects of man's selection; and in this case the individual differences given by nature, which man for some object selects, must of necessity first occur. Others have objected that the term selection implies conscious choice in the animals which become modified; and it has even been urged that, as plants have no volition, natural selection is not applicable to them! In the literal sense of the word, no doubt, natural selection is a false term; but who ever objected to chemists speaking of the elective affinities of the various elements? – and yet an acid cannot strictly be said to elect the base with which it in preference combines. It has been said that I speak of natural selection as an active power or Deity; but who objects to an author speaking of the attraction of gravity as ruling the movements of the planets? Every one knows what is meant and is implied by such metaphorical expressions; and they are almost necessary for brevity. So again it is difficult to avoid personifying the word Nature; but I mean by Nature, only the aggregate action and product of many natural laws, and by laws the sequence of events as ascertained by us. With a little familiarity such superficial objections will be forgotten.

We shall best understand the probable course of natural selection by taking the case of a country undergoing some slight physical change, for instance, of climate. The proportional numbers of its inhabitants will almost immediately undergo a change, and some species will probably become extinct. We may conclude, from what we have seen of the intimate and complex manner in which the inhabitants of each country are bound together, that any change in the numerical proportions of the inhabitants, independently of the change of climate itself, would seriously affect the others. If the country were open on its borders, new forms would certainly immigrate, and this would likewise seriously disturb the relations of some of the former inhabitants. Let it be remembered how powerful the influence of a single introduced tree or mammal has been shown to be. But in the case of an island, or of a country

partly surrounded by barriers, into which new and better adapted forms could not freely enter, we should then have places in the economy of nature which would assuredly be better filled up if some of the original inhabitants were in some manner modified; for, had the area been open to immigration, these same places would have been seized on by intruders. In such cases, slight modifications, which in any way favoured the individuals of any species, by better adapting them to their altered conditions, would tend to be preserved; and natural selection would have free scope for the work of improvement.

We have good reason to believe, as shown in the first chapter, that changes in the conditions of life give a tendency to increased variability; and in the foregoing cases the conditions have changed, and this would manifestly be favourable to natural selection, by affording a better chance of the occurrence of profitable variations. Unless such occur, natural selection can do nothing. Under the term of 'variations,' it must never be forgotten that mere individual differences are included. As man can produce a great result with his domestic animals and plants by adding up in any given direction individual differences, so could natural selection, but far more easily from having incomparably longer time for action. Nor do I believe that any great physical change, as of climate, or any unusual degree of isolation, to check immigration, is necessary in order that new and unoccupied places should be left for natural selection to fill up by improving some of the varying inhabitants. For as all the inhabitants of each country are struggling together with nicely balanced forces, extremely slight modifications in the structure or habits of one species would often give it an advantage over others; and still further modifications of the same kind would often still further increase the advantage, as long as the species continued under the same conditions of life and profited by similar means of subsistence and defence. No country can be named in which all the native inhabitants are now so perfectly adapted to each other and to the physical conditions under which they live, that none of them could be still better adapted or improved; for in all countries, the natives have been so far conquered by naturalised productions that they have allowed some foreigners to take firm possession of the land. And as foreigners have thus in every country beaten some of the natives, we may safely conclude that the natives might have been modified with advantage, so as to have better resisted the intruders.

As man can produce, and certainly has produced, a great result by his methodical and unconscious means of selection, what may not natural selection effect? Man can act only on external and visible characters: Nature, if I may be allowed to personify the natural preservation or survival of the fittest, cares nothing for appearances, except in so far as they are useful to any being. She can act on every internal organ, on every shade of constitutional difference, on the whole machinery of life. Man selects only for his own good; Nature only for that of the being which she tends. Every selected character is fully exercised by her, as is implied by the fact of their selection. Man keeps the natives of many climates in the same country. He seldom exercises each selected character in some peculiar and fitting manner; he feeds a long and a short-beaked pigeon on the same food; he does not exercise a long-backed or long-legged quadruped in any peculiar manner; he exposes sheep with long and short wool to the same climate; does not allow the most vigorous males to struggle for the females; he does not rigidly destroy all inferior animals, but protects during each varying season, as far as lies in his power, all his productions. He often begins his selection by some half-monstrous form, or at least by some modification prominent enough to catch the eye or to be plainly useful to him. Under nature, the slightest differences of structure or constitution may well turn the nicely balanced scale in the struggle for life, and so be preserved. How fleeting are the wishes and efforts of man!

how short his time! and consequently how poor will be his results, compared with those accumulated by Nature during whole geological periods? Can we wonder, then, that Nature's productions should be far 'truer' in character than man's productions; that they should be infinitely better adapted to the most complex conditions of life, and should plainly bear the stamp of far higher workmanship?

It may metaphorically be said that natural selection is daily and hourly scrutinising, throughout the world, the slightest variations; rejecting those that are bad, preserving and adding up all that are good; silently and insensibly working, *whenever and wherever opportunity offers*, at the improvement of each organic being in relation to its organic and inorganic conditions of life. We see nothing of these slow changes in progress, until the hand of time has marked the long lapse of ages, and then so imperfect is our view into long-past geological ages that we see only that the forms of life are now different from what they formerly were.

In order that any great amount of modification should be effected in a species, a variety, when once formed must again, perhaps after a long interval of time, vary or present individual differences of the same favourable nature as before; and these must again be preserved, and so onward, step by step. Seeing that individual differences of the same kind perpetually recur, this can hardly be considered as an unwarrantable assumption. But whether it is true, we can judge only by seeing how far the hypothesis accords with and explains the general phenomena of nature. On the other hand, the ordinary belief that the amount of possible variation is a strictly limited quantity, is likewise a simple assumption.

Although natural selection can act only through and for the good of each being, yet characters and structures, which we are apt to consider as of very trifling importance, may thus be acted on. When we see leaf-eating insects green, and bark-feeders mottled-grey; the alpine ptarmigan white in winter, the red grouse the colour of heather, we must believe that these tints are of service to these birds and insects in preserving them from danger. Grouse, if not destroyed at some period of their lives, would increase in countless numbers; they are known to suffer largely from birds of prey; and hawks are guided by eyesight to their prey – so much so, that on parts of the continent persons are warned not to keep white pigeons, as being the most liable to destruction. Hence natural selection might be effective in giving the proper colour to each kind of grouse, and in keeping that colour, when once acquired, true and constant. Nor ought we to think that the occasional destruction of an animal of any particular colour would produce little effect; we should remember how essential it is in a flock of white sheep to destroy a lamb with the faintest trace of black. We have seen how the colour of hogs, which feed on the 'paint root' in Virginia, determines whether they shall live or die. In plants, the down on the fruit and the colour of the flesh are considered by botanists as characters of the most trifling importance; yet we hear from an excellent horticulturist, Downing, that in the United States smooth-skinned fruits suffer far more from a beetle, a Curculio, than those with down; that purple plums suffer far more from a certain disease than yellow plums; whereas another disease attacks yellow-fleshed peaches far more than those with other coloured flesh. If, with all the aids of art, these slight differences make a great difference in cultivating the several varieties, assuredly, in a state of nature, where the trees would have to struggle with other trees and with a host of enemies, such differences would effectually settle which variety, whether a smooth or downy, a yellow or a purple fleshed fruit, should succeed.

In looking at many small points of difference between species, which, as far as our ignorance permits us to judge, seem quite unimportant, we must not forget that climate, food, etc., have no

The Silkworm Moth at different stages of its life:
A caterpillar *B* female moth *C* cocoon

doubt produced some direct effect. It is also necessary to bear in mind that, owing to the law of correlation, when one part varies and the variations are accumulated through natural selection, other modifications, often of the most unexpected nature, will ensue.

As we see that those variations which, under domestication, appear at any particular period of life, tend to reappear in the offspring at the same period; for instance, in the shape, size and flavour of the seeds of the many varieties of our culinary and agricultural plants; in the caterpillar and cocoon stages of the varieties of the silkworm; in the eggs of poultry, and in the colour of the down of their chickens; in the horns of our sheep and cattle when nearly adult; so in a state of nature natural selection will be enabled to act on and modify organic beings at any age, by the accumulation of variations profitable at that age, and by their inheritance at a corresponding age. If it profit a plant to have its seeds more and more widely disseminated by the wind, I can see no greater difficulty in this being effected through natural selection, than in the cotton planter increasing and improving by selection the down in the pods on his cotton trees. Natural selection may modify and adapt the larva of an insect to a score of contingencies, wholly different from those which concern the mature insect; and these modifications may affect, through correlation, the structure of the adult. So, conversely, modifications in the adult may affect the structure of the larva; but in all cases natural selection will ensure that they shall not be injurious: for if they were so, the species would become extinct.

Natural selection will modify the structure of the young in relation to the parent and of the parent in relation to the young. In social animals it will adapt the structure of each individual for the benefit of the whole community; if the community profits by the selected change. What natural selection cannot do, is to modify the structure of one species, without giving it any advantage, for the good of another species; and though statements to this effect may be found in works of natural history, I cannot find one case which will bear investigation. A structure used only once in an animal's life, if of high importance to it, might be modified to any extent by natural selection; for instance, the great jaws possessed by certain insects, used exclusively for opening the cocoon – or the hard tip to the beak of unhatched birds, used for breaking the eggs. It has been asserted, that of the best short-beaked tumbler pigeons a greater number perish in the egg than are able to get out of it; so that fanciers assist in the act of

hatching. Now, if nature had to make the beak of a full-grown pigeon very short for the bird's own advantage, the process of modification would be very slow, and there would be simultaneously the most rigorous selection of all the young birds within the egg, which had the most powerful and hardest beaks, for all with weak beaks would inevitably perish: or, more delicate and more easily broken shells might be selected, the thickness of the shell being known to vary like every other structure.

It may be well here to remark that with all beings there must be much fortuitous destruction, which can have little or no influence on the course of natural selection. For instance, a vast number of eggs or seeds are annually devoured, and these could be modified through natural selection only if they varied in some manner which protected them from their enemies. Yet many of these eggs or seeds would perhaps, if not destroyed, have yielded individuals better adapted to their conditions of life than any of those which happened to survive. So again a vast number of mature animals and plants, whether or not they be the best adapted to their conditions, must be annually destroyed by accidental causes, which would not be in the least degree mitigated by certain changes of structure or constitution which would in other ways be beneficial to the species. But let the destruction of the adults be ever so heavy, if the number which can exist in any district be not wholly kept down by such causes, – or again let the destruction of eggs or seeds be so great that only a hundredth or a thousandth part are developed, – yet of those which do survive, the best adapted individuals, supposing that there is any variability in a favourable direction, will tend to propagate their kind in larger numbers than the less well adapted. If the numbers be wholly kept down by the causes just indicated, as will often have been the case, natural selection will be powerless in certain beneficial directions; but this is no valid objection to its efficiency at other times and in other ways; for we are far from having any reason to suppose that many species ever undergo modification and improvement at the same time in the same area.

SEXUAL SELECTION

Inasmuch as peculiarities often appear under domestication in one sex and become hereditarily attached to that sex, so no doubt it will be under nature. Thus it is rendered possible for the two sexes to be modified through natural selection in relation to different habits of life, as is sometimes the case; or for one sex to be modified in relation to the other sex, as commonly occurs. This leads me to say a few words on what I have called Sexual Selection. This form of selection depends, not on a struggle for existence in relation to other organic beings or to external conditions, but on a struggle between the individuals of one sex, generally the males, for the possession of the other sex. The result is not death to the unsuccessful competitor, but few or no offspring. Sexual selection is, therefore, less rigorous than natural selection. Generally, the most vigorous males, those which are best fitted for their places in nature, will leave most progeny. But in many cases victory depends not so much on general vigour, but on having special weapons, confined to the male sex. A hornless stag or spurless cock would have a poor chance of leaving numerous offspring. Sexual selection, by always allowing the victor to breed, might surely give indomitable courage, length of spur, and strength to the wing to strike in the spurred leg, in nearly the same manner as does the brutal cockfighter by the careful selection of his best cocks. How low in the scale of nature the law of battle descends I know not; male alligators have been described as fighting, bellowing, and whirling round, like Indians in a war dance, for the possession of the females; male salmons have been observed fighting all day long; male stag beetles sometimes bear wounds from the huge mandibles of other males; the males of certain hymenopterous insects have been frequently seen by that inimitable observer M. Fabre, fighting for a particular female who sits by, an apparently unconcerned beholder of the struggle, and then retires with the conqueror. The war is, perhaps, severest between the males of polygamous animals, and these seem oftenest provided with special weapons. The males of carnivorous animals are already well armed; though to them and to others, special means of defence may be given through means of sexual selection, as the mane of the lion, and the hooked jaw to the male salmon; for the shield may be as important for victory as the sword or spear.

Amongst birds, the contest is often of a more peaceful character. All those who have attended to the subject, believe that there is the severest rivalry between the males of many species to attract, by singing, the females. The rock thrush of Guiana, birds of paradise, and some others, congregate, and successive males display with the most elaborate care, and show off in the best manner, their gorgeous plumage; they likewise perform strange antics before the females, which, standing by as spectators, at last choose the most attractive partner. Those who have closely attended to birds in confinement well know that they often take individual preferences and dislikes: thus Sir R. Heron has described how a pied peacock was eminently attractive to all his hen birds. I cannot here enter on the necessary details; but if man can in a short time give beauty and an elegant carriage to his bantams, according to his standard of beauty, I can see no good reason to doubt that female birds, by selecting, during thousands of generations, the most melodious or beautiful males, according to their standard of beauty, might produce a marked effect. Some well-known laws, with respect to the plumage of male and female birds, in comparison with the plumage of the young, can partly be explained through the action of sexual selection on variations occurring at different ages, and transmitted to the males alone or to both sexes at corresponding ages; but I have not space here to enter on this subject.

Thus it is, as I believe, that when the males and females of any animal have the same general habits of life, but differ in structure, colour, or ornament, such differences have been mainly caused by sexual selection: that is, by individual males having had, in successive generations, some slight advantage over other males, in their weapons, means of defence, or charms; which they have transmitted to their male offspring alone. Yet, I would not wish to attribute all sexual differences to this agency: for we see in our domestic animals peculiarities arising and becoming attached to the male sex, which apparently have not been augmented through selection by man. The tuft of hair on the breast of the wild turkey cock cannot be of any use, and it is doubtful whether it can be ornamental in the eyes of the female bird; indeed, had the tuft appeared under domestication it would have been called a monstrosity.

ILLUSTRATIONS OF THE ACTION OF NATURAL SELECTION, OR THE SURVIVAL OF THE FITTEST

In order to make it clear how, as I believe, natural selection acts, I must beg permission to give one or two imaginary illustrations. Let us take the case of a wolf, which preys on various animals, securing some by craft, some by strength, and some by fleetness; and let us suppose that the fleetest prey, a deer for instance, had from any change in the country increased in numbers, or that other prey had decreased in numbers, during that season of the year when the wolf was hardest pressed for food. Under such circumstances the swiftest and slimmest wolves have the best chance of surviving, and so be preserved or selected, – provided always that they retained strength to master their prey at this or some other period of the year, when they were compelled to prey on other animals. I can see no more reason to doubt that this would be the result, than that man should be able to improve the fleetness of his greyhounds by careful and methodical selection, or by that kind of

unconscious selection which follows from each man trying to keep the best dogs without any thought of modifying the breed. I may add that, according to Mr. Pierce, there are two varieties of the wolf inhabiting the Catskill Mountains, in the United States, one with a light greyhound-like form, which pursues deer, and the other more bulky, with shorter legs, which more frequently attacks the shepherd's flocks.

It should be observed that in the above illustration, I speak of the slimmest individual wolves, and not of any single strongly marked variation having been preserved. In former editions of this work I sometimes spoke as if this latter alternative had frequently occurred. I saw the great importance of individual differences, and this led me fully to discuss the results of unconscious selection by man, which depends on the preservation of all the more or less valuable individuals, and on the destruction of the worst. I saw, also, that the preservation in a state of nature of any occasional deviation of structure, such as a monstrosity, would be a rare event; and that, if at first preserved, it would generally be lost by subsequent intercrossing with ordinary individuals. Nevertheless, until reading an able and valuable article in the 'North British Review' (1867), I did not appreciate how rarely single variations, whether slight or strongly marked, could be perpetuated. The author takes the case of a pair of animals, producing during their lifetime two hundred offspring, of which, from various causes of destruction, only two on an average survive to pro-create their kind. This is rather an extreme estimate for most of the higher animals, but by no means so for many of the lower organisms. He then shows that if a single individual were born, which varied in some manner, giving it twice as good a chance of life as that of the other individuals, yet the chances would be strongly against its survival. Supposing it to survive and to breed, and that half its young inherited the favourable variation; still, as the reviewer goes on to show, the young would have only a slightly better chance of surviving and breeding; and this chance would go on decreasing in the succeeding generations. The justice of these remarks cannot, I think, be disputed. If, for instance, a bird of some kind could procure its food more easily by having its beak curved, and if one were born with its beak strongly curved, and which consequently flourished, nevertheless there would be a very poor chance of this one individual perpetuating its kind to the exclusion of the common form; but there can hardly be a doubt, judging by what we see taking place under domestication, that this result would follow from the preservation during many generations of a large number of individuals with more or less strongly curved beaks, and from the destruction of a still larger number with the straightest beaks.

It should not, however, be overlooked that certain rather strongly marked variations, which no one would rank as mere individual differences, frequently recur owing to a similar organisation being similarly acted on – of which fact numerous instances could be given with our domestic productions. In such cases, if the varying individual did not actually transmit to its offspring its newly acquired character, it would undoubtedly transmit to them, as long as the existing conditions remained the same, a still stronger tendency to vary in the same manner. There can also be little doubt that the tendency to vary in the same manner has often been so strong that all the individuals of the same species have been similarly modified without the aid of any form of selection. Or only a third, fifth, or tenth part of the individuals may have been thus affected, of which fact several instances could be given. Thus Graba estimates that about one-fifth of the guillemots in the Faroe Islands consist of a variety so well marked, that it was formerly ranked as a distinct species under the name of Uria lacrymans. In cases of this kind, if the variation were of a beneficial nature, the original form would soon be supplanted by the modified form, through the survival of the fittest.

To the effects of intercrossing in eliminating variations of all kinds, I shall have to recur; but it may be here remarked that most animals and plants keep to their proper homes, and do not needlessly wander about; we see this even with migratory birds, which almost always return to the same spot. Consequently each newly-formed variety would generally be at first local, as seems to be the common rule with varieties in a state of nature; so that similarly modified individuals would soon exist in a small body together, and would often breed together. If the new variety were successful in its battle for life, it would slowly spread from a central district, competing with and conquering the unchanged individuals on the margins of an ever-increasing circle.

It may be worth while to give another and more complex illustration of the action of natural selection. Certain plants excrete sweet juice, apparently for the sake of eliminating something injurious from the sap: this is effected, for instance, by glands at the base of the stipules in some Leguminosae, and at the backs of the leaves of the common laurel. This juice, though small in quantity, is greedily sought by insects; but their visits do not in any way benefit the plant. Now, let us suppose that the juice or nectar was excreted from the inside of the flowers of a certain number of plants of any species. Insects in seeking the nectar would get dusted with pollen, and would often transport it from one flower to another. The flowers of two distinct individuals of the same species would thus get crossed; and the act of crossing, as can be fully proved, gives rise to vigorous seedlings, which consequently would have the best chance of flourishing and surviving. The plants which produced flowers with the largest glands or nectaries, excreting most nectar, would oftenest be visited by insects, and would oftenest be crossed; and so in the long run would gain the upper hand and form a local variety. The flowers, also, which had their stamens and pistils placed, in relation to the size and habits of the particular insect which visited them, so as to

Orchids, honeysuckle and honeybush flowers – whose nectar is highly attractive – rely upon a variety of creatures such as flies, moths and birds for cross-pollination to strengthen the species favour in any degree the transportal of the pollen, would likewise be favoured. We might have taken the case of insects visiting flowers for the sake of collecting pollen instead of nectar; and as pollen is formed for the sole purpose of fertilisation, its destruction appears to be a simple loss to the plant; yet if a little pollen were carried, at first occasionally and then habitually, by the pollen-devouring insects from flower to flower, and a cross thus effected, although nine-tenths of the pollen were destroyed it might still be a great gain to the plant to be thus robbed; and the individuals which produced more and more pollen, and had larger anthers, would be selected.

When our plant, by the above process long continued, had been rendered highly attractive to insects, they would, unintentionally on their part, regularly carry pollen from flower to flower; and that they do this effectually I could easily show by many striking facts. I will give only one, as likewise illustrating one step in the separation of the sexes of plants. Some holly trees bear only male flowers, which have four stamens producing a rather small quantity of pollen, and a rudimentary pistil; other holly trees bear only female flowers; these have a full-sized pistil, and four stamens with shrivelled anthers, in which not a grain of pollen can be detected. Having found a female tree exactly sixty yards from a male tree, I put the stigmas of twenty flowers, taken from different branches, under the microscope, and on all, without exception, there were a few pollen grains, and on some a profusion. As the wind had set for several days from the female to the male tree, the pollen could not thus have been carried. The weather had been cold and boisterous and therefore not favourable to bees, nevertheless every female flower which I examined had been effectually fertilised by the bees, which had flown from tree to tree in search of nectar. But to return to our imaginary case; as soon as the plant had been rendered so highly attractive to insects that pollen was regularly carried from flower to flower, another process might commence. No

naturalist doubts the advantage of what has been called the 'physiological division of labour;' hence we may believe that it would be advantageous to a plant to produce stamens alone in one flower or on one whole plant, and pistils alone in another flower or on another plant. In plants under culture and placed under new conditions of life, sometimes the male organs and sometimes the female organs become more or less impotent; now if we suppose this to occur in ever so slight a degree under nature, then, as pollen is already carried regularly from flower to flower, and as a more complete separation of the sexes of our plant would be advantageous on the principle of the division of labour, individuals with this tendency more and more increased, would be continually favoured or selected, until at last a complete separation of the sexes might be effected. It would take up too much space to show the various steps, through dimorphism and other means, by which the separation of the sexes in plants of various kinds is apparently now in progress; but I may add that some of the species of holly in North America are, according to Asa Gray, in an exactly intermediate condition, or, as he expresses it, are more or less dioeciously polygamous.

Let us now turn to the nectar-feeding insects; we may suppose the plant of which we have been slowly increasing the nectar by continued selection, to be a common plant; and that certain insects depended in main part on its nectar for food. I could give many facts showing how anxious bees are to save time: for instance, their habit of cutting holes and sucking the nectar at the bases of certain flowers, which with a very little more trouble they can enter by the mouth. Bearing such facts in mind, it may be believed that under certain circumstances individual differences in the curvature or length of the proboscis, etc., too slight to be appreciated by us, might profit a bee or other insect, so that certain individuals would be able to obtain their food more quickly than others; and thus the communities to which they belonged would flourish and throw off many swarms inheriting the same peculiarities. The tubes of the corolla of the common red or incarnate clovers (Trifolium pratense and incarnatum) do not on a hasty glance appear to differ in length; yet the hive bee can easily suck the nectar out of the incarnate clover, but not out of the common red clover, which is visited by humblebees alone; so that whole fields of the red clover offer in vain an abundant supply of precious nectar to the hive bee. That this nectar is much liked by the hive bee is certain; for I have repeatedly seen, but only in the autumn, many hive bees sucking the flowers through holes bitten in the base of the tube by humblebees. The difference in the length of the corolla in the two kinds of clover, which determines the visits of the hive bee, must be very trifling; for I have been assured that when red clover has been mown, the flowers of the second crop are somewhat smaller, and that these are visited by many hive bees. I do not know whether this statement is accurate; nor whether another published statement can be trusted, namely, that the Ligurian bee, which is generally considered a mere variety of the common hive bee, and which freely crosses with it, is able to reach and suck the nectar of the red clover. Thus, in a country where this kind of clover abounded, it might be a great advantage to the hive bee to have a slightly longer or differently constructed proboscis. On the other hand, as the fertility of this clover absolutely depends on bees visiting the flowers, if humblebees were to become rare in any country, it might be a great advantage to the plant to have a shorter or more deeply divided corolla, so that the hive bees should be enabled to suck its flowers. Thus I can understand how a flower and a bee might slowly become, either simultaneously or one after the other, modified and adapted to each other in the most perfect manner, by the continued preservation of all the individuals which presented slight deviations of structure mutually favourable to each other.

I am well aware that this doctrine of natural selection, exemplified in the above imaginary instances, is open to the same objections which were first urged against Sir Charles Lyell's noble views on 'the modern changes of the earth, as illustrative of geology'; but we now seldom hear the agencies which we see still at work, spoken of as trifling and insignificant, when used in explaining the excavation of the deepest valleys or the formation of long lines of inland cliffs. Natural selection acts only by the preservation and accumulation of small inherited modifications, each profitable to the preserved being; and as modern geology has almost banished such views as the excavation of a great valley by a single diluvial wave, so will natural selection banish the belief of the continued creation of new organic beings, or of any great and sudden modification in their structure.

ON THE INTERCROSSING OF INDIVIDUALS

I must here introduce a short digression. In the case of animals and plants with separated sexes, it is of course obvious that two individuals must always (with the exception of the curious and not well understood cases of parthenogenesis) unite for each birth; but in the case of hermaphrodites this is far from obvious. Nevertheless there is reason to believe that with all hermaphrodites two individuals, either occasionally or habitually, concur for the reproduction of their kind. This view was long ago doubtfully suggested by Sprengel, Knight and Kölreuter. We shall presently see its importance; but I must here treat the subject with extreme brevity, though I have the materials prepared for an ample discussion. All vertebrate animals, all insects and some other large groups of animals, pair for each birth. Modern research has much diminished the number of supposed hermaphrodites, and of real hermaphrodites a large number pair; that is, two individuals regularly unite for reproduction, which is all that concerns us. But still there are many hermaphrodite animals which certainly do not habitually pair, and a vast majority of plants are hermaphrodites. What reason, it may be asked, is there for supposing in these cases that two individuals ever concur in reproduction? As it is impossible here to enter on details, I must trust to some general considerations alone.

In the first place, I have collected so large a body of facts, and made so many experiments, showing, in accordance with the almost universal belief of breeders, that with animals and plants a cross between different varieties, or between individuals of the same variety but of another strain, gives vigour and fertility to the offspring; and on the other hand, that *close* interbreeding diminishes vigour and fertility; that these facts alone incline me to believe that it is a general law of nature that no organic being fertilises itself for a perpetuity of generations; but that a cross with another individual is occasionally – perhaps at long intervals of time – indispensable.

On the belief that this is a law of nature, we can, I think, understand several large classes of facts, such as the following, which on any other view are inexplicable. Every hybridizer knows how unfavourable exposure to wet is to the fertilisation of a flower, yet what a multitude of flowers have their anthers and stigmas fully exposed to the weather! If an occasional cross be indispensable, notwithstanding that the plant's own anthers and pistil stand so near each other as almost to ensure self-fertilisation, the fullest freedom for the entrance of pollen from another individual will explain the above state of exposure of the organs. Many flowers, on the other hand, have their organs of fructification closely enclosed, as in the great papilionaceous or pea-family; but these almost invariably present beautiful and curious adaptations in relation to the visits of insects. So necessary are the visits of bees to many papilionaceous flowers, that their fertility is greatly diminished if these visits be prevented. Now, it is scarcely possible for insects to fly from flower to flower, and not to carry pollen from one to the other, to the great good of the plant. Insects act

like a camel-hair pencil, and it is sufficient, to ensure fertilisation, just to touch with the same brush the anthers of one flower and then the stigma of another; but it must not be supposed that bees would thus produce a multitude of hybrids between distinct species; for if a plant's own pollen and that from another species are placed on the same stigma, the former is so prepotent that it invariably and completely destroys, as has been shown by Gärtner, the influence of the foreign pollen.

When the stamens of a flower suddenly spring towards the pistil, or slowly move one after the other towards it, the contrivance seems adapted solely to ensure self-fertilisation; and no doubt it is useful for this end: but the agency of insects is often required to cause the stamens to spring forward, as Kölreuter has shown to be the case with the barberry; and in this very genus, which seems to have a special contrivance for self-fertilisation, it is well known that, if closely-allied forms or varieties are planted near each other, it is hardly possible to raise pure seedlings, so largely do they naturally cross. In numerous other cases, far from self-fertilisation being favoured, there are special contrivances which effectually prevent the stigma receiving pollen from its own flower, as I could show from the works of Sprengel and others, as well as from my own observations: for instance, in Lobelia fulgens, there is a really beautiful and elaborate contrivance by which all the infinitely numerous pollen-granules are swept out of the conjoined anthers of each flower, before the stigma of that individual flower is ready to receive them; and as this flower is never visited, at least in my garden, by insects, it never sets a seed, though by placing pollen from one flower on the stigma of another, I raise plenty of seedlings. Another species of Lobelia, which is visited by bees, seeds freely in my garden. In very many other cases, though there is no special mechanical contrivance to prevent the stigma receiving pollen from the same flower, yet, as Sprengel, and more recently Hildebrand and others have shown, and as I can confirm, either the anthers burst before the stigma is ready for fertilisation, or the stigma is ready before the pollen of that flower is ready, so that these so-named dichogamous plants have in fact separated sexes, and must habitually be crossed. So it is with the reciprocally dimorphic and trimorphic plants previously alluded to. How strange are these facts! How strange that the pollen and stigmatic surface of the same flower, though placed so close together, as if for the very purpose of self-fertilisation, should be in so many cases mutually useless to each other? How simply are these facts explained on the view of an occasional cross with a distinct individual being advantageous or indispensable!

If several varieties of the cabbage, radish, onion, and of some other plants, be allowed to seed near each other, a large majority of the seedlings thus raised turn out, as I found, mongrels: for instance, I raised 233 seedling cabbages from some plants of different varieties growing near each other, and of these only 78 were true to their kind, and some even of these were not perfectly true. Yet the pistil of each cabbage-flower is surrounded not only by its own six stamens but by those of the many other flowers on the same plant; and the pollen of each flower readily gets on its stigma without insect agency; for I have found that plants carefully protected from insects produce the full number of pods. How, then, comes it that such a vast number of the seedlings are mongrelized? It must arise from the pollen of a distinct *variety* having a prepotent effect over the flower's own pollen; and that this is part of the general law of good being derived from the intercrossing of distinct individuals of the same species. When distinct *species* are crossed the case is reversed, for a plant's own pollen is always prepotent over foreign pollen; but to this subject we shall return in a future chapter.

In the case of a large tree covered with innumerable flowers, it may be objected that pollen could seldom be carried from tree to tree, and at most only from flower to flower on the same tree; and flowers on the same tree can be considered as distinct individuals only in a limited sense. I believe this objection to be valid, but that nature has largely provided against it by giving to trees a strong tendency to bear flowers with separated sexes. When the sexes are separated, although the male and female flowers may be produced on the same tree, pollen must be regularly carried from flower to flower; and this will give a better chance of pollen being occasionally carried from tree to tree. That trees belonging to all orders have their sexes more often separated than other plants, I find to be the case in this country; and at my request Dr. Hooker tabulated the trees of New Zealand, and Dr. Asa Gray those of the United States, and the result was as I anticipated. On the other hand, Dr. Hooker informs me that the rule does not hold good in Australia: but if most of the Australian trees are dichogamous, the same result would follow as if they bore flowers with separated sexes. I have made these few remarks on trees simply to call attention to the subject.

Turning for a brief space to animals: various terrestrial species are hermaphrodites, such as the land Mollusca and earthworms; but these all pair. As yet I have not found a single terrestrial animal which can fertilise itself. This remarkable fact, which offers so strong a contrast with terrestrial plants, is intelligible on the view of an occasional cross being indispensable; for owing to the nature of the fertilising element there are no means, analogous to the action of insects and of the wind with plants, by which an occasional cross could be effected with terrestrial animals without the concurrence of two individuals. Of aquatic animals, there are many self-fertilising hermaphrodites; but here the currents of water offer an obvious means for an occasional cross. As in the case of flowers, I have as yet failed, after consultation with one of the highest authorities, namely, Professor Huxley, to discover a single hermaphrodite animal with the organs of reproduction so perfectly enclosed that access from without, and the occasional influence of a distinct individual, can be shown to be physically impossible. Cirripedes long appeared to me to present, under this point of view, a case of great difficulty; but I have been enabled, by a fortunate chance, to prove that two individuals, though both are self-fertilising hermaphrodites, do sometimes cross.

It must have struck most naturalists as a strange anomaly that, both with animals and plants, some species of the same family and even of the same genus, though agreeing closely with each other in their whole organisation, are hermaphrodites, and some unisexual. But if, in fact, all hermaphrodites do occasionally intercross, the difference between them and unisexual species is, as far as function is concerned, very small.

From these several considerations and from the many special facts which I have collected, but which I am unable here to give, it appears that with animals and plants an occasional intercross between distinct individuals is a very general, if not universal, law of nature.

CIRCUMSTANCES FAVOURABLE FOR THE PRODUCTION OF NEW FORMS THROUGH NATURAL SELECTION

This is an extremely intricate subject. A great amount of variability, under which term individual differences are always included, will evidently be favourable. A large number of individuals, by giving a better chance within any given period for the appearance of profitable variations, will compensate for a lesser amount of variability in each individual, and is, I believe, a highly important element of success. Though Nature grants long periods of time for the work of natural selection, she does not grant an indefinite period; for as all organic beings are striving to seize on each place in the economy of nature, if any one species does not become modified and improved in a corresponding degree with its competitors it will be exterminated. Unless favourable variations be inherited by some at least of the offspring,

nothing can be effected by natural selection. The tendency to reversion may often check or prevent the work; but as this tendency has not prevented man from forming by selection numerous domestic races, why should it prevail against natural selection?

In the case of methodical selection, a breeder selects for some definite object, and if the individuals be allowed freely to intercross, his work will completely fail. But when many men, without intending to alter the breed, have a nearly common standard of perfection, and all try to procure and breed from the best animals, improvement surely but slowly follows from this unconscious process of selection, notwithstanding that there is no separation of selected individuals. Thus it will be under nature; for within a confined area, with some place in the natural polity not perfectly occupied, all the individuals varying in the right direction, though in different degrees, will tend to be preserved. But if the area be large, its several districts will almost certainly present different conditions of life; and then, if the same species undergoes modification in different districts, the newly formed varieties will intercross on the confines of each. But we shall see in the sixth chapter that intermediate varieties, inhabiting intermediate districts, will in the long run generally be supplanted by one of the adjoining varieties. Intercrossing will chiefly affect those animals which unite for each birth and wander much, and which do not breed at a very quick rate. Hence with animals of this nature, for instance birds, varieties will generally be confined to separated countries; and this I find to be the case. With hermaphrodite organisms which cross only occasionally, and likewise with animals which unite for each birth, but which wander little and can increase at a rapid rate, a new and improved variety might be quickly formed on any one spot, and might there maintain itself in a body and afterward spread, so that the individuals of the new variety would chiefly cross together. On this principle, nurserymen always prefer saving seed from a large body of plants, as the chance of intercrossing is thus lessened.

Even with animals which unite for each birth, and which do not propagate rapidly, we must not assume that free intercrossing would always eliminate the effects of natural selection; for I can bring forward a considerable body of facts showing that within the same area two varieties of the same animal may long remain distinct, from haunting different stations, from breeding at slightly different seasons, or from the individuals of each variety preferring to pair together.

Intercrossing plays a very important part in nature by keeping the individuals of the same species, or of the same variety, true and uniform in character. It will obviously thus act far more efficiently with those animals which unite for each birth; but, as already stated, we have reason to believe that occasional intercrosses take place with all animals and plants. Even if these take place only at long intervals of time, the young thus produced will gain so much in vigour and fertility over the offspring from long-continued self-fertilisation, that they will have a better chance of surviving and propagating their kind; and thus in the long run the influence of crosses, even at rare intervals, will be great. With respect to organic beings extremely low in the scale, which do not propagate sexually, nor conjugate, and which cannot possibly intercross, uniformity of character can be retained by them under the same conditions of life, only through the principle of inheritance, and through natural selection which will destroy any individuals departing from the proper type. If the conditions of life change and the form undergoes modification, uniformity of character can be given to the modified offspring, solely by natural selection preserving similar favourable variations.

Isolation also is an important element in the modification of species through natural selection. In a confined or isolated area, if not very large, the organic and inorganic conditions of life will generally be almost uniform; so that natural selection will tend to modify all the varying individuals of the same species in the same manner. Intercrossing with the inhabitants of the surrounding districts, will also be thus prevented. Moritz Wagner has lately published an interesting essay on this subject, and has shown that the service rendered by isolation in preventing crosses between newly-formed varieties is probably greater even than I supposed. But from reasons already assigned I can by no means agree with this naturalist, that migration and isolation are necessary elements for the formation of new species. The importance of isolation is likewise great in preventing, after any physical change in the conditions, such as of climate, elevation of the land, etc., the immigration of better adapted organisms; and thus new places in the natural economy of the district will be left open to be filled up by the modification of the old inhabitants. Lastly, isolation will give time for a new variety to be improved at a slow rate; and this may sometimes be of much importance. If, however, an isolated area be very small, either from being surrounded by barriers, or from having very peculiar physical conditions, the total number of the inhabitants will be small; and this will retard the production of new species through natural selection, by decreasing the chances of favourable variations arising.

The mere lapse of time by itself does nothing, either for or against natural selection. I state this because it has been erroneously asserted that the element of time has been assumed by me to play an all-important part in modifying species, as if all the forms of life were necessarily undergoing change through some innate law. Lapse of time is only so far important, and its importance in this respect is great, that it gives a better chance of beneficial variations arising and of their being selected, accumulated, and fixed. It likewise tends to increase the direct action of the physical conditions of life, in relation to the constitution of each organism.

If we turn to nature to test the truth of these remarks, and look at any small isolated area, such as an oceanic island, although the number of the species inhabiting it is small, as we shall see in our chapter on Geographical Distribution; yet of these species a very large proportion are endemic, – that is, have been produced there and nowhere else in the world. Hence an oceanic island at first sight seems to have been highly favourable for the production of new species. But we may thus deceive ourselves, for to ascertain whether a small isolated area, or a large open area like a continent, has been most favourable for the production of new organic forms, we ought to make the comparison within equal times; and this we are incapable of doing.

Although isolation is of great importance in the production of new species, on the whole I am inclined to believe that largeness of area is still more important, especially for the production of species which shall prove capable of enduring for a long period, and of spreading widely. Throughout a great and open area, not only will there be a better chance of favourable variations, arising from the large number of individuals of the same species there supported, but the conditions of life are much more complex from the large number of already existing species; and if some of these many species become modified and improved, others will have to be improved in a corresponding degree, or they will be exterminated. Each new form, also, as soon as it has been much improved, will be able to spread over the open and continuous area, and will thus come into competition with many other forms. Moreover, great areas, though now continuous, will often, owing to former oscillations of level, have existed in a broken condition, so that the good effects of isolation will generally, to a certain extent, have concurred. Finally, I conclude that, although small isolated areas have been in some respects highly favourable for the production of new species, yet that the course of modification will generally have been more rapid on large areas; and what is more

important, that the new forms produced on large areas, which already have been victorious over many competitors, will be those that will spread most widely, and will give rise to the greatest number of new varieties and species. They will thus play a more important part in the changing history of the organic world.

In accordance with this view, we can, perhaps, understand some facts which will be again alluded to in our chapter on Geographical Distribution; for instance, the fact of the productions of the smaller continent of Australia now yielding before those of the larger Europaeo-Asiatic area. Thus, also, it is that continental productions have everywhere become so largely naturalised on islands. On a small island, the race for life will have been less severe, and there will have been less modification and less extermination. Hence, we can understand how it is that the flora of Madeira, according to Oswald Heer, resembles to a certain extent the extinct tertiary flora of Europe. All fresh water basins, taken together, make a small area compared with that of the sea or of the land. Consequently, the competition between fresh water productions will have been less severe than elsewhere; new forms will have been more slowly produced, and old forms more slowly exterminated. And it is in fresh water basins that we find seven genera of Ganoid fishes, remnants of a once preponderant order: and in fresh water we find some of the most anomalous forms now known in the world, as the Ornithorhynchus and Lepidosiren, which, like fossils, connect to a certain extent orders at present widely separated in the natural scale. These anomalous forms may be called living fossils; they have endured to the present day, from having inhabited a confined area, and from having been exposed to less varied, and therefore less severe, competition.

To sum up, as far as the extreme intricacy of the subject permits, the circumstances favourable and unfavourable for the production of new species through natural selection. I conclude that

Ganoid fish:
A Bichier *B* Reed fish *C* Gar pike *D* Common sturgeon *E* Shovelnose sturgeon
F Slender-beaked sturgeon *G* Spoonbill sturgeon *H* Bow-fin

for terrestrial productions a large continental area, which has undergone many oscillations of level, will have been the most favourable for the production of many new forms of life, fitted to endure for a long time and to spread widely. Whilst the area existed as a continent, the inhabitants will have been numerous in individuals and kinds, and will have been subjected to severe competition. When converted by subsidence into large separate islands, there will still have existed many individuals of the same species on each island: intercrossing on the confines of the range of each new species will have been checked: after physical changes of any kind, immigration will have been prevented, so that new places in the polity of each island will have had to be filled up by the modification of the old inhabitants; and time will have been allowed for the varieties in each to become well modified and perfected. When, by renewed elevation, the islands were reconverted into a continental area, there will again have been very severe competition; the most favoured or improved varieties will have been enabled to spread; there will have been much extinction of the less improved forms, and the relative proportional numbers of the various inhabitants of the reunited continent will again have been changed; and again there will have been a fair field for natural selection to improve still further the inhabitants, and thus to produce new species.

That natural selection generally acts with extreme slowness I fully admit. It can act only when there are places in the natural polity of a district which can be better occupied by the modification of some of its existing inhabitants. The occurrence of such places will often depend on physical changes, which generally take place very slowly, and on the immigration of better adapted forms being prevented. As some few of the old inhabitants become modified, the mutual relations of others will often be disturbed; and this will create new places, ready to be filled up by better adapted forms; but all this will take place very slowly. Although all the individuals of the same species differ in some slight degree from each other, it would often be long before differences of the right nature in various parts of the organisation might occur. The result would often be greatly retarded by free intercrossing. Many will exclaim that these several causes are amply sufficient to neutralise the power of natural selection. I do not believe so. But I do believe that natural selection will generally act very slowly, only at long intervals of time, and only on a few of the inhabitants of the same region. I further believe that these slow, intermittent results accord well with what geology tells us of the rate and manner at which the inhabitants of the world have changed.

Slow though the process of selection may be, if feeble man can do much by artificial selection, I can see no limit to the amount of change, to the beauty and complexity of the co-adaptations between all organic beings, one with another and with their physical conditions of life, which may have been effected in the long course of time through nature's power of selection, that is by the survival of the fittest.

EXTINCTION CAUSED BY NATURAL SELECTION

This subject will be more fully discussed in our chapter on Geology; but it must here be alluded to from being intimately connected with natural selection. Natural selection acts solely through the preservation of variations in some way advantageous, which consequently endure. Owing to the high geometrical rate of increase of all organic beings, each area is already fully stocked with inhabitants, and it follows from this, that as the favoured forms increase in number, so, generally, will the less favoured decrease and become rare. Rarity, as geology tells us, is the precursor to extinction. We can see that any form which is represented by few individuals will run a good chance of utter extinction, during great fluctuations in the nature of the seasons, or from a temporary increase in the number of its enemies. But we may go further than this; for as new forms are produced, unless we admit that specific forms can go on indefinitely increasing in number, many old forms must become extinct. That the number of specific forms has not indefinitely increased, geology plainly tells us; and we shall presently attempt to show why it is that the number of species throughout the world has not become immeasurably great.

We have seen that the species which are most numerous in individuals have the best chance of producing favourable variations within any given period. We have evidence of this, in the facts stated in the second chapter, showing that it is the common and diffused or dominant species which offer the greatest number of recorded varieties. Hence, rare species will be less quickly modified or improved within any given period; they will consequently be beaten in the race for life by the modified and improved descendants of the commoner species.

From these several considerations I think it inevitably follows, that as new species in the course of time are formed through natural selection, others will become rarer and rarer, and finally extinct. The forms which stand in closest competition with those undergoing modification and improvement, will naturally suffer most. And we have seen in the chapter on the Struggle for Existence that it is the most closely-allied forms, – varieties of the same species, and species of the same genus or related genera, – which, from having nearly the same structure, constitution and habits, generally come into the severest competition with each other. Consequently, each new variety or species, during the progress of its formation, will generally press hardest on its nearest kindred, and tend to exterminate them. We see the same process of extermination amongst our domesticated productions, through the selection of improved forms by man. Many curious instances could be given showing how quickly new breeds of cattle, sheep and other animals, and varieties of flowers, take the place of older and inferior kinds. In Yorkshire, it is historically known that the ancient black cattle were displaced by the longhorns, and that these 'were swept away by the shorthorns' (I quote the words of an agricultural writer) 'as if by some murderous pestilence.'

DIVERGENCE OF CHARACTER

The principle, which I have designated by this term, is of high importance, and explains, as I believe, several important facts. In the first place, varieties, even strongly marked ones, though having somewhat of the character of species – as is shown by the hopeless doubts in many cases how to rank them – yet certainly differ far less from each other than do good and distinct species. Nevertheless, according to my view, varieties are species in the process of formation, or are, as I have called them, incipient species. How, then, does the lesser difference between varieties become augmented into the greater difference between species? That this does habitually happen, we must infer from most of the innumerable species throughout nature presenting well-marked differences; whereas varieties, the supposed prototypes and parents of future well-marked species, present slight and ill-defined differences. Mere chance, as we may call it, might cause one variety to differ in some character from its parents, and the offspring of this variety again to differ from its parent in the very same character and in a greater degree; but this alone would never account for so habitual and large a degree of difference as that between the species of the same genus.

As has always been my practice, I have sought light on this head from our domestic productions. We shall here find something analogous. It will be admitted that the production of races so different as shorthorn and Hereford cattle, race and cart horses, the several breeds of pigeons, etc., could never have been effected by the mere chance accumulation of similar variations during many successive generations. In practice, a fancier is, for instance, struck by a pigeon having a slightly shorter beak; another fancier is struck by a pigeon having a rather longer beak; and on the acknowledged principle that 'fanciers do not and will not admire a medium standard, but like extremes,' they both go on (as has actually occurred with the sub-breeds of the tumbler pigeon) choosing and breeding from birds with longer and longer beaks, or with shorter and shorter beaks. Again, we may suppose that at an early period of history, the men of one nation or district required swifter horses, whilst those of another required stronger and bulkier horses. The early differences would be very slight; but, in the course of time, from the continued selection of swifter horses in the one case, and of stronger ones in the other, the differences would become greater, and would be noted as forming two sub-breeds. Ultimately, after the lapse of centuries, these sub-breeds would become converted into two well-established and distinct breeds. As the differences became greater, the inferior animals with intermediate characters, being neither very swift nor very strong, would not have been used for breeding, and will thus have tended to disappear. Here, then, we see in man's productions the action of what may be called the principle of divergence, causing differences, at first barely appreciable, steadily to increase, and the breeds to diverge in character, both from each other and from their common parent.

But how, it may be asked, can any analogous principle apply in nature? I believe it can and does apply most efficiently (though it was a long time before I saw how), from the simple circumstance that the more diversified the descendants from any one species become in structure, constitution, and habits, by so much will they be better enabled to seize on many and widely diversified places in the polity of nature, and so be enabled to increase in numbers.

We can clearly discern this in the case of animals with simple habits. Take the case of a carnivorous quadruped, of which the number that can be supported in any country has long ago arrived at its full average. If its natural power of increase be allowed to act, it can succeed in increasing (the country not undergoing any change in conditions) only by its varying descendants seizing on places at present occupied by other animals: some of them, for instance, being enabled to feed on new kinds of prey, either dead or alive; some inhabiting new stations, climbing trees, frequenting water, and some perhaps becoming less carnivorous. The more diversified in habits and structure the descendants of our carnivorous animals become, the more places they will be enabled to occupy. What applies to one animal will apply throughout all time to all animals – that is, if they vary – for otherwise natural selection can effect nothing. So it will be with plants. It has been experimentally proved, that if a plot of ground be sown with one species of grass, and a similar plot be sown with several distinct genera of grasses, a greater number of plants and a greater weight of dry herbage can be raised in the latter than in the former case. The same has been found to hold good when one variety and several mixed varieties of wheat have been sown on equal spaces of ground. Hence, if any one species of grass were to go on varying, and the varieties were continually selected which differed from each other in the same manner, though in a very slight degree, as do the distinct species and genera of grasses, a greater number of individual plants of this species, including its modified descendants, would succeed in living on the same piece of ground. And we know that each species and each variety of grass is annually sowing almost countless seeds; and is thus striving, as it may be said, to the utmost to increase in number. Consequently, in the course of many thousand generations, the most distinct varieties of any one species of grass would have the best chance of succeeding and of increasing in numbers, and thus of supplanting the less distinct varieties; and varieties, when rendered very distinct from each other, take the rank of species.

The truth of the principle that the greatest amount of life can be supported by great diversification of structure, is seen under many natural circumstances. In an extremely small area, especially if freely open to immigration, and where the contest between individual and individual must be very severe, we always find great diversity in its inhabitants. For instance, I found that a piece of turf, three feet by four in size, which had been exposed for many years to exactly the same conditions, supported twenty species of plants, and these belonged to eighteen genera and to eight orders, which shows how much these plants differed from each other. So it is with the plants and insects on small and uniform islets: also in small ponds of fresh water. Farmers find that they can raise more food by a rotation of plants belonging to the most different orders: nature follows what may be called a simultaneous rotation. Most of the animals and plants which live close round any small piece of ground, could live on it (supposing its nature not to be in any way peculiar), and may be said to be striving to the utmost to live there; but, it is seen, that where they come into the closest competition, the advantages of diversification of structure, with the accompanying differences of habit and constitution, determine that the inhabitants, which thus jostle each other most closely, shall, as a general rule, belong to what we call different genera and orders.

The same principle is seen in the naturalisation of plants through man's agency in foreign lands. It might have been expected that the plants which would succeed in becoming naturalised in any land would generally have been closely allied to the indigenes; for these are commonly looked at as specially created and adapted for their own country. It might also, perhaps, have been expected that naturalised plants would have belonged to a few groups more especially adapted to certain stations in their new homes. But the case is very different; and Alph. de Candolle has well remarked, in his great and admirable work, that floras gain by naturalisation, proportionally with the number of the native genera and species, far more in new genera than in new species. To give a single instance: in the last edition of Dr. Asa Gray's 'Manual of the Flora of the Northern United States,' 260 naturalised plants are enumerated, and these belong to 162 genera. We thus see that these naturalised plants are of a highly diversified nature. They differ, moreover, to a large extent, from the indigenes, for out of the 162 naturalised genera, no less than 100 genera are not there indigenous, and thus a large proportional addition is made to the genera now living in the United States.

By considering the nature of the plants or animals which have in any country struggled successfully with the indigenes, and have there become naturalised, we may gain some crude idea in what manner some of the natives would have had to be modified in order to gain an advantage over their compatriots; and we may at least infer that diversification of structure, amounting to new generic differences, would be profitable to them.

The advantage of diversification of structure in the inhabitants of the same region is, in fact, the same as that of the physiological division of labour in the organs of the same individual body – a subject so well elucidated by Milne Edwards. No physiologist doubts that a stomach by being adapted to digest vegetable matter alone, or flesh alone, draws most nutriment from these substances. So in the general economy of any land, the more widely and perfectly the animals and plants are diversified for different habits of life, so will a greater number of individuals be capable of there supporting themselves. A set of animals, with their organisation but little diversified, could hardly compete with a set more perfectly diversified in structure. It may be doubted, for instance, whether the Australian marsupials, which are divided into groups differing but little from each other, and feebly representing, as Mr. Waterhouse and others have remarked, our carnivorous, ruminant, and rodent mammals, could successfully compete with these well-developed orders. In the Australian mammals, we see the process of diversification in an early and incomplete stage of development.

THE PROBABLE EFFECTS OF THE ACTION OF NATURAL SELECTION THROUGH DIVERGENCE OF CHARACTER AND EXTINCTION, ON THE DESCENDANTS OF A COMMON ANCESTOR

After the foregoing discussion, which has been much compressed, we may assume that the modified descendants of any one species will succeed so much the better as they become more diversified in structure, and are thus enabled to encroach on places occupied by other beings. Now let us see how this principle of benefit being derived from divergence of character, combined with the principles of natural selection and of extinction, tends to act.

The accompanying diagram [see overleaf] will aid us in understanding this rather perplexing subject. Let A to L represent the species of a genus large in its own country; these species are supposed to resemble each other in unequal degrees, as is so generally the case in nature, and as is represented in the diagram by the letters standing at unequal distances. I have said a large genus, because as we saw in the second chapter, on an average more species vary in large genera than in small genera; and the varying species of the large genera present a greater number of varieties. We have, also, seen that the species, which are the commonest and most widely-diffused, vary more than do the rare and restricted species. Let (A) be a common, widely-diffused, and varying species, belonging to a genus large in its own country. The branching and diverging dotted lines of unequal lengths proceeding from (A), may represent its varying offspring. The variations are supposed to be extremely slight, but of the most diversified nature; they are not supposed all to appear simultaneously, but often after long intervals of time; nor are they all

supposed to endure for equal periods. Only those variations which are in some way profitable will be preserved or naturally selected. And here the importance of the principle of benefit derived from divergence of character comes in; for this will generally lead to the most different or divergent variations (represented by the outer dotted lines) being preserved and accumulated by natural selection. When a dotted line reaches one of the horizontal lines, and is there marked by a small numbered letter, a sufficient amount of variation is supposed to have been accumulated to form it into a fairly well-marked variety, such as would be thought worthy of record in a systematic work.

The intervals between the horizontal lines in the diagram, may represent each a thousand or more generations. After a thousand generations, species (A) is supposed to have produced two fairly well-marked varieties, namely a^1 and m^1. These two varieties will generally still be exposed to the same conditions which made their parents variable, and the tendency to variability is in itself hereditary; consequently they will likewise tend to vary, and commonly in nearly the same manner as did their parents. Moreover, these two varieties, being only slightly modified forms, will tend to inherit those advantages which made their parent (A) more numerous than most of the other inhabitants of the same country; they will also partake of those more general advantages which made the genus to which the parent-species belonged, a large genus in its own country. And all these circumstances are favourable to the production of new varieties.

If, then, these two varieties be variable, the most divergent of their variations will generally be preserved during the next thousand generations. And after this interval, variety a^1 is supposed in the diagram to have produced variety a^2, which will, owing to the principle of divergence, differ more from (A) than did variety a^1. Variety m^1 is supposed to have produced two varieties, namely m^2 and s^2, differing from each other, and more considerably from their common

parent (A). We may continue the process by
similar steps for any length of time; some of the
varieties, after each thousand generations,
producing only a single variety, but in a more and
more modified condition, some producing two or
three varieties, and some failing to produce any.
Thus the varieties or modified descendants of the
common parent (A), will generally go on
increasing in number and diverging in character.
In the diagram the process is represented up to
the ten-thousandth generation, and under a
condensed and simplified form up to the
fourteen-thousandth generation.

But I must here remark that I do not suppose
that the process ever goes on so regularly as is
represented in the diagram, though in itself made
somewhat irregular, nor that it goes on
continuously; it is far more probable that each
form remains for long periods unaltered, and then
again undergoes modification. Nor do I suppose
that the most divergent varieties are invariably
preserved: a medium form may often long endure,
and may or may not produce more than one
modified descendant; for natural selection will
always act according to the nature of the places
which are either unoccupied or not perfectly
occupied by other beings; and this will depend on
infinitely complex relations. But as a general rule,
the more diversified in structure the descendants
from any one species can be rendered, the more
places they will be enabled to seize on, and the
more their modified progeny will increase. In our
diagram the line of succession is broken at regular
intervals by small numbered letters marking the
successive forms which have become sufficiently
distinct to be recorded as varieties. But these
breaks are imaginary, and might have been
inserted anywhere, after intervals long enough to
allow the accumulation of a considerable amount
of divergent variation.

As all the modified descendants from a
common and widely-diffused species, belonging
to a large genus, will tend to partake of the same
advantages which made their parent successful in

life, they will generally go on multiplying in
number as well as diverging in character: this is
represented in the diagram by the several
divergent branches proceeding from (A). The
modified offspring from the later and more highly
improved branches in the lines of descent, will, it
is probable, often take the place of, and so
destroy, the earlier and less improved branches:
this is represented in the diagram by some of the
lower branches not reaching to the upper
horizontal lines. In some cases, no doubt, the
process of modification will be confined to a
single line of descent, and the number of
modified descendants will not be increased;
although the amount of divergent modification
may have been augmented. This case would be
represented in the diagram, if all the lines
proceeding from (A) were removed, excepting
that from a^1 to a^{10}. In the same way the English
race horse and English pointer have apparently
both gone on slowly diverging in character from
their original stocks, without either having given
off any fresh branches or races.

After ten thousand generations, species (A) is
supposed to have produced three forms, a^{10}, f^{10},
and m^{10}, which, from having diverged in character
during the successive generations, will have come
to differ largely, but perhaps unequally, from each
other and from their common parent. If we
suppose the amount of change between each
horizontal line in our diagram to be excessively
small, these three forms may still be only well-
marked varieties; but we have only to suppose the
steps in the process of modification to be more
numerous or greater in amount, to convert these
three forms into doubtful or at least into well-
defined species. Thus the diagram illustrates the
steps by which the small differences distinguishing
varieties are increased into the larger differences
distinguishing species. By continuing the same
process for a greater number of generations (as
shown in the diagram in a condensed and
simplified manner), we get eight species, marked
by the letters between a^{14} and m^{14}, all descended

from (A). Thus, as I believe, species are multiplied and genera are formed.

In a large genus it is probable that more than one species would vary. In the diagram I have assumed that a second species (I) has produced, by analogous steps, after ten thousand generations, either two well-marked varieties (w^{10} and z^{10}) or two species, according to the amount of change supposed to be represented between the horizontal lines. After fourteen thousand generations, six new species, marked by the letters n^{14} to z^{14}, are supposed to have been produced. In any genus, the species which are already very different in character from each other, will generally tend to produce the greatest number of modified descendants; for these will have the best chance of seizing on new and widely different places in the polity of nature: hence in the diagram I have chosen the extreme species (A), and the nearly extreme species (I), as those which have largely varied, and have given rise to new varieties and species. The other nine species (marked by capital letters) of our original genus, may for long but unequal periods continue to transmit unaltered descendants; and this is shown in the diagram by the dotted lines unequally prolonged upwards.

But during the process of modification, represented in the diagram, another of our principles, namely that of extinction, will have played an important part. As in each fully stocked country natural selection necessarily acts by the selected form having some advantage in the struggle for life over other forms, there will be a constant tendency in the improved descendants of any one species to supplant and exterminate in each stage of descent their predecessors and their original progenitor. For it should be remembered that the competition will generally be most severe between those forms which are most nearly related to each other in habits, constitution and structure. Hence all the intermediate forms between the earlier and later states, that is between the less and more improved states of the same species, as well

as the original parent-species itself, will generally tend to become extinct. So it probably will be with many whole collateral lines of descent, which will be conquered by later and improved lines. If, however, the modified offspring of a species get into some distinct country, or become quickly adapted to some quite new station, in which offspring and progenitor do not come into competition, both may continue to exist.

If, then, our diagram be assumed to represent a considerable amount of modification, species (A) and all the earlier varieties will have become extinct, being replaced by eight new species (a^{14} to m^{14}); and species (I) will be replaced by six (n^{14} to z^{14}) new species.

But we may go further than this. The original species of our genus were supposed to resemble each other in unequal degrees, as is so generally the case in nature; species (A) being more nearly related to B, C, and D than to the other species; and species (I) more to G, H, K, L, than to the others. These two species (A) and (I), were also supposed to be very common and widely diffused species, so that they must originally have had some advantage over most of the other species of the genus. Their modified descendants, fourteen in number at the fourteen-thousandth generation, will probably have inherited some of the same advantages: they have also been modified and improved in a diversified manner at each stage of descent, so as to have become adapted to many related places in the natural economy of their country. It seems, therefore, extremely probable that they will have taken the places of, and thus exterminated, not only their parents (A) and (I), but likewise some of the original species which were most nearly related to their parents. Hence very few of the original species will have transmitted offspring to the fourteen-thousandth generation. We may suppose that only one (F) of the two species (E and F) which were least closely related to the other nine original species, has transmitted descendants to this late stage of descent.

The new species in our diagram, descended from the original eleven species, will now be fifteen in number. Owing to the divergent tendency of natural selection, the extreme amount of difference in character between species a^{14} and z^{14} will be much greater than that between the most distinct of the original eleven species. The new species, moreover, will be allied to each other in a widely different manner. Of the eight descendants from (A) the three marked a^{14}, q^{14}, p^{14}, will be nearly related from having recently branched off from a^{10}; b^{14} and f^{14}, from having diverged at an earlier period from a^5, will be in some degree distinct from the three first-named species; and lastly, o^{14}, e^{14}, and m^{14}, will be nearly related one to the other, but, from having diverged at the first commencement of the process of modification, will be widely different from the other five species, and may constitute a sub-genus or a distinct genus.

The six descendants from (I) will form two sub-genera or genera. But as the original species (I) differed largely from (A), standing nearly at the extreme end of the original genus, the six descendants from (I) will, owing to inheritance alone, differ considerably from the eight descendants from (A); the two groups, moreover, are supposed to have gone on diverging in different directions. The intermediate species, also (and this is a very important consideration), which connected the original species (A) and (I), have all become, except (F), extinct, and have left no descendants. Hence the six new species descended from (I), and the eight descendants from (A), will have to be ranked as very distinct genera, or even as distinct sub-families.

Thus it is, as I believe, that two or more genera are produced by descent with modification, from two or more species of the same genus. And the two or more parent-species are supposed to be descended from some one species of an earlier genus. In our diagram this is indicated by the broken lines beneath the capital letters, converging in sub-branches downwards towards a single point; this point represents a species, the supposed progenitor of our several new sub-genera and genera.

It is worth while to reflect for a moment on the character of the new species F^{14}, which is supposed not to have diverged much in character, but to have retained the form of (F), either unaltered or altered only in a slight degree. In this case, its affinities to the other fourteen new species will be of a curious and circuitous nature. Being descended from a form that stood between the parent-species (A) and (I), now supposed to be extinct and unknown, it will be in some degree intermediate in character between the two groups descended from these two species. But as these two groups have gone on diverging in character from the type of their parents, the new species (F14) will not be directly intermediate between them, but rather between types of the two groups; and every naturalist will be able to call such cases before his mind.

In the diagram, each horizontal line has hitherto been supposed to represent a thousand generations, but each may represent a million or more generations; it may also represent a section of the successive strata of the earth's crust including extinct remains. We shall, when we come to our chapter on geology, have to refer again to this subject, and I think we shall then see that the diagram throws light on the affinities of extinct beings, which, though generally belonging to the same orders, families, or genera, with those now living, yet are often, in some degree, intermediate in character between existing groups; and we can understand this fact, for the extinct species lived at various remote epochs when the branching lines of descent had diverged less.

I see no reason to limit the process of modification, as now explained, to the formation of genera alone. If, in the diagram, we suppose the amount of change represented by each successive group of diverging dotted lines to be great, the forms marked a^{14} to p^{14}, those marked b^{14} and f^{14}, and those marked o^{14} to m^{14}, will form three very distinct genera. We shall also have two very distinct genera descended from (I), differing widely from the descendants of (A). Those two groups of genera will thus form two distinct families, or orders, according to the amount of divergent modification supposed to be represented in the diagram. And the two new families, or orders, are descended from two species of the original genus; and these are supposed to be descended from some still more ancient and unknown form.

We have seen that in each country it is the species belonging to the larger genera which oftenest present varieties or incipient species. This, indeed, might have been expected; for as natural selection acts through one form having some advantage over other forms in the struggle for existence, it will chiefly act on those which already have some advantage; and the largeness of any group shows that its species have inherited from a common ancestor some advantage in common. Hence, the struggle for the production of new and modified descendants will mainly lie between the larger groups, which are all trying to increase in number. One large group will slowly conquer another large group, reduce its number, and thus lessen its chance of further variation and improvement. Within the same large group, the later and more highly perfected sub-groups, from branching out and seizing on many new places in the polity of nature, will constantly tend to supplant and destroy the earlier and less improved sub-groups. Small and broken groups and sub-groups will finally disappear. Looking to the future, we can predict that the groups of organic beings which are now large and triumphant, and which are least broken up, that is, which have as yet suffered least extinction, will, for a long period, continue to increase. But which groups will ultimately prevail, no man can predict; for we know that many groups, formerly most extensively developed, have now become extinct. Looking still more remotely to the future, we may predict that, owing to the continued and steady increase of the

larger groups, a multitude of smaller groups will become utterly extinct, and leave no modified descendants; and consequently that, of the species living at any one period, extremely few will transmit descendants to a remote futurity. I shall have to return to this subject in the chapter on Classification, but I may add that as, according to this view, extremely few of the more ancient species have transmitted descendants to the present day, and, as all the descendants of the same species form a class, we can understand how it is that there exist so few classes in each main division of the animal and vegetable kingdoms. Although few of the most ancient species have left modified descendants, yet, at remote geological periods, the earth may have been almost as well peopled with species of many genera, families, orders and classes, as at the present day.

ON THE DEGREE TO WHICH ORGANISATION TENDS TO ADVANCE

Natural selection acts exclusively by the preservation and accumulation of variations, which are beneficial under the organic and inorganic conditions to which each creature is exposed at all periods of life. The ultimate result is that each creature tends to become more and more improved in relation to its conditions. This improvement inevitably leads to the gradual advancement of the organisation of the greater number of living beings throughout the world. But here we enter on a very intricate subject, for naturalists have not defined to each other's satisfaction what is meant by an advance in organisation. Amongst the vertebrata the degree of intellect and an approach in structure to man clearly come into play. It might be thought that the amount of change which the various parts and organs pass through in their development from embryo to maturity would suffice as a standard of comparison; but there are cases, as with certain parasitic crustaceans, in which several parts of the structure become less perfect, so that the mature animal cannot be called higher than its larva. Von

Baer's standard seems the most widely applicable and the best, namely, the amount of differentiation of the parts of the same organic being, in the adult state, as I should be inclined to add, and their specialisation for different functions; or, as Milne Edwards would express it, the completeness of the division of physiological labour. But we shall see how obscure this subject is if we look, for instance, to fishes, amongst which some naturalists rank those as highest which, like the sharks, approach nearest to amphibians; whilst other naturalists rank the common bony or teleostean fishes as the highest, inasmuch as they are most strictly fish-like, and differ most from the other vertebrate classes. We see still more plainly the obscurity of the subject by turning to plants, amongst which the standard of intellect is of course quite excluded; and here some botanists rank those plants as highest which have every organ, as sepals, petals, stamens, and pistils, fully developed in each flower; whereas other botanists, probably with more truth, look at the plants which have their several organs much modified and reduced in number as the highest.

If we take as the standard of high organisation, the amount of differentiation and specialisation of the several organs in each being when adult (and this will include the advancement of the brain for intellectual purposes), natural selection clearly leads towards this standard: for all physiologists admit that the specialisation of organs, inasmuch as in this state they perform their functions better, is an advantage to each being; and hence the accumulation of variations tending towards specialisation is within the scope of natural selection. On the other hand, we can see, bearing in mind that all organic beings are striving to increase at a high ratio and to seize on every unoccupied or less well occupied place in the economy of nature, that it is quite possible for natural selection gradually to fit a being to a situation in which several organs would be superfluous or useless: in such cases there would be retrogression in the scale of organisation.

Whether organisation on the whole has actually advanced from the remotest geological periods to the present day will be more conveniently discussed in our chapter on Geological Succession.

But it may be objected that if all organic beings thus tend to rise in the scale, how is it that throughout the world a multitude of the lowest forms still exist; and how is it that in each great class some forms are far more highly developed than others? Why have not the more highly developed forms everywhere supplanted and exterminated the lower? Lamarck, who believed in an innate and inevitable tendency towards perfection in all organic beings, seems to have felt this difficulty so strongly that he was led to suppose that new and simple forms are continually being produced by spontaneous generation. Science has not as yet proved the truth of this belief, whatever the future may reveal. On our theory the continued existence of lowly organisms offers no difficulty; for natural selection, or the Survival of the Fittest, does not necessarily include progressive development – it only takes advantage of such variations as arise and are beneficial to each creature under its complex relations of life. And it may be asked what advantage, as far as we can see, would it be to an infusorian animalcule – to an intestinal worm – or even to an earth-worm, to be highly organised. If it were no advantage, these forms would be left, by natural selection, unimproved or but little improved, and might remain for indefinite ages in their present lowly condition. And geology tells us that some of the lowest forms, as the infusoria and rhizopods, have remained for an enormous period in nearly their present state. But to suppose that most of the many now existing low forms have not in the least advanced since the first dawn of life would be extremely rash; for every naturalist who has dissected some of the beings now ranked as very low in the scale, must have been struck with their really wondrous and beautiful organisation.

Nearly the same remarks are applicable, if we look to the different grades of organisation within

Chrondral (cartilaginous) and boned fish including
2 Hammerhead shark 5 Sting ray 6 Sturgeon 8 Blenny

mammals when inhabiting the water lie under a disadvantage in having to come continually to the surface to breathe. With fishes, members of the shark family would not tend to supplant the lancelet; for the lancelet, as I hear from Fritz Müller, has as sole companion and competitor on the barren sandy shore of South Brazil, an anomalous annelid. The three lowest orders of mammals, namely, marsupials, edentata, and rodents, co-exist in South America in the same region with numerous monkeys, and probably interfere little with each other. Although organisation, on the whole, may have advanced and be still advancing throughout the world, yet the scale will always present many degrees of perfection; for the high advancement of certain whole classes, or of certain members of each class, does not at all necessarily lead to the extinction of those groups with which they do not enter into close competition. In some cases, as we shall hereafter see, lowly organised forms appear to have been preserved to the present day, from inhabiting confined or peculiar stations, where they have been subjected to less severe competition, and where their scanty numbers have retarded the chance of favourable variations arising.

Finally, I believe that many lowly organised forms now exist throughout the world, from various causes. In some cases variations or individual differences of a favourable nature may never have arisen for natural selection to act on and accumulate. In no case, probably, has time sufficed for the utmost possible amount of development. In some few cases there has been what we must call retrogression of organisation. But the main cause lies in the fact that under very simple conditions of life a high organisation would be of no service – possibly would be of actual disservice, as being of a more delicate nature, and more liable to be put out of order and injured.

Looking to the first dawn of life, when all organic beings, as we may believe, presented the simplest structure, how, it has been asked, could

the same great group; for instance, in the Vertebrata, to the co-existence of mammals and fish – amongst Mammalia, to the co-existence of man and the ornithorhynchus – amongst fishes, to the co-existence of the shark and the lancelet (Amphioxus), which latter fish in the extreme simplicity of its structure approaches the invertebrate classes. But mammals and fish hardly come into competition with each other; the advancement of the whole class of mammals, or of certain members in this class, to the highest grade would not lead to their taking the place of fishes. Physiologists believe that the brain must be bathed by warm blood to be highly active, and this requires aerial respiration; so that warm-blooded the first step in the advancement or differentiation of parts have arisen? Mr. Herbert Spencer would probably answer that, as soon as simple unicellular organisms came by growth or division to be compounded of several cells, or became attached to any supporting surface, his law 'that homologous units of any order become differentiated in proportion as their relations to incident forces become different' would come into action. But as we have no facts to guide us, speculation on the subject is almost useless. It is, however, an error to suppose that there would be no struggle for existence, and, consequently, no natural selection, until many forms had been produced: variations in a single species inhabiting an isolated station might be beneficial, and thus the whole mass of individuals might be modified, or two distinct forms might arise. But, as I remarked towards the close of the introduction, no one ought to feel surprise at much remaining as yet unexplained on the origin of species, if we make due allowance for our profound ignorance on the mutual relations of the inhabitants of the world at the present time, and still more so during past ages.

CONVERGENCE OF CHARACTER

Mr. H.C. Watson thinks that I have overrated the importance of divergence of character (in which, however, he apparently believes), and that convergence, as it may be called, has likewise played a part. If two species belonging to two distinct though allied genera, had both produced a large number of new and divergent forms, it is conceivable that these might approach each other so closely that they would have all to be classed under the same genus; and thus the descendants of two distinct genera would converge into one. But it would in most cases be extremely rash to attribute to convergence a close and general similarity of structure in the modified descendants of widely distinct forms. The shape of a crystal is determined solely by the molecular forces, and it is not surprising that dissimilar substances should

sometimes assume the same form; but with organic beings we should bear in mind that the form of each depends on an infinitude of complex relations, namely on the variations which have arisen, these being due to causes far too intricate to be followed out – on the nature of the variations which have been preserved or selected, and this depends on the surrounding physical conditions, and in a still higher degree on the surrounding organisms with which each being has come into competition – and lastly, on inheritance (in itself a fluctuating element) from innumerable progenitors, all of which have had their forms determined through equally complex relations. It is incredible that the descendants of two organisms, which had originally differed in a marked manner, should ever afterwards converge so closely as to lead to a near approach to identity throughout their whole organisation. If this had occurred, we should meet with the same form, independently of genetic connection, recurring in widely separated geological formations; and the balance of evidence is opposed to any such an admission.

Mr. Watson has also objected that the continued action of natural selection, together with divergence of character, would tend to make an indefinite number of specific forms. As far as mere inorganic conditions are concerned, it seems probable that a sufficient number of species would soon become adapted to all considerable diversities of heat, moisture, etc.; but I fully admit that the mutual relations of organic beings are more important; and as the number of species in any country goes on increasing, the organic conditions of life must become more and more complex. Consequently there seems at first no limit to the amount of profitable diversification of structure, and therefore no limit to the number of species which might be produced. We do not know that even the most prolific area is fully stocked with specific forms: at the Cape of Good Hope and in Australia, which support such an astonishing number of species, many European plants have become naturalised. But geology shows us, that from an early part of the tertiary period the number of species of shells, and that from the middle part of this same period, the number of mammals, has not greatly or at all increased. What then checks an indefinite increase in the number of species? The amount of life (I do not mean the number of specific forms) supported on an area must have a limit, depending so largely as it does on physical conditions; therefore, if an area be inhabited by very many species, each or nearly each species will be represented by few individuals; and such species will be liable to extermination from accidental fluctuations in the nature of the seasons or in the number of their enemies. The process of extermination in such cases would be rapid, whereas the production of new species must always be slow. Imagine the extreme case of as many species as individuals in England, and the first severe winter or very dry summer would exterminate thousands on thousands of species. Rare species, and each species will become rare if the number of species in any country becomes indefinitely increased, will, on the principal often explained, present within a given period few favourable variations;

Antlers of a Red Deer:
br brow-tine *bez* bez-tine *a.r.* antler royal *s.r.* sur-royal

consequently, the process of giving birth to new specific forms would thus be retarded. When any species becomes very rare, close interbreeding will help to exterminate it; authors have thought that this comes into play in accounting for the deterioration of the aurochs in Lithuania, of red deer in Scotland, and of bears in Norway, etc. Lastly, and this I am inclined to think is the most important element, a dominant species, which has already beaten many competitors in its own home, will tend to spread and supplant many others. Alph. de Candolle has shown that those species which spread widely tend generally to spread *very* widely, consequently they will tend to supplant and exterminate several species in several areas, and thus check the inordinate increase of specific forms throughout the world. Dr. Hooker has recently shown that in the southeast corner of Australia, where, apparently, there are many invaders from different quarters of the globe, the endemic Australian species have been greatly reduced in number. How much weight to attribute to these several considerations I will not pretend to say; but conjointly they must limit in each country the tendency to an indefinite augmentation of specific forms.

SUMMARY OF CHAPTER

If under changing conditions of life organic beings present individual differences in almost every part of their structure, and this cannot be disputed; if there be, owing to their geometrical rate of increase, a severe struggle for life at some age, season or year, and this certainly cannot be disputed; then, considering the infinite complexity of the relations of all organic beings to each other and to their conditions of life, causing an infinite diversity in structure, constitution, and habits, to be advantageous to them, it would be a most extraordinary fact if no variations had ever occurred useful to each being's own welfare, in the same manner as so many variations have occurred useful to man. But if variations useful to any organic being ever do occur, assuredly individuals

CERVUS CANADENSIS, RAY.
AMERICAN ELK - WAPITI DEER.
½ Natural Size
MALE AND FEMALE

Male and female American Elk (Wapiti Deer). The Native American word 'waapiti' means 'white rump'.
This species is very close to the European Red Deer

thus characterised will have the best chance of being preserved in the struggle for life; and from the strong principle of inheritance, these will tend to produce offspring similarly characterised. This principle of preservation, or the Survival of the Fittest, I have called Natural Selection. It leads to the improvement of each creature in relation to its organic and inorganic conditions of life; and consequently, in most cases, to what must be regarded as an advance in organisation. Nevertheless, low and simple forms will long endure if well fitted for their simple conditions of life.

Natural selection, on the principle of qualities being inherited at corresponding ages, can modify the egg, seed, or young as easily as the adult. Amongst many animals, sexual selection will have given its aid to ordinary selection by assuring to the most vigorous and best adapted males the greatest number of offspring. Sexual selection will also give characters useful to the males alone in their struggles or rivalry with other males; and these characters will be transmitted to one sex or to both sexes, according to the form of inheritance which prevails.

Whether natural selection has really thus acted in adapting the various forms of life to their several conditions and stations, must be judged by the general tenor and balance of evidence given in the following chapters. But we have already seen how it entails extinction; and how largely extinction has acted in the world's history, geology plainly declares. Natural selection, also, leads to divergence of character; for the more organic beings diverge in structure, habits, and constitution, by so much the more can a large number be supported on the area, – of which we see proof by looking to the inhabitants of any small spot, and to the productions naturalised in foreign lands. Therefore, during the modification of the descendants of any one species, and during the incessant struggle of all species to increase in numbers, the more diversified the descendants become, the better will be their chance of success in the battle for life. Thus the small differences distinguishing varieties of the same species, steadily tend to increase, till they equal the greater differences between species of the same genus, or even of distinct genera.

We have seen that it is the common, the widely diffused, and widely ranging species, belonging to the larger genera within each class, which vary most; and these tend to transmit to their modified offspring that superiority which now makes them dominant in their own countries. Natural selection, as has just been remarked, leads to divergence of character and to much extinction of the less improved and intermediate forms of life. On these principles, the nature of the affinities, and the generally well defined distinctions between the innumerable organic beings in each class throughout the world, may be explained. It is a truly wonderful fact – the wonder of which we are apt to overlook from familiarity – that all animals and all plants throughout all time and space should be related to each other in groups, subordinate to groups, in the manner which we everywhere behold – namely, varieties of the same species most closely related, species of the same genus less closely and unequally related, forming sections and sub-genera, species of distinct genera much less closely related, and genera related in different degrees, forming sub-families, families, orders, sub-classes, and classes. The several subordinate groups in any class cannot be ranked in a single file, but seem clustered round points, and these round other points, and so on in almost endless cycles. If species had been independently created, no explanation would have been possible of this kind of classification; but it is explained through inheritance and the complex action of natural selection, entailing extinction and divergence of character, as we have seen illustrated in the diagram.

The affinities of all the beings of the same class have sometimes been represented by a great tree. I believe this simile largely speaks the truth.

The green and budding twigs may represent existing species; and those produced during former years may represent the long succession of extinct species. At each period of growth all the growing twigs have tried to branch out on all sides, and to overtop and kill the surrounding twigs and branches, in the same manner as species and groups of species have at all times overmastered other species in the great battle for life. The limbs divided into great branches, and these into lesser and lesser branches, were themselves once, when the tree was young, budding twigs; and this connexion of the former and present buds by ramifying branches may well represent the classification of all extinct and living species in groups subordinate to groups. Of the many twigs which flourished when the tree was a mere bush, only two or three, now grown into great branches, yet survive and bear the other branches; so with the species which lived during long-past geological periods, very few have left living and modified descendants. From the first growth of the tree, many a limb and branch has decayed and dropped off; and these fallen branches of various sizes may represent those whole orders, families, and genera which have now no living representatives, and which are known to us only in a fossil state. As we here and there see a thin, straggling branch springing from a fork low down in a tree, and which by some chance has been favoured and is still alive on its summit, so we occasionally see an animal like the Ornithorhynchus or Lepidosiren, which in some small degree connects by its affinities two large branches of life, and which has apparently been saved from fatal competition by having inhabited a protected station. As buds give rise by growth to fresh buds, and these, if vigorous, branch out and overtop on all sides many a feebler branch, so by generation I believe it has been with the great Tree of Life, which fills with its dead and broken branches the crust of the earth, and covers the surface with its ever-branching and beautiful ramifications.

Chapter V

LAWS OF VARIATION

Effects of changed conditions–Use and disuse, combined with natural selection; organs of flight and of vision–Acclimatisation–Correlated variation–Compensation and economy of growth–False correlations–Multiple, rudimentary, and lowly organised structures variable–Parts developed in an unusual manner are highly variable; specific characters more variable than generic: secondary sexual characters variable–Species of the same genus vary in an analogous manner–Reversions to long-lost characters–Summary.

I have hitherto sometimes spoken as if the variations – so common and multiform with organic beings under domestication, and in a lesser degree with those under nature – were due to chance. This, of course, is a wholly incorrect expression, but it serves to acknowledge plainly our ignorance of the cause of each particular variation. Some authors believe it to be as much the function of the reproductive system to produce individual differences, or slight deviations of structure, as to make the child like its parents. But the fact of variations and monstrosities occurring much more frequently under domestication than under nature, and the greater variability of species having wide ranges than of those with restricted ranges, lead to the conclusion that variability is generally related to the conditions of life to which each species has been exposed during several successive generations. In the first chapter I attempted to show that changed conditions act in two ways, directly on the whole organisation or on certain parts alone, and indirectly through the reproductive system. In all cases there are two factors, the nature of the organism, which is much the most important of the two, and the nature of the conditions. The direct action of changed conditions leads to definite or indefinite results. In the latter case the organisation seems

to become plastic, and we have much fluctuating variability. In the former case the nature of the organism is such that it yields readily, when subjected to certain conditions, and all, or nearly all the individuals become modified in the same way.

It is very difficult to decide how far changed conditions, such as of climate, food, etc., have acted in a definite manner. There is reason to believe that in the course of time the effects have been greater than can be proved by clear evidence. But we may safely conclude that the innumerable complex coadaptations of structure, which we see throughout nature between various organic beings, cannot be attributed simply to such action. In the following cases the conditions seem to have produced some slight definite effect: E. Forbes asserts that shells at their southern limit, and when living in shallow water, are more brightly coloured than those of the same species from further north or from a greater depth; but this certainly does not always hold good. Mr. Gould believes that birds of the same species are more brightly coloured under a clear atmosphere, than when living near the coast or on islands; and Wollaston is convinced that residence near the sea affects the colours of insects. Moquin-Tandon gives a list of plants which, when growing near the seashore, have their leaves in some degree fleshy, though not elsewhere fleshy. These slightly varying organisms are interesting in as far as they present characters analogous to those possessed by the species which are confined to similar conditions.

When a variation is of the slightest use to any being, we cannot tell how much to attribute to the accumulative action of natural selection, and how much to the definite action of the conditions of life. Thus, it is well known to furriers that animals of the same species have thicker and better fur the further north they live; but who can tell how much of this difference may be due to the warmest clad individuals having been favoured and preserved during many generations, and how much to the action of the severe climate? For it would appear that climate has some direct action on the hair of our domestic quadrupeds.

Instances could be given of similar varieties being produced from the same species under external conditions of life as different as can well be conceived; and, on the other hand, of dissimilar varieties being produced under apparently the same external conditions. Again, innumerable instances are known to every naturalist, of species keeping true, or not varying at all, although living under the most opposite climates. Such considerations as these incline me to lay less weight on the direct action of the surrounding conditions, than on a tendency to vary, due to causes of which we are quite ignorant.

In one sense the conditions of life may be said, not only to cause variability, either directly or indirectly, but likewise to include natural selection, for the conditions determine whether this or that variety shall survive. But when man is the selecting agent, we clearly see that the two elements of change are distinct; variability is in some manner excited, but it is the will of man which accumulates the variations in certain directions; and it is this latter agency which answers to the survival of the fittest under nature.

EFFECTS OF THE INCREASED USE AND DISUSE OF PARTS, AS CONTROLLED BY NATURAL SELECTION

From the facts alluded to in the first chapter, I think there can be no doubt that use in our domestic animals has strengthened and enlarged certain parts, and disuse diminished them; and that such modifications are inherited. Under free nature, we have no standard of comparison, by which to judge of the effects of long-continued use or disuse, for we know not the parent-forms; but many animals possess structures which can be best explained by the effects of disuse. As Professor Owen has remarked, there is no greater anomaly in nature than a bird that cannot fly; yet there are several in this state. The logger-headed duck of South America can only flap along the

Left leg of an African Ostrich: lateral view. It has only two clawed toes but these enable the bird to reach a running speed of 40 mph

Male American Bustard displaying its inflated throat pouch

surface of the water, and has its wings in nearly the same condition as the domestic Aylesbury duck: it is a remarkable fact that the young birds, according to Mr. Cunningham, can fly, whilst the adults have lost this power. As the larger ground-feeding birds seldom take flight except to escape danger, it is probable that the nearly wingless condition of several birds, now inhabiting or which lately inhabited several oceanic islands, tenanted by no beasts of prey, has been caused by disuse. The ostrich indeed inhabits continents, and is exposed to danger from which it cannot escape by flight, but it can defend itself, by kicking its enemies, as efficiently as many quadrupeds. We may believe that the progenitor of the ostrich genus had habits like those of the bustard, and that, as the size and weight of its body were increased during successive generations, its legs were used more and its wings less, until they became incapable of flight.

Kirby has remarked (and I have observed the same fact) that the anterior tarsi, or feet, of many male dung-feeding beetles are often broken off; he examined seventeen specimens in his own collection, and not one had even a relic left. In the Onites apelles the tarsi are so habitually lost that the insect has been described as not having them. In some other genera they are present, but in a rudimentary condition. In the Ateuchus or sacred beetle of the Egyptians, they are totally deficient. The evidence that accidental mutilations can be inherited is at present not decisive; but the remarkable cases observed by Brown-Séquard in guinea-pigs, of the inherited effects of operations, should make us cautious in denying this tendency. Hence, it will perhaps be safest to look at the entire absence of the anterior tarsi in Ateuchus, and their rudimentary condition in some other genera, not as cases of inherited mutilations, but as due to the effects of long-continued disuse; for as many dung-feeding beetles are generally found with their tarsi lost, this must happen early in life; therefore the tarsi cannot be of much importance or be much used by these insects.

In some cases we might easily put down to disuse modifications of structure which are wholly, or mainly, due to natural selection. Mr. Wollaston has discovered the remarkable fact that 200 beetles, out of the 550 species (but more are now known) inhabiting Madeira, are so far deficient in wings that they cannot fly; and that, of the twenty-nine endemic genera, no less than twenty-three have all their species in this condition! Several facts, – namely, that beetles in many parts of the world are very frequently blown to sea and perish; that the beetles in Madeira, as observed by Mr. Wollaston, lie much concealed, until the wind lulls and the sun shines; that the proportion of wingless beetles is larger on the exposed Desertas than in Madeira itself; and especially the extraordinary fact, so strongly insisted on by Mr. Wollaston, that certain large groups of beetles, elsewhere excessively numerous, which absolutely require the use of their wings, are here almost entirely absent. These several considerations make me believe that the wingless condition of so many Madeira beetles is mainly due to the action of natural selection, combined probably with disuse. For during many successive generations each individual beetle which flew least, either from its wings having been ever so little less perfectly developed or from indolent habit, will have had the best chance of surviving from not being blown out to sea; and, on the other hand, those beetles which most readily took to flight would oftenest have been blown to sea, and thus destroyed.

The insects in Madeira which are not ground-feeders, and which, as certain flower-feeding coleoptera and lepidoptera, must habitually use their wings to gain their subsistence, have, as Mr. Wollaston suspects, their wings not at all reduced, but even enlarged. This is quite compatible with the action of natural selection. For when a new insect first arrived on the island, the tendency of natural selection to enlarge or to reduce the wings, would depend on whether a greater number of individuals were saved by successfully battling with the winds, or by giving

Six insectivores:
1 Hedgehog *2* Mole *3* Elephant shrew *4* Shrew *5* Tanrec, a native of Madagascar
6 Banxring, a squirrel-like East Indian mammal

up the attempt and rarely or never flying. As with mariners shipwrecked near a coast, it would have been better for the good swimmers if they had been able to swim still further, whereas it would have been better for the bad swimmers if they had not been able to swim at all and had stuck to the wreck.

The eyes of moles and of some burrowing rodents are rudimentary in size, and in some cases are quite covered by skin and fur. This state of the eyes is probably due to gradual reduction from disuse, but aided perhaps by natural selection. In South America, a burrowing rodent, the tucutucu, or Ctenomys, is even more subterranean in its habits than the mole; and I was assured by a Spaniard, who had often caught them, that they were frequently blind. One which I kept alive was certainly in this condition, the cause, as appeared on dissection, having been inflammation of the nictitating membrane. As frequent inflammation of the eyes must be injurious to any animal, and as eyes are certainly not necessary to animals having subterranean habits, a reduction in their size, with the adhesion of the eyelids and growth of fur over them, might in such case be an advantage; and if so, natural selection would aid the effects of disuse.

It is well known that several animals, belonging to the most different classes, which inhabit the caves of Carniola and Kentucky, are blind. In some of the crabs the foot-stalk for the eye remains, though the eye is gone; the stand for the telescope is there, though the telescope with its glasses has been lost. As it is difficult to imagine that eyes, though useless, could be in any way injurious to animals living in darkness, their loss may be attributed to disuse. In one of the blind animals, namely, the cave rat (Neotoma), two of which were captured by Professor Silliman at above half a mile distance from the mouth of the cave, and therefore not in the profoundest depths, the eyes were lustrous and of large size; and these animals, as I am informed by Professor Silliman, after having been exposed for about a month to a graduated light, acquired a dim perception of objects.

It is difficult to imagine conditions of life more similar than deep limestone caverns under a nearly similar climate; so that, in accordance with the old view of the blind animals having been separately created for the American and European caverns, very close similarity in their organisation and affinities might have been expected. This is certainly not the case if we look at the two whole faunas; with respect to the insects alone, Schiödte has remarked: 'We are accordingly prevented from considering the entire phenomenon in any other light than something purely local, and the similarity which is exhibited in a few forms between the Mammoth Cave (in Kentucky) and the caves in Carniola, otherwise than as a very plain expression of that analogy which subsists generally between the fauna of Europe and of North America.' On my view we must suppose that American animals, having in most cases ordinary powers of vision, slowly migrated by successive generations from the outer world into the deeper and deeper recesses of the Kentucky caves, as did European animals into the caves of Europe. We have some evidence of this gradation of habit; for, as Schiödte remarks: 'We accordingly look upon the subterranean faunas as small ramifications which have penetrated into the earth from the geographically limited faunas of the adjacent tracts, and which, as they extended themselves into darkness, have been accommodated to surrounding circumstances. Animals not far remote from ordinary forms, prepare the transition from light to darkness. Next follow those that are constructed for twilight; and, last of all, those destined for total darkness, and whose formation is quite peculiar.' These remarks of Schiödte's it should be understood, apply not to the same, but to distinct species. By the time that an animal had reached, after numberless generations, the deepest recesses, disuse will on this view have more or less perfectly obliterated its eyes, and natural selection will often have effected other changes, such as an increase in the length of the antennae or palpi, as a compensation for blindness. Notwithstanding such modifications, we might expect still to see in the cave-animals of America, affinities to the other inhabitants of that continent, and in those of Europe to the inhabitants of the European continent. And this is the case with some of the American cave animals, as I hear from Professor Dana; and some of the European cave insects are very closely allied to those of the surrounding country. It would be difficult to give any rational explanation of the affinities of the blind cave animals to the other inhabitants of the two continents on the ordinary view of their independent creation. That several of the inhabitants of the caves of the Old and New Worlds should be closely related, we might expect from the well-known relationship of most of their other productions. As a blind species of Bathyscia is found in abundance on shady rocks far from caves, the loss of vision in the cave species of this one genus has probably had no relation to its dark habitation; for it is natural that an insect already deprived of vision should readily become adapted to dark caverns. Another blind genus (Anophthalmus) offers this remarkable peculiarity, that the species, as Mr. Murray observes, have not as yet been found anywhere except in caves; yet those which inhabit the several caves of Europe and America are distinct; but it is possible that the progenitors of these several species, whilst they were furnished with eyes, may formerly have ranged over both continents, and then have become extinct, excepting in their present secluded abodes. Far from feeling surprise that some of the cave animals should be very anomalous, as Agassiz has remarked in regard to the blind fish, the Amblyopsis, and as is the case with the blind Proteus, with reference to the reptiles of Europe, I am only surprised that more wrecks of ancient life have not been preserved, owing to the less severe competition to which the scanty inhabitants of these dark abodes will have been exposed.

ACCLIMATISATION

Habit is hereditary with plants, as in the period of flowering, in the time of sleep, in the amount of rain requisite for seeds to germinate, etc., and this leads me to say a few words on acclimatisation. As it is extremely common for distinct species belonging to the same genus to inhabit hot and cold countries, if it be true that all the species of the same genus are descended from a single parent-form, acclimatisation must be readily effected during a long course of descent. It is notorious that each species is adapted to the climate of its own home: species from an arctic or even from a temperate region cannot endure a tropical climate, or conversely. So again, many succulent plants cannot endure a damp climate. But the degree of adaptation of species to the climates under which they live is often overrated. We may infer this from our frequent inability to predict whether or not an imported plant will endure our climate, and from the number of plants and animals brought from different countries which are here perfectly healthy. We have reason to believe that species in a state of nature are closely limited in their ranges by the competition of other organic beings quite as much as, or more than, by adaptation to particular climates. But whether or not this adaptation is in most cases very close, we have evidence with some few plants, of their becoming, to a certain extent, naturally habituated to different temperatures; that is, they become acclimatised: thus the pines and rhododendrons, raised from seed collected by Dr. Hooker from the same species growing at different heights on the Himalayas, were found to possess in this country different constitutional powers of resisting cold. Mr. Thwaites informs me that he has observed similar facts in Ceylon; analogous observations have been made by Mr. H.C. Watson on European species of plants brought from the Azores to England; and I could give other cases. In regard to animals, several authentic instances could be adduced of species having largely extended, within historical times, their range from warmer to colder latitudes, and conversely; but we do not positively know that these animals were strictly adapted to their native climate, though in all ordinary cases we assume such to be the case; nor do we know that they have subsequently become specially acclimatised to their new homes, so as to be better fitted for them than they were at first.

As we may infer that our domestic animals were originally chosen by uncivilised man because they were useful and because they bred readily under confinement, and not because they were subsequently found capable of far-extended transportation, the common and extraordinary capacity in our domestic animals of not only withstanding the most different climates, but of being perfectly fertile (a far severer test) under them, may be used as an argument that a large proportion of other animals now in a state of nature could easily be brought to bear widely different climates. We must not, however, push the foregoing argument too far, on account of the probable origin of some of our domestic animals from several wild stocks; the blood, for instance, of a tropical and arctic wolf may perhaps be mingled in our domestic breeds. The rat and mouse cannot be considered as domestic animals, but they have been transported by man to many parts of the world, and now have a far wider range than any other rodent; for they live under the cold climate of Faroë in the north and of the Falklands in the south, and on many an island in the torrid zones. Hence adaptation to any special climate may be looked at as a quality readily grafted on an innate wide flexibility of constitution, common to most animals. On this view, the capacity of enduring the most different climates by man himself and by his domestic animals, and the fact of the extinct elephant and rhinoceros having formerly endured a glacial climate, whereas the living species are now all tropical or sub-tropical in their habits, ought not to be looked at as anomalies, but as examples of a very common flexibility of constitution, brought, under peculiar circumstances, into action.

How much of the acclimatisation of species to any peculiar climate is due to mere habit, and how much to the natural selection of varieties having different innate constitutions, and how much to both means combined, is an obscure question. That habit or custom has some influence, I must believe, both from analogy and from the incessant advice given in agricultural works, even in the ancient encyclopaedias of China, to be very cautious in transporting animals from one district to another. And as it is not likely that man should have succeeded in selecting so many breeds and sub-breeds with constitutions specially fitted for their own districts, the result must, I think, be due to habit. On the other hand, natural selection would inevitably tend to preserve those individuals which were born with constitutions best adapted to any country which they inhabited. In treatises on many kinds of cultivated plants, certain varieties are said to withstand certain climates better than others; this is strikingly shown in works on fruit trees published in the United States, in which certain varieties are habitually recommended for the northern and others for the southern States; and as most of these varieties are of recent origin, they cannot owe their constitutional differences to habit. The case of the Jerusalem artichoke, which is never propagated in England by seed, and of which, consequently, new varieties have not been produced, has even been advanced, as proving that acclimatisation cannot be effected, for it is now as tender as ever it was! The case, also, of the kidney bean has been often cited for a similar purpose, and with much greater weight; but until some one will sow, during a score of generations, his kidney beans so early that a very large proportion are destroyed by frost, and then collect seed from the few survivors, with care to prevent accidental crosses, and then again get seed from these seedlings, with the same precautions, the experiment cannot be said to have been even tried. Nor let it be supposed that differences in the constitution of seedling kidney

Top: The Long-haired or Plague rat of Australia
Bottom: Darwin's Mouse

beans never appear, for an account has been published how much more hardy some seedlings are than others; and of this fact I have myself observed striking instances.

On the whole, we may conclude that habit, or use and disuse, have, in some cases, played a considerable part in the modification of the constitution and structure; but that the effects have often been largely combined with, and sometimes overmastered by, the natural selection of innate variations.

CORRELATED VARIATION

I mean by this expression that the whole organisation is so tied together, during its growth and development, that when slight variations in any one part occur and are accumulated through natural selection, other parts become modified. This is a very important subject, most imperfectly understood, and no doubt wholly different classes of facts may be here easily confounded together. We shall presently see that simple inheritance often gives the false appearance of correlation. One of the most obvious real cases is, that variations of structure arising in the young or larvae naturally tend to affect the structure of the mature animal. The several parts which are homologous, and which, at an early embryonic period, are identical in structure, and which are necessarily exposed to similar conditions, seem eminently liable to vary in a like manner: we see this in the right and left sides of the body varying in the same manner; in the front and hind legs, and even in the jaws and limbs, varying together, for the lower jaw is believed by some anatomists to be homologous with the limbs. These tendencies, I do not doubt, may be mastered more or less completely by natural selection: thus a family of stags once existed with an antler only on one side; and if this had been of any great use to the breed, it might probably have been rendered permanent by natural selection.

Homologous parts, as has been remarked by some authors, tend to cohere; this is often seen in monstrous plants: and nothing is more common than the union of homologous parts in normal structures, as in the union of the petals into a tube. Hard parts seem to affect the form of adjoining soft parts; it is believed by some authors that with birds the diversity in the shape of the pelvis causes the remarkable diversity in the shape of the kidneys. Others believe that the shape of the pelvis in the human mother influences by pressure the shape of the head of the child. In snakes, according to Schlegel, the shape of the body and the manner of swallowing determine the position and form of several of the most important viscera.

The nature of the bond is frequently quite obscure. M. Is. Geoffroy St. Hilaire has forcibly remarked that certain malconformations frequently, and that others rarely, co-exist, without our being able to assign any reason. What can be more singular than the relation in cats between complete whiteness and blue eyes with deafness, or between the tortoise-shell colour and the female sex; or in pigeons, between their feathered feet and skin betwixt the outer toes, or between the presence of more or less down on the young pigeon when first hatched, with the future colour of its plumage; or, again, the relation between the hair and the teeth in the naked Turkish dog, though here no doubt homology comes into play? With respect to this latter case of correlation, I think it can hardly be accidental that the two orders of mammals which are most abnormal in their dermal covering, viz., Cetacea (whales) and Edentata (armadilloes, scaly ant-eaters, etc.), are likewise on the whole the most abnormal in their teeth, but there are so many exceptions to this rule, as Mr. Mivart has remarked, that it has little value.

I know of no case better adapted to show the importance of the laws of correlation and variation, independently of utility, and therefore of natural selection, than that of the difference between the outer and inner flowers in some Compositous and Umbelliferous plants. Everyone is familiar with the difference between the ray and

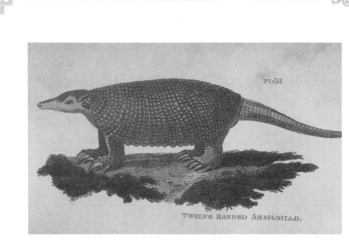

The Twelve-banded Armadillo, the largest of the
20 species of armadillo

central florets of, for instance, the daisy, and this difference is often accompanied with the partial or complete abortion of the reproductive organs. But in some of these plants the seeds also differ in shape and sculpture. These differences have sometimes been attributed to the pressure of the involucra on the florets, or to their mutual pressure, and the shape of the seeds in the ray-florets of some Compositae countenances this idea; but with the Umbelliferae it is by no means, as Dr. Hooker informs me, the species with the densest heads which most frequently differ in their inner and outer flowers. It might have been thought that the development of the ray-petals, by drawing nourishment from the reproductive organs causes their abortion; but this can hardly be the sole case, for in some Compositae the seeds of the outer and inner florets differ, without any difference in the corolla. Possibly these several differences may be connected with the different flow of nutriment towards the central and external flowers. We know, at least, that with irregular flowers, those nearest to the axis are most subject to peloria, that is to become abnormally symmetrical. I may add, as an instance of this fact, and as a striking case of correlation, that in many pelargoniums, the two upper petals in the central flower of the truss often lose their patches of darker colour; and when this occurs, the adherent nectary is quite aborted, the central

flower thus becoming peloric or regular. When the colour is absent from only one of the two upper petals, the nectary is not quite aborted but is much shortened.

With respect to the development of the corolla, Sprengel's idea that the ray-florets serve to attract insects, whose agency is highly advantageous or necessary for the fertilisation of these plants, is highly probable; and if so, natural selection may have come into play. But with respect to the seeds, it seems impossible that their differences in shape, which are not always correlated with any difference in the corolla, can be in any way beneficial; yet in the Umbelliferae these differences are of such apparent importance – the seeds being sometimes orthospermous in the exterior flowers and coelospermous in the central flowers – that the elder De Candolle founded his main divisions in the order on such characters. Hence modifications of structure, viewed by systematists as of high value, may be wholly due to the laws of variation and correlation, without being, as far as we can judge, of the slightest service to the species.

We may often falsely attribute to correlated variation structures which are common to whole groups of species, and which in truth are simply due to inheritance; for an ancient progenitor may have acquired through natural selection some one modification in structure, and, after thousands of generations, some other and independent modification; and these two modifications, having been transmitted to a whole group of descendants with diverse habits, would naturally be thought to be in some necessary manner correlated. Some other correlations are apparently due to the manner in which natural selection can alone act. For instance, Alph. De Candolle has remarked that winged seeds are never found in fruits which do not open; I should explain this rule by the impossibility of seeds gradually becoming winged through natural selection, unless the capsules were open; for in this case alone could the seeds, which were a little better adapted to be wafted by the wind, gain an advantage over others less well fitted for wide dispersal.

COMPENSATION AND ECONOMY OF GROWTH

The elder Geoffroy and Goethe propounded, at about the same time, their law of compensation or balancement of growth; or, as Goethe expressed it, 'in order to spend on one side, nature is forced to economise on the other side.' I think this holds true to a certain extent with our domestic productions: if nourishment flows to one part or organ in excess, it rarely flows, at least in excess, to another part; thus it is difficult to get a cow to give much milk and to fatten readily. The same varieties of the cabbage do not yield abundant and nutritious foliage and a copious supply of oil-bearing seeds. When the seeds in our fruits become atrophied, the fruit itself gains largely in size and quality. In our poultry, a large tuft of feathers on the head is generally accompanied by a diminished comb, and a large beard by diminished wattles. With species in a state of nature it can hardly be maintained that the law is of universal application; but many good observers, more especially botanists, believe in its truth. I will not, however, here give any instances, for I see hardly any way of distinguishing between the effects, on the one hand, of a part being largely developed through natural selection and another and adjoining part being reduced by the same process or by disuse, and, on the other hand, the actual withdrawal of nutriment from one part owing to the excess of growth in another and adjoining part.

I suspect, also, that some of the cases of compensation which have been advanced, and likewise some other facts, may be merged under a more general principle, namely, that natural selection is continually trying to economise in every part of the organisation. If under changed conditions of life a structure, before useful, becomes less useful, its diminution will be favoured, for it will profit the individual not to have its nutriment wasted in building up a useless structure. I can thus only understand a fact with which I was much struck when examining cirripedes, and of which many other instances could be given: namely, that when a cirripede is parasitic within another cirripede and is thus protected, it loses more or less completely its own shell or carapace. This is the case with the male Ibla, and in a truly extraordinary manner with the Proteolepas: for the carapace in all other cirripedes consists of the three highly important anterior segments of the head enormously developed, and furnished with great nerves and muscles; but in the parasitic and protected Proteolepas, the whole anterior part of the head is reduced to the merest rudiment attached to the bases of the prehensile antennae. Now the saving of a large and complex structure, when rendered superfluous, would be a decided advantage to each successive individual of the species; for in the struggle for life to which every animal is exposed, each would have a better chance of supporting itself, by less nutriment being wasted.

Thus, as I believe, natural selection will tend in the long run to reduce any part of the organisation, as soon as it becomes, through changed habits, superfluous, without by any means causing some other part to be largely developed in a corresponding degree. And conversely, that natural selection may perfectly well succeed in largely developing an organ without requiring as a necessary compensation the reduction of some adjoining part.

MULTIPLE, RUDIMENTARY, AND LOWLY-ORGANISED STRUCTURES ARE VARIABLE

It seems to be a rule, as remarked by Is. Geoffroy St. Hilaire, both with varieties and species, that when any part or organ is repeated many times in the same individual (as the vertebrae in snakes, and the stamens in polyandrous flowers) the number is variable; whereas the number of the same part or organ, when it occurs in lesser numbers, is constant. The same author as well as some botanists, have further remarked that multiple parts are extremely liable to vary in structure. As 'vegetative repetition,' to use Professor Owen's expression, is a sign of low organisation; the foregoing statements accord with the common opinion of naturalists, that beings which stand low in the scale of nature are more variable than those which are higher. I presume that lowness here means that the several parts of the organisation have been but little specialised for particular functions; and as long as the same part has to perform diversified work, we can perhaps see why it should remain variable, that is, why natural selection should not have preserved or rejected each little deviation of form so carefully as when the part has to serve for some one special purpose. In the same way that a knife which has to cut all sorts of things may be of almost any shape; whilst a tool for some particular purpose must be of some particular shape. Natural selection, it should never be forgotten, can act solely through and for the advantage of each being.

Rudimentary parts, as is generally admitted, are apt to be highly variable. We shall have to recur to this subject; and I will here only add that their variability seems to result from their uselessness, and consequently from natural selection having had no power to check deviations in their structure.

A PART DEVELOPED IN ANY SPECIES IN AN EXTRAORDINARY DEGREE OR MANNER, IN COMPARISON WITH THE SAME PART IN ALLIED SPECIES, TENDS TO BE HIGHLY VARIABLE

Several years ago I was much struck by a remark, to the above effect, made by Mr. Waterhouse. Professor Owen, also, seems to have come to a nearly similar conclusion. It is hopeless to attempt to convince any one of the truth of the above proposition without giving the long array of facts which I have collected, and which cannot possibly be here introduced. I can only state my conviction that it is a rule of high generality. I am aware of

several causes of error, but I hope that I have made due allowances for them. It should be understood that the rule by no means applies to any part, however unusually developed, unless it be unusually developed in one species or in a few species in comparison with the same part in many closely allied species. Thus, the wing of the bat is a most abnormal structure in the class of mammals; but the rule would not apply here, because the whole group of bats possesses wings; it would apply only if some one species had wings developed in a remarkable manner in comparison with the other species of the same genus. The rule applies very strongly in the case of secondary sexual characters, when displayed in any unusual manner. The term, secondary sexual characters, used by Hunter, relates to characters which are attached to one sex, but are not directly connected with the act of reproduction. The rule applies to males and females; but more rarely to females, as they seldom offer remarkable secondary sexual characters. The rule being so plainly applicable in the case of secondary sexual characters, may be due to the great variability of these characters, whether or not displayed in any unusual manner – of which fact I think there can be little doubt. But that our rule is not confined to secondary sexual characters is clearly shown in the case of hermaphrodite cirripedes; I particularly attended to Mr. Waterhouse's remark, whilst investigating this order, and I am fully convinced that the rule almost always holds good. I shall, in a future work, give a list of all the more remarkable cases. I will here give only one, as it illustrates the rule in its largest application. The opercular valves of sessile cirripedes (rock barnacles) are, in every sense of the word, very important structures, and they differ extremely little even in distinct genera; but in the several species of one genus, Pyrgoma, these valves present a marvellous amount of diversification; the homologous valves in the different species being sometimes wholly unlike in shape; and the amount of variation in the individuals of the same

The Vampire Bat of South America, published in *The Zoology of the Voyage of H.M.S. Beagle*

species is so great that it is no exaggeration to state that the varieties of the same species differ more from each other in the characters derived from these important organs, than do the species belonging to other distinct genera.

As with birds the individuals of the same species, inhabiting the same country, vary extremely little, I have particularly attended to them; and the rule certainly seems to hold good in this class. I cannot make out that it applies to plants, and this would have seriously shaken my belief in its truth, had not the great variability in plants made it particularly difficult to compare their relative degrees of variability.

When we see any part or organ developed in a remarkable degree or manner in a species, the fair presumption is that it is of high importance to that species: nevertheless it is in this case eminently liable to variation. Why should this be so? On the view that each species has been independently created, with all its parts as we now see them, I can see no explanation. But on the view that groups of species are descended from some other species, and have been modified through natural selection, I think we can obtain some light. First let me make some preliminary remarks. If, in our domestic animals, any part or the whole animal be neglected, and no selection be applied, that part (for instance, the comb in the Dorking fowl) or the whole breed will cease to have a uniform character: and the breed may be said to be degenerating. In rudimentary organs, and in those which have been but little specialised for any particular purpose, and perhaps in polymorphic groups, we see a nearly parallel case; for in such cases natural selection either has not or cannot come into full play, and thus the organisation is left in a fluctuating condition. But what here more particularly concerns us is, that those points in our domestic animals, which at the present time are undergoing rapid change by continued selection, are also eminently liable to variation. Look at the individuals of the same breed of the pigeon; and

see what a prodigious amount of difference there is in the beak of tumblers, in the beak and wattle of carriers, in the carriage and tail of fantails, etc., these being the points now mainly attended to by English fanciers. Even in the same sub-breed, as in that of the short-faced tumbler, it is notoriously difficult to breed nearly perfect birds, many departing widely from the standard. There may truly be said to be a constant struggle going on between, on the one hand, the tendency to reversion to a less perfect state, as well as an innate tendency to new variations, and, on the other hand, the power of steady selection to keep the breed true. In the long run selection gains the day, and we do not expect to fail so completely as to breed a bird as coarse as a common tumbler pigeon from a good short-faced strain. But as long as selection is rapidly going on, much variability in the parts undergoing modification may always be expected.

Now let us turn to nature. When a part has been developed in an extraordinary manner in any one species, compared with the other species of the same genus, we may conclude that this part has undergone an extraordinary amount of modification since the period when the several species branched off from the common progenitor of the genus. This period will seldom be remote in any extreme degree, as species rarely endure for more than one geological period. An extraordinary amount of modification implies an unusually large and long-continued amount of variability, which has continually been accumulated by natural selection for the benefit of the species. But as the variability of the extraordinarily developed part or organ has been so great and long-continued within a period not excessively remote, we might, as a general rule, still expect to find more variability in such parts than in other parts of the organisation which have remained for a much longer period nearly constant. And this, I am convinced, is the case. That the struggle between natural selection on the one hand, and the tendency to reversion and variability on the other hand, will in the course of time cease; and that the most abnormally developed organs may be made constant, I see no reason to doubt. Hence, when an organ, however abnormal it may be, has been transmitted in approximately the same condition to many modified descendants, as in the case of the wing of the bat, it must have existed, according to our theory, for an immense period in nearly the same state; and thus it has come not to be more variable than any other structure. It is only in those cases in which the modification has been comparatively recent and extraordinarily great that we ought to find the *generative variability*, as it may be called, still present in a high degree. For in this case the variability will seldom as yet have been fixed by the continued selection of the individuals varying in the required manner and degree, and by the continued rejection of those tending to revert to a former and less modified condition.

SPECIFIC CHARACTERS MORE VARIABLE THAN GENERIC CHARACTERS

The principle discussed under the last heading may be applied to our present subject. It is notorious that specific characters are more variable than generic. To explain by a simple example what is meant: if in a large genus of plants some species had blue flowers and some had red, the colour would be only a specific character, and no one would be surprised at one of the blue species varying into red, or conversely; but if all the species had blue flowers, the colour would become a generic character, and its variation would be a more unusual circumstance. I have chosen this example because the explanation which most naturalists would advance is not here applicable, namely, that specific characters are more variable than generic, because they are taken from parts of less physiological importance than those commonly used for classing genera. I believe this explanation is partly, yet only indirectly, true; I shall, however, have to return to this point in the chapter on

Classification. It would be almost superfluous to adduce evidence in support of the statement, that ordinary specific characters are more variable than generic; but with respect to important characters, I have repeatedly noticed in works on natural history, that when an author remarks with surprise that some important organ or part, which is generally very constant throughout a large group of species, *differs* considerably in closely-allied species, it is often *variable* in the individuals of the same species. And this fact shows that a character, which is generally of generic value, when it sinks in value and becomes only of specific value, often becomes variable, though its physiological importance may remain the same. Something of the same kind applies to monstrosities: at least Is. Geoffroy St. Hilaire apparently entertains no doubt, that the more an organ normally differs in the different species of the same group, the more subject it is to anomalies in the individuals.

On the ordinary view of each species having been independently created, why should that part of the structure, which differs from the same part in other independently created species of the same genus, be more variable than those parts which are closely alike in the several species? I do not see that any explanation can be given. But on the view that species are only strongly marked and fixed varieties, we might expect often to find them still continuing to vary in those parts of their structure which have varied within a moderately recent period, and which have thus come to differ. Or to state the case in another manner: the points in which all the species of a genus resemble each other, and in which they differ from allied genera, are called generic characters; and these characters may be attributed to inheritance from a common progenitor, for it can rarely have happened that natural selection will have modified several distinct species, fitted to more or less widely different habits, in exactly the same manner: and as these so-called generic characters have been inherited from before the period when the several species first branched off from their common progenitor, and subsequently have not varied or come to differ in any degree, or only in a slight degree, it is not probable that they should vary at the present day. On the other hand, the points in which species differ from other species of the same genus are called specific characters; and as these specific characters have varied and come to differ since the period when the species branched off from a common progenitor, it is probable that they should still often be in some degree variable – at least more variable than those parts of the organisation which have for a very long period remained constant.

SECONDARY SEXUAL CHARACTERS VARIABLE

I think it will be admitted by naturalists, without my entering on details, that secondary sexual characters are highly variable. It will also be admitted that species of the same group differ from each other more widely in their secondary sexual characters, than in other parts of their organisation; compare, for instance, the amount of difference between the males of gallinaceous birds, in which secondary sexual characters are strongly displayed, with the amount of difference between the females. The cause of the original variability of these characters is not manifest; but we can see why they should not have been rendered as constant and uniform as others, for they are accumulated by sexual selection, which is less rigid in its action than ordinary selection, as it does not entail death, but only gives fewer offspring to the less favoured males. Whatever the cause may be of the variability of secondary sexual characters, as they are highly variable, sexual selection will have had a wide scope for action, and may thus have succeeded in giving to the species of the same group a greater amount of difference in these than in other respects.

It is a remarkable fact, that the secondary differences between the two sexes of the same species are generally displayed in the very same parts of the organisation in which the species of the same genus differ from each other. Of this fact I will give in illustration the first two instances which happen to stand on my list; and as the differences in these cases are of a very unusual nature, the relation can hardly be accidental. The same number of joints in the tarsi is a character common to very large groups of beetles, but in the Engidae, as Westwood has remarked, the number varies greatly and the number likewise differs in the two sexes of the same species. Again in the fossorial hymenoptera, the neuration of the wings is a character of the highest importance, because common to large groups; but in certain genera the neuration differs in the different species, and likewise in the two sexes of the same species. Sir J. Lubbock has recently remarked, that several minute crustaceans offer excellent illustrations of this law. 'In Pontella, for instance, the sexual characters are afforded mainly by the anterior antennae and by the fifth pair of legs: the specific differences also are principally given by these organs.' This relation has a clear meaning on my view: I look at all the species of the same genus as having as

The Giant Wood Wasp is not a true wasp but a Sawfly. The large 'sting' is, in fact, its ovipositor

certainly descended from the same progenitor, as have the two sexes of any one species. Consequently, whatever part of the structure of the common progenitor, or of its early descendants, became variable; variations of this part would, it is highly probable, be taken advantage of by natural and sexual selection, in order to fit the several places in the economy of nature, and likewise to fit the two sexes of the same species to each other, or to fit the males to struggle with other males for the possession of the females.

Finally, then, I conclude that the greater variability of specific characters, or those which distinguish species from species, than of generic characters, or those which are possessed by all the species; that the frequent extreme variability of any part which is developed in a species in an extraordinary manner in comparison with the same part in its congeners; and the slight degree of variability in a part, however extraordinarily it may be developed, if it be common to a whole group of species; that the great variability of secondary sexual characters and their great difference in closely allied species; that secondary sexual and ordinary specific differences are generally displayed in the same parts of the organisation, are all principles closely connected together. All being mainly due to the species of the same group being the descendants of a common progenitor, from whom they have inherited much in common, to parts which have recently and largely varied being more likely still to go on varying than parts which have long been inherited and have not varied, to natural selection having more or less completely, according to the lapse of time, overmastered the tendency to reversion and to further variability, to sexual selection being less rigid than ordinary selection, and to variations in the same parts having been accumulated by natural and sexual selection, and thus having been adapted for secondary sexual, and for ordinary purposes.

DISTINCT SPECIES PRESENT ANALOGOUS VARIATIONS, SO THAT A VARIETY OF ONE SPECIES OFTEN ASSUMES A CHARACTER PROPER TO AN ALLIED SPECIES, OR REVERTS TO SOME OF THE CHARACTERS OF AN EARLY PROGENITOR

These propositions will be most readily understood by looking to our domestic races. The most distinct breeds of the pigeon, in countries widely apart, present sub-varieties with reversed feathers on the head, and with feathers on the feet, characters not possessed by the aboriginal rock-pigeon; these then are analogous variations in two or more distinct races. The frequent presence of fourteen or even sixteen tail-feathers in the pouter may be considered as a variation representing the normal structure of another race, the fantail. I presume that no one will doubt that all such analogous variations are due to the several races of the pigeon having inherited from a common parent the same constitution and tendency to variation, when acted on by similar unknown influences. In the vegetable kingdom we have a case of analogous variation, in the enlarged stems, or as commonly called roots, of the Swedish turnip and ruta-baga, plants which several botanists rank as varieties produced by cultivation from a common parent: if this be not so, the case will then be one of analogous variation in two so-called distinct species; and to these a third may be added, namely, the common turnip. According to the ordinary view of each species having been independently created, we should have to attribute this similarity in the enlarged stems of these three plants, not to the *vera causa* of community of descent, and a consequent tendency to vary in a like manner, but to three separate yet closely related acts of creation. Many similar cases of analogous variation have been observed by Naudin in the great gourd family, and by various authors in our cereals. Similar cases occurring with insects under natural conditions have lately been discussed with much ability by Mr. Walsh, who has grouped them under his law of Equable Variability.

With pigeons, however, we have another case, namely, the occasional appearance in all the breeds, of slaty-blue birds with two black bars on the wings, white loins, a bar at the end of the tail, with the outer feathers externally edged near their bases with white. As all these marks are characteristic of the parent rock-pigeon, I presume that no one will doubt that this is a case of reversion, and not of a new yet analogous variation appearing in the several breeds. We may, I think, confidently come to this conclusion, because, as we have seen, these coloured marks are eminently liable to appear in the crossed offspring of two distinct and differently coloured breeds; and in this case there is nothing in the external conditions of life to cause the reappearance of the slaty-blue, with the several marks, beyond the influence of the mere act of crossing on the laws of inheritance.

No doubt it is a very surprising fact that characters should reappear after having been lost for many, probably for hundreds of generations. But when a breed has been crossed only once by some other breed, the offspring occasionally show for many generations a tendency to revert in character to the foreign breed – some say, for a dozen or even a score of generations. After twelve generations, the proportion of blood, to use a common expression, from one ancestor, is only one in 2048; and yet, as we see, it is generally believed that a tendency to reversion is retained by this remnant of foreign blood. In a breed which has not been crossed, but in which *both* parents have lost some character which their progenitor possessed, the tendency, whether strong or weak, to reproduce the lost character might, as was formerly remarked, for all that we can see to the contrary, be transmitted for almost any number of generations. When a character which has been lost in a breed, reappears after a great number of generations, the most probable hypothesis is, not that one individual suddenly takes after an ancestor removed by some hundred generations, but that in each successive generation the

character in question has been lying latent, and at last, under unknown favourable conditions, is developed. With the barb pigeon, for instance, which very rarely produces a blue bird, it is probable that there is a latent tendency in each generation to produce blue plumage. The abstract improbability of such a tendency being transmitted through a vast number of generations, is not greater than that of quite useless or rudimentary organs being similarly transmitted. A mere tendency to produce a rudiment is indeed sometimes thus inherited.

As all the species of the same genus are supposed to be descended from a common progenitor, it might be expected that they would occasionally vary in an analogous manner; so that the varieties of two or more species would resemble each other, or that a variety of one species would resemble in certain characters another and distinct species, this other species being, according to our view, only a well-marked and permanent variety. But characters exclusively due to analogous variation would probably be of an unimportant nature, for the preservation of all functionally important characters will have been determined through natural selection, in accordance with the different habits of the species. It might further be expected that the species of the same genus would occasionally exhibit reversions to long-lost characters. As, however, we do not know the common ancestor of any natural group, we cannot distinguish between reversionary and analogous characters. If, for instance, we did not know that the parent rock-pigeon was not feather-footed or turn-crowned, we could not have told, whether such characters in our domestic breeds were reversions or only analogous variations; but we might have inferred that the blue colour was a case of reversion from the number of the markings, which are correlated with this tint, and which would not probably have all appeared together from simple variation. More especially we might have inferred this from the blue colour and the several marks so often appearing when differently coloured breeds are crossed. Hence, although under nature it must generally be left doubtful, what cases are reversions to formerly existing characters, and what are new but analogous variations, yet we ought, on our theory, sometimes to find the varying offspring of a species assuming characters which are already present in other members of the same group. And this undoubtedly is the case.

The difficulty in distinguishing variable species is largely due to the varieties mocking, as it were, other species of the same genus. A considerable catalogue, also, could be given of forms intermediate between two other forms, which themselves can only doubtfully be ranked as species; and this shows, unless all these closely allied forms be considered as independently created species, that they have in varying assumed some of the characters of the others. But the best evidence of analogous variations is afforded by parts or organs which are generally constant in character, but which occasionally vary so as to resemble, in some degree, the same part or organ in an allied species. I have collected a long list of such cases; but here, as before, I lie under the great disadvantage of not being able to give them. I can only repeat that such cases certainly occur, and seem to me very remarkable.

I will, however, give one curious and complex case, not indeed as affecting any important character, but from occurring in several species of the same genus, partly under domestication and partly under nature. It is a case almost certainly of reversion. The ass sometimes has very distinct transverse bars on its legs, like those on the legs of a zebra. It has been asserted that these are plainest in the foal, and from inquiries which I have made, I believe this to be true. The stripe on the shoulder is sometimes double, and is very variable in length and outline. A white ass, but *not* an albino, has been described without either spinal or shoulder stripe; and these stripes are sometimes very obscure, or actually quite lost, in dark-coloured asses. The koulan of Pallas is said to have

been seen with a double shoulder stripe. Mr. Blyth has seen a specimen of the hemionus with a distinct shoulder stripe, though it properly has none; and I have been informed by Colonel Poole that foals of this species are generally striped on the legs and faintly on the shoulder. The quagga, though so plainly barred like a zebra over the body, is without bars on the legs; but Dr. Gray has figured one specimen with very distinct zebra-like bars on the hocks.

With respect to the horse, I have collected cases in England of the spinal stripe in horses of the most distinct breeds, and of *all* colours; transverse bars on the legs are not rare in duns, mouse-duns, and in one instance in a chestnut; a faint shoulder stripe may sometimes be seen in duns, and I have seen a trace in a bay horse. My son made a careful examination and sketch for me of a dun Belgian cart horse with a double stripe on each shoulder and with leg stripes. I have myself seen a dun Devonshire pony, and a small dun Welsh pony has been carefully described to me, both with *three* parallel stripes on each shoulder.

In the northwest part of India the Kattywar breed of horses is so generally striped, that, as I hear from Colonel Poole, who examined this breed for the Indian Government, a horse without stripes is not considered as purely bred. The spine is always striped; the legs are generally barred; and the shoulder stripe, which is sometimes double and sometimes treble, is common; the side of the face, moreover, is sometimes striped. The stripes are often plainest in the foal; and sometimes quite disappear in old horses. Colonel Poole has seen both gray and bay Kattywar horses striped when first foaled. I have also reason to suspect, from information given me by Mr. W.W. Edwards, that with the English race horse the spinal stripe is much commoner in the foal than in the full-grown animal. I have myself recently bred a foal from a bay mare (offspring of a Turkoman horse and a Flemish mare) by a bay English race horse. This foal, when a week old, was marked on its hinder quarters and on its forehead with numerous very

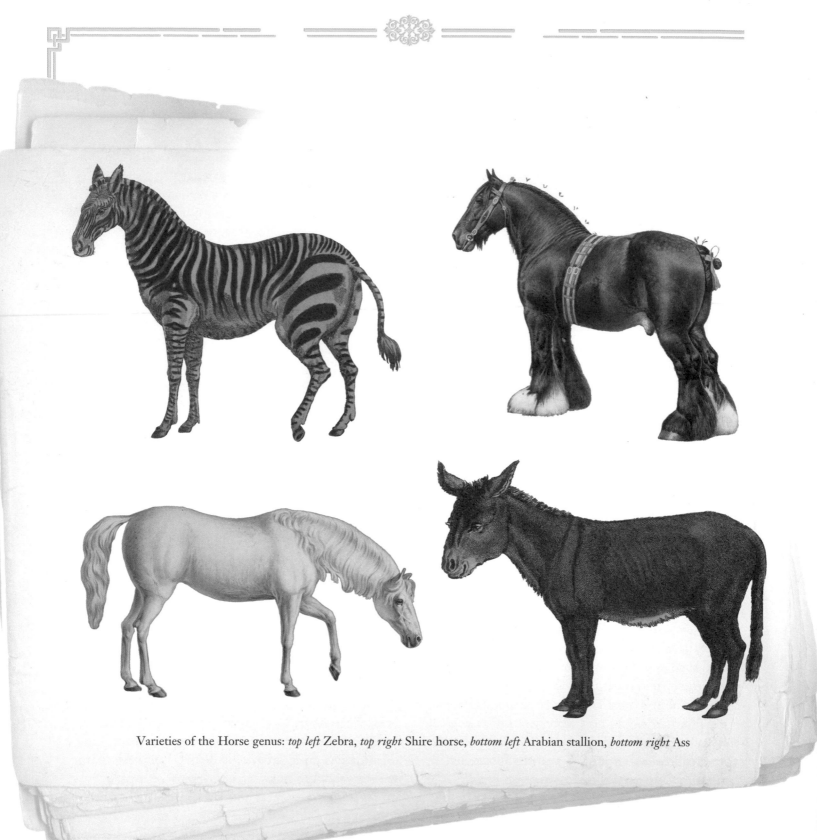

Varieties of the Horse genus: *top left* Zebra, *top right* Shire horse, *bottom left* Arabian stallion, *bottom right* Ass narrow, dark, zebra-like bars, and its legs were feebly striped. All the stripes soon disappeared completely. Without here entering on further details I may state that I have collected cases of leg and shoulder stripes in horses of very different breeds in various countries from Britain to Eastern China; and from Norway in the north to the Malay Archipelago in the south. In all parts of the world these stripes occur far oftenest in duns and mouse-duns; by the term dun a large range of colour is included, from one between brown and black to a close approach to cream colour.

I am aware that Colonel Hamilton Smith, who has written on this subject, believes that the several breeds of the horse are descended from several aboriginal species, one of which, the dun, was striped; and that the above-described appearances are all due to ancient crosses with the dun stock. But this view may be safely rejected, for it is highly improbable that the heavy Belgian cart horse, Welsh ponies, Norwegian cobs, the lanky Kattywar race, etc., inhabiting the most distant parts of the world, should have all have been crossed with one supposed aboriginal stock.

Now let us turn to the effects of crossing the several species of the horse genus. Rollin asserts, that the common mule from the ass and horse is particularly apt to have bars on its legs; according to Mr. Gosse, in certain parts of the United States, about nine out of ten mules have striped legs. I once saw a mule with its legs so much striped that any one might have thought that it was a hybrid zebra; and Mr. W.C. Martin, in his excellent treatise on the horse, has given a figure of a similar mule. In four coloured drawings, which I have seen, of hybrids between the ass and zebra, the legs were much more plainly barred than the rest of the body; and in one of them there was a double shoulder stripe. In Lord Morton's famous hybrid, from a chestnut mare and male quagga, the hybrid and even the pure offspring subsequently produced from the same mare by a black Arabian sire, were much more plainly barred across the legs than is even the pure

quagga. Lastly, and this is another most remarkable case, a hybrid has been figured by Dr. Gray (and he informs me that he knows of a second case) from the ass and the hemionus; and this hybrid, though the ass only occasionally has stripes on his legs and the hemionus has none and has not even a shoulder stripe, nevertheless had all four legs barred, and had three short shoulder stripes, like those on the dun Devonshire and Welsh ponies, and even had some zebra-like stripes on the sides of its face. With respect to this last fact, I was so convinced that not even a stripe of colour appears from what is commonly called chance, that I was led solely from the occurrence of the face stripes on this hybrid from the ass and hemionus to ask Colonel Poole whether such face stripes ever occurred in the eminently striped Kattywar breed of horses, and was, as we have seen, answered in the affirmative.

What now are we to say to these several facts? We see several distinct species of the horse genus becoming, by simple variation, striped on the legs like a zebra, or striped on the shoulders like an ass. In the horse we see this tendency strong whenever a dun tint appears – a tint which approaches to that of the general colouring of the other species of the genus. The appearance of the stripes is not accompanied by any change of form, or by any other new character. We see this tendency to become striped most strongly displayed in hybrids from between several of the most distinct species. Now observe the case of the several breeds of pigeons: they are descended from a pigeon (including two or three sub-species or geographical races) of a bluish colour, with certain bars and other marks; and when any breed assumes by simple variation a bluish tint, these bars and other marks invariably reappear; but without any other change of form or character. When the oldest and truest breeds of various colours are crossed, we see a strong tendency for the blue tint and bars and marks to reappear in the mongrels. I have stated that the most probable hypothesis to account for the reappearance of very ancient characters, is – that there is a *tendency* in the young of each successive generation to produce the long-lost character, and that this tendency, from unknown causes, sometimes prevails. And we have just seen that in several species of the horse genus the stripes are either plainer or appear more commonly in the young than in the old. Call the breeds of pigeons, some of which have bred true for centuries, species; and how exactly parallel is the case with that of the species of the horse genus! For myself, I venture confidently to look back thousands on thousands of generations, and I see an animal striped like a zebra, but perhaps otherwise very differently constructed, the common parent of our domestic horse (whether or not it be descended from one or more wild stocks) of the ass, the hemionus, quagga, and zebra.

He who believes that each equine species was independently created, will, I presume, assert that each species has been created with a tendency to vary, both under nature and under domestication, in this particular manner, so as often to become striped like the other species of the genus; and that each has been created with a strong tendency, when crossed with species inhabiting distant quarters of the world, to produce hybrids resembling in their stripes, not their own parents, but other species of the genus. To admit this view is, as it seems to me, to reject a real for an unreal, or at least for an unknown cause. It makes the works of God a mere mockery and deception; I would almost as soon believe with the old and ignorant cosmogonists, that fossil shells had never lived, but had been created in stone so as to mock the shells now living on the seashore.

SUMMARY

Our ignorance of the laws of variation is profound. Not in one case out of a hundred can we pretend to assign any reason why this or that part has varied. But whenever we have the means of instituting a comparison, the same laws appear to have acted in producing the lesser differences between varieties of the same species, and the greater differences between species of the same genus. Changed conditions generally induce mere fluctuating variability, but sometimes they cause direct and definite effects; and these may become strongly marked in the course of time, though we have not sufficient evidence on this head. Habit in producing constitutional peculiarities, and use in strengthening, and disuse in weakening and diminishing organs, appear in many cases to have been potent in their effects. Homologous parts tend to vary in the same manner, and homologous parts tend to cohere. Modifications in hard parts and in external parts sometimes affect softer and internal parts. When one part is largely developed, perhaps it tends to draw nourishment from the adjoining parts; and every part of the structure which can be saved without detriment will be saved. Changes of structure at an early age may affect parts subsequently developed; and many cases of correlated variation, the nature of which we are unable to understand, undoubtedly occur. Multiple parts are variable in number and in structure, perhaps arising from such parts not having been closely specialised for any particular function, so that their modifications have not been closely checked by natural selection. It follows probably from this same cause, that organic beings low in the scale are more variable than those standing higher in the scale, and which have their whole organisation more specialised. Rudimentary organs, from being useless, are not regulated by natural selection, and hence are variable. Specific characters – that is, the characters which have come to differ since the several species of the same genus branched off from a common parent – are more variable than generic characters, or those which have long been inherited, and have not differed within this same period. In these remarks we have referred to special parts or organs being still variable, because they have recently varied and thus come to differ; but we have also seen in the second chapter that the same principle applies to the whole individual; for in a district where many species of a genus are found – that is, where there has been much former variation and differentiation, or where the manufactory of new specific forms has been actively at work – in that district and amongst these species, we now find, on an average, most varieties. Secondary sexual characters are highly variable, and such characters differ much in the species of the same group. Variability in the same parts of the organisation has generally been taken advantage of in giving secondary sexual differences to the two sexes of the same species, and specific differences to the several species of the same genus. Any part or organ developed to an extraordinary size or in an extraordinary manner, in comparison with the same part or organ in the allied species, must have gone through an extraordinary amount of modification since the genus arose; and thus we can understand why it should often still be variable in a much higher degree than other parts; for variation is a long-continued and slow process, and natural selection will in such cases not as yet have had time to overcome the tendency to further variability and to reversion to a less

modified state. But when a species with an extraordinarily developed organ has become the parent of many modified descendants – which on our view must be a very slow process, requiring a long lapse of time – in this case, natural selection has succeeded in giving a fixed character to the organ, in however extraordinary a manner it may have been developed. Species inheriting nearly the same constitution from a common parent, and exposed to similar influences, naturally tend to present analogous variations, or these same species may occasionally revert to some of the characters of their ancient progenitors. Although new and important modifications may not arise from reversion and analogous variation, such modifications will add to the beautiful and harmonious diversity of nature.

Whatever the cause may be of each slight difference between the offspring and their parents – and a cause for each must exist – we have reason to believe that it is the steady accumulation of beneficial differences which has given rise to all the more important modifications of structure in relation to the habits of each species.

Chapter VI

DIFFICULTIES OF THE THEORY

Difficulties of the theory of descent with modification–Absence or rarity of transitional varieties–Transitions in habits of life–Diversified habits in the same species–Species with habits widely different from those of their allies–Organs of extreme perfection–Modes of transition–Cases of difficulty–Natura non facit saltum–Organs of small importance–Organs not in all cases absolutely perfect–The law of Unity of Type and of the Conditions of Existence embraced by the theory of Natural Selection.

Long before the reader has arrived at this part of my work, a crowd of difficulties will have occurred to him. Some of them are so serious that to this day I can hardly reflect on them without being in some degree staggered; but, to the best of my judgment, the greater number are only apparent, and those that are real are not, I think, fatal to the theory.

These difficulties and objections may be classed under the following heads: – First, why, if species have descended from other species by fine gradations, do we not everywhere see innumerable transitional forms? Why is not all nature in confusion, instead of the species being, as we see them, well defined?

Secondly, is it possible that an animal having, for instance, the structure and habits of a bat, could have been formed by the modification of some other animal with widely different habits and structure? Can we believe that natural selection could produce, on the one hand, an organ of trifling importance, such as the tail of a giraffe, which serves as a fly-flapper, and, on the other hand, an organ so wonderful as the eye?

Thirdly, can instincts be acquired and modified through natural selection? What shall we say to the instinct which leads the bee to make cells, and which has practically anticipated the discoveries of profound mathematicians?

Fourthly, how can we account for species, when crossed, being sterile and producing sterile offspring, whereas, when varieties are crossed, their fertility is unimpaired?

The two first heads will be here discussed; some miscellaneous objections in the following chapter; Instinct and Hybridism in the two succeeding chapters.

ON THE ABSENCE OR RARITY OF TRANSITIONAL VARIETIES

As natural selection acts solely by the preservation of profitable modifications, each new form will tend in a fully-stocked country to take the place of, and finally to exterminate, its own less improved parent-form and other less-favoured forms with which it comes into competition. Thus extinction and natural selection go hand in hand. Hence, if we look at each species as descended from some unknown form, both the parent and all the transitional varieties will generally have been exterminated by the very process of the formation and perfection of the new form.

But, as by this theory innumerable transitional forms must have existed, why do we not find them embedded in countless numbers in the crust of the earth? It will be more convenient to discuss this question in the chapter on the Imperfection of the Geological Record; and I will here only state that I believe the answer mainly lies in the record being incomparably less perfect than is generally supposed. The crust of the earth is a vast museum; but the natural collections have been imperfectly made, and only at long intervals of time.

But it may be urged that when several closely-allied species inhabit the same territory, we surely ought to find at the present time many transitional forms. Let us take a simple case: in travelling from north to south over a continent, we generally meet at successive intervals with closely allied or representative species, evidently filling nearly the same place in the natural economy of the land. These representative species often meet and interlock; and as the one becomes rarer and rarer, the other becomes more and more frequent, till the one replaces the other. But if we compare these species where they intermingle, they are generally as absolutely distinct from each other in every detail of structure as are specimens taken from the metropolis inhabited by each. By my theory these allied species are descended from a common parent; and during the process of modification, each has become adapted to the conditions of life of its own region, and has supplanted and exterminated its original parent-form and all the transitional varieties between its past and present states. Hence we ought not to expect at the present time to meet with numerous transitional varieties in each region, though they must have existed there, and may be embedded there in a fossil condition. But in the intermediate region, having intermediate conditions of life, why do we not now find closely-linking intermediate varieties? This difficulty for a long time quite confounded me. But I think it can be in large part explained.

In the first place we should be extremely cautious in inferring, because an area is now continuous, that it has been continuous during a long period. Geology would lead us to believe that most continents have been broken up into islands even during the later tertiary periods; and in such

islands distinct species might have been separately formed without the possibility of intermediate varieties existing in the intermediate zones. By changes in the form of the land and of climate, marine areas now continuous must often have existed within recent times in a far less continuous and uniform condition than at present. But I will pass over this way of escaping from the difficulty; for I believe that many perfectly defined species have been formed on strictly continuous areas; though I do not doubt that the formerly broken condition of areas now continuous, has played an important part in the formation of new species, more especially with freely-crossing and wandering animals.

In looking at species as they are now distributed over a wide area, we generally find them tolerably numerous over a large territory, then becoming somewhat abruptly rarer and rarer on the confines, and finally disappearing. Hence the neutral territory between two representative species is generally narrow in comparison with the territory proper to each. We see the same fact in ascending mountains, and sometimes it is quite remarkable how abruptly, as Alph. De Candolle has observed, a common alpine species disappears. The same fact has been noticed by E. Forbes in sounding the depths of the sea with the dredge. To those who look at climate and the physical conditions of life as the all-important elements of distribution, these facts ought to cause surprise, as climate and height or depth graduate away insensibly. But when we bear in mind that almost every species, even in its metropolis, would increase immensely in numbers, were it not for other competing species; that nearly all either prey on or serve as prey for others; in short, that each organic being is either directly or indirectly related in the most important manner to other organic beings – we see that the range of the inhabitants of any country by no means exclusively depends on insensibly changing physical conditions, but in large part on the presence of other species, on which it lives, or by which it is destroyed, or with which it comes into competition; and as these species are already defined objects, not blending one into another by insensible gradations, the range of any one species, depending as it does on the range of others, will tend to be sharply defined. Moreover, each species on the confines of its range, where it exists in lessened numbers, will, during fluctuations in the number of its enemies or of its prey, or in the nature of the seasons, be extremely liable to utter extermination; and thus its geographical range will come to be still more sharply defined.

As allied or representative species, when inhabiting a continuous area, are generally distributed in such a manner that each has a wide range, with a comparatively narrow neutral territory between them, in which they become rather suddenly rarer and rarer; then, as varieties do not essentially differ from species, the same rule will probably apply to both; and if we take a varying species inhabiting a very large area, we shall have to adapt two varieties to two large areas, and a third variety to a narrow intermediate zone. The intermediate variety, consequently, will exist in lesser numbers from inhabiting a narrow and lesser area; and practically, as far as I can make out, this rule holds good with varieties in a state of nature. I have met with striking instances of the rule in the case of varieties intermediate between well-marked varieties in the genus Balanus. And it would appear from information given me by Mr. Watson, Dr. Asa Gray, and Mr. Wollaston, that generally, when varieties intermediate between two other forms occur, they are much rarer numerically than the forms which they connect. Now, if we may trust these facts and inferences, and conclude that varieties linking two other varieties together generally have existed in lesser numbers than the forms which they connect, then we can understand why intermediate varieties should not endure for very long periods: – why, as a general rule, they should be exterminated and disappear, sooner than the forms which they originally linked together.

For any form existing in lesser numbers would, as already remarked, run a greater chance of being exterminated than one existing in large numbers; and in this particular case the intermediate form would be eminently liable to the inroads of closely allied forms existing on both sides of it. But it is a far more important consideration, that during the process of further modification, by which two varieties are supposed to be converted and perfected into two distinct species, the two which exist in larger numbers, from inhabiting larger areas, will have a great advantage over the intermediate variety, which exists in smaller numbers in a narrow and intermediate zone. For forms existing in larger numbers will have a better chance, within any given period, of presenting further favourable variations for natural selection to seize on, than will the rarer forms which exist in lesser numbers. Hence, the more common forms, in the race for life, will tend to beat and supplant the less common forms, for these will be more slowly modified and improved. It is the same principle which, as I believe, accounts for the common species in each country, as shown in the second chapter, presenting on an average a greater number of well-marked varieties than do the rarer species. I may illustrate what I mean by supposing three varieties of sheep to be kept, one adapted to an extensive mountainous region; a second to a comparatively narrow, hilly tract; and a third to the wide plains at the base; and that the inhabitants are all trying with equal steadiness and skill to improve their stocks by selection; the chances in this case will be strongly in favour of the great holders on the mountains or on the plains improving their breeds more quickly than the small holders on the intermediate narrow, hilly tract; and consequently the improved mountain or plain breed will soon take the place of the less improved hill breed; and thus the two breeds, which originally existed in greater numbers, will come into close contact with each other, without the interposition of the supplanted, intermediate hill variety.

The Old Lincoln Breed of sheep, which has been extensively cross-bred to enhance its size and long wool

To sum up, I believe that species come to be tolerably well-defined objects, and do not at any one period present an inextricable chaos of varying and intermediate links: first, because new varieties are very slowly formed, for variation is a slow process, and natural selection can do nothing until favourable individual differences or variations occur, and until a place in the natural polity of the country can be better filled by some modification of some one or more of its inhabitants. And such new places will depend on slow changes of climate, or on the occasional immigration of new inhabitants, and, probably, in a still more important degree, on some of the old inhabitants becoming slowly modified, with the new forms thus produced and the old ones acting and reacting on each other. So that, in any one region and at any one time, we ought to see only a few species presenting slight modifications of structure in some degree permanent; and this assuredly we do see.

Secondly, areas now continuous must often have existed within the recent period as isolated portions, in which many forms, more especially amongst the classes which unite for each birth and wander much, may have separately been rendered sufficiently distinct to rank as representative species. In this case, intermediate varieties between the several representative species and their common parent, must formerly have existed within each isolated portion of the land, but these links during the process of natural selection will have been supplanted and exterminated, so that they will no longer be found in a living state.

Thirdly, when two or more varieties have been formed in different portions of a strictly continuous area, intermediate varieties will, it is probable, at first have been formed in the intermediate zones, but they will generally have had a short duration. For these intermediate varieties will, from reasons already assigned (namely from what we know of the actual distribution of closely allied or representative species, and likewise of acknowledged varieties),

exist in the intermediate zones in lesser numbers than the varieties which they tend to connect. From this cause alone the intermediate varieties will be liable to accidental extermination; and during the process of further modification through natural selection, they will almost certainly be beaten and supplanted by the forms which they connect; for these, from existing in greater numbers will, in the aggregate, present more varieties, and thus be further improved through natural selection and gain further advantages.

Lastly, looking not to any one time, but at all time, if my theory be true, numberless intermediate varieties, linking closely together all the species of the same group, must assuredly have existed; but the very process of natural selection constantly tends, as has been so often remarked, to exterminate the parent forms and the intermediate links. Consequently evidence of their former existence could be found only amongst fossil remains, which are preserved, as we shall attempt to show in a future chapter, in an extremely imperfect and intermittent record.

ON THE ORIGIN AND TRANSITIONS OF ORGANIC BEINGS WITH PECULIAR HABITS AND STRUCTURE

It has been asked by the opponents of such views as I hold, how, for instance, could a land carnivorous animal have been converted into one with aquatic habits; for how could the animal in its transitional state have subsisted? It would be easy to show that there now exist carnivorous animals presenting close intermediate grades from strictly terrestrial to aquatic habits; and as each exists by a struggle for life, it is clear that each must be well adapted to its place in nature. Look at the Mustela vison of North America, which has webbed feet, and which resembles an otter in its fur, short legs, and form of tail; during summer this animal dives for and preys on fish, but during the long winter it leaves the frozen waters, and preys, like other polecats, on mice and land animals. If a different case had been taken, and it had been asked how an

The American Mink (Mustela vison), cousin of weasels, otters, polecats and ferrets insectivorous quadruped could possibly have been converted into a flying bat, the question would have been far more difficult to answer. Yet I think such difficulties have little weight.

Here, as on other occasions, I lie under a heavy disadvantage, for, out of the many striking cases which I have collected, I can give only one or two instances of transitional habits and structures in allied species; and of diversified habits, either constant or occasional, in the same species. And it seems to me that nothing less than a long list of such cases is sufficient to lessen the difficulty in any particular case like that of the bat.

Look at the family of squirrels; here we have the finest gradation from animals with their tails only slightly flattened, and from others, as Sir J. Richardson has remarked, with the posterior part of their bodies rather wide and with the skin on their flanks rather full, to the so-called flying squirrels; and flying squirrels have their limbs and even the base of the tail united by a broad expanse of skin, which serves as a parachute and allows them to glide through the air to an astonishing distance from tree to tree. We cannot doubt that each structure is of use to each kind of squirrel in its own country, by enabling it to escape birds or beasts of prey, or to collect food more quickly, or, as there is reason to believe, to lessen the danger from occasional falls. But it does not follow from this fact that the structure of each squirrel is the best that it is possible to conceive under all possible conditions. Let the climate and vegetation change, let other competing rodents or new beasts of prey immigrate, or old ones become modified, and all analogy would lead us to believe that some, at least, of the squirrels would decrease in numbers or become exterminated, unless they also become modified and improved in structure in a corresponding manner. Therefore, I can see no difficulty, more especially under changing conditions of life, in the continued preservation of individuals with fuller and fuller flank-membranes, each modification being useful, each being propagated, until, by the accumulated effects of

this process of natural selection, a perfect so-called flying squirrel was produced.

Now look at the Galeopithecus or so-called flying lemur, which was formerly ranked amongst bats, but is now believed to belong to the Insectivora. An extremely wide flank-membrane stretches from the corners of the jaw to the tail, and includes the limbs with the elongated fingers. This flank-membrane is furnished with an extensor muscle. Although no graduated links of structure, fitted for gliding through the air, now connect the Galeopithecus with the other Insectivora, yet there is no difficulty in supposing that such links formerly existed, and that each was developed in the same manner as with the less perfectly gliding squirrels; each grade of structure having been useful to its possessor. Nor can I see any insuperable difficulty in further believing it possible that the membrane-connected fingers and fore-arm of the Galeopithecus might have been greatly lengthened by natural selection; and this, as far as the organs of flight are concerned, would have converted the animal into a bat. In certain bats in which the wing-membrane extends from the top of the shoulder to the tail and includes the hind-legs, we perhaps see traces of an apparatus originally fitted for gliding through the air rather than for flight.

If about a dozen genera of birds were to become extinct, who would have ventured to surmise that birds might have existed which used their wings solely as flappers, like the logger-headed duck (Micropterus of Eyton); as fins in the water and as front legs on the land, like the penguin; as sails, like the ostrich; and functionally for no purpose, like the apteryx? Yet the structure of each of these birds is good for it, under the conditions of life to which it is exposed, for each has to live by a struggle: but it is not necessarily the best possible under all possible conditions. It must not be inferred from these remarks that any of the grades of wing-structure here alluded to, which perhaps may all be the result of disuse, indicate the steps by which birds actually acquired

The hand of a Brown Lemur from Madagascar. It shows a padded semi-opposable thumb, which makes it a very agile climber, unlike the Colugo

The Flying Lemur or Colugo, of the Southeast Asian rainforest. The muscled membrane stretching from jaw to tail enables it to glide up to 200 ft from one tree to another

their perfect power of flight; but they serve to show what diversified means of transition are at least possible.

Seeing that a few members of such water-breathing classes as the Crustacea and Mollusca are adapted to live on the land; and seeing that we have flying birds and mammals, flying insects of the most diversified types, and formerly had flying reptiles, it is conceivable that flying-fish, which now glide far through the air, slightly rising and turning by the aid of their fluttering fins, might have been modified into perfectly winged animals. If this had been effected, who would have ever imagined that in an early transitional state they had been inhabitants of the open ocean, and had used their incipient organs of flight exclusively, so far as we know, to escape being devoured by other fish?

When we see any structure highly perfected for any particular habit, as the wings of a bird for flight, we should bear in mind that animals displaying early transitional grades of the structure will seldom have survived to the present day, for they will have been supplanted by their successors, which were gradually rendered more perfect through natural selection. Furthermore, we may conclude that transitional states between structures fitted for very different habits of life will rarely have been developed at an early period in great numbers and under many subordinate forms. Thus, to return to our imaginary illustration of the flying-fish, it does not seem probable that fishes capable of true flight would have been developed under many subordinate forms, for taking prey of many kinds in many ways, on the land and in the water, until their organs of flight had come to a high stage of perfection, so as to have given them a decided advantage over other animals in the battle for life. Hence the chance of discovering species with transitional grades of structure in a fossil condition will always be less, from their having existed in lesser numbers, than in the case of species with fully developed structures.

I will now give two or three instances, both of diversified and of changed habits, in the individuals of the same species. In either case it would be easy for natural selection to adapt the structure of the animal to its changed habits, or exclusively to one of its several habits. It is, however, difficult to decide and immaterial for us, whether habits generally change first and structure afterwards; or whether slight modifications of structure lead to changed habits; both probably often occurring almost simultaneously. Of cases of changed habits it will suffice merely to allude to that of the many British insects which now feed on exotic plants, or exclusively on artificial substances. Of diversified habits innumerable instances could be given: I have often watched a tyrant flycatcher (Saurophagus sulphuratus) in South America, hovering over one spot and then proceeding to another, like a kestrel, and at other times standing stationary on the margin of water, and then dashing into it like a kingfisher at a fish. In our own country the larger titmouse (Parus major) may be seen climbing branches, almost like a creeper; it sometimes, like a shrike, kills small birds by blows on the head; and I have many times seen and heard it hammering the seeds of the yew on a branch, and thus breaking them like a nuthatch. In North America the black bear was seen by Hearne swimming for hours with widely open mouth, thus catching, almost like a whale, insects in the water.

As we sometimes see individuals following habits different from those proper to their species and to the other species of the same genus, we might expect that such individuals would occasionally give rise to new species, having anomalous habits, and with their structure either slightly or considerably modified from that of their type. And such instances occur in nature. Can a more striking instance of adaptation be given than that of a woodpecker for climbing trees and seizing insects in the chinks of the bark? Yet in North America there are woodpeckers which feed largely on fruit, and others with elongated wings which chase insects on the wing. On the plains of La Plata, where hardly a tree grows, there is a woodpecker (Colaptes campestris) which has two toes before and two behind, a long-pointed tongue, pointed tail-feathers, sufficiently stiff to support the bird in a vertical position on a post, but not so stiff as in the typical wood-peckers, and a straight, strong beak. The beak, however, is not so straight or so strong as in the typical woodpeckers but it is strong enough to bore into wood. Hence this Colaptes, in all the essential parts of its structure, is a woodpecker. Even in such trifling characters as the colouring, the harsh tone of the voice, and undulatory flight, its close blood-relationship to our common woodpecker is plainly declared; yet, as I can assert, not only from my own observations, but from those of the accurate Azara, in certain large districts it does not climb trees, and it makes its nest in holes in banks! In certain other districts, however, this same woodpecker, as Mr. Hudson states, frequents trees, and bores holes in the trunk for its nest. I may mention as another illustration of the varied habits of this genus, that a Mexican Colaptes has been described by De Saussure as boring holes into hard wood in order to lay up a store of acorns.

Petrels are the most aerial and oceanic of birds, but, in the quiet sounds of Tierra del Fuego, the Puffinuria berardi, in its general habits, in its astonishing power of diving, in its manner of swimming and of flying when made to take flight, would be mistaken by any one for an auk or a grebe; nevertheless, it is essentially a petrel, but with many parts of its organisation profoundly modified in relation to its new habits of life; whereas the woodpecker of La Plata has had its structure only slightly modified. In the case of the water-ouzel, the acutest observer, by examining its dead body, would never have suspected its sub-aquatic habits; yet this bird, which is allied to the thrush family, subsists by diving – using its wings under water and grasping stones with its feet. All the members of the great order of Hymenopterous

Agriornis micropterus.

The Grey-bellied Shrike-Tyrant of South America

Specialized climbing feet of the Green Woodpecker (left) compared with those of the African Plantain-eater (right)

insects are terrestrial, excepting the genus Proctotrupes, which Sir John Lubbock has discovered to be aquatic in its habits; it often enters the water and dives about by the use not of its legs but of its wings, and remains as long as four hours beneath the surface; yet it exhibits no modification in structure in accordance with its abnormal habits.

He who believes that each being has been created as we now see it, must occasionally have felt surprise when he has met with an animal having habits and structure not in agreement. What can be plainer than that the webbed feet of ducks and geese are formed for swimming? Yet there are upland geese with webbed feet which rarely go near the water; and no one except Audubon, has seen the frigate-bird, which has all its four toes webbed, alight on the surface of the ocean. On the other hand, grebes and coots are eminently aquatic, although their toes are only bordered by membrane. What seems plainer than that the long toes, not furnished with membrane, of the Grallatores, are formed for walking over swamps and floating plants? – the water-hen and

landrail are members of this order, yet the first is nearly as aquatic as the coot, and the second is nearly as terrestrial as the quail or partridge. In such cases, and many others could be given, habits have changed without a corresponding change of structure. The webbed feet of the upland goose may be said to have become almost rudimentary in function, though not in structure. In the frigate-bird, the deeply scooped membrane between the toes shows that structure has begun to change.

He who believes in separate and innumerable acts of creation may say, that in these cases it has pleased the Creator to cause a being of one type to take the place of one belonging to another type; but this seems to me only restating the fact in dignified language. He who believes in the struggle for existence and in the principle of natural selection, will acknowledge that every organic being is constantly endeavouring to increase in numbers; and that if any one being varies ever so little, either in habits or structure, and thus gains an advantage over some other inhabitant of the same country, it will seize on the place of that inhabitant, however different that may be from its own place. Hence it will cause him no surprise that there should be geese and frigate-birds with webbed feet, living on the dry land and rarely alighting on the water, that there should be long-toed corncrakes, living in meadows instead of in swamps; that there should be woodpeckers where hardly a tree grows; that there should be diving thrushes and diving Hymenoptera, and petrels with the habits of auks.

ORGANS OF EXTREME PERFECTION AND COMPLICATION

To suppose that the eye with all its inimitable contrivances for adjusting the focus to different distances, for admitting different amounts of light, and for the correction of spherical and chromatic aberration, could have been formed by natural selection, seems, I freely confess, absurd in the highest degree. When it was first said that the sun stood still and the world turned round, the common sense of mankind declared the doctrine false; but the old saying of *Vox populi, vox Dei*, as every philosopher knows, cannot be trusted in science. Reason tells me, that if numerous gradations from a simple and imperfect eye to one complex and perfect can be shown to exist, each grade being useful to its possessor, as is certainly the case; if further, the eye ever varies and the variations be inherited, as is likewise certainly the case; and if such variations should be useful to any animal under changing conditions of life, then the difficulty of believing that a perfect and complex eye could be formed by natural selection, though insuperable by our imagination, should not be considered as subversive of the theory. How a nerve comes to be sensitive to light, hardly concerns us more than how life itself originated; but I may remark that, as some of the lowest organisms in which nerves cannot be detected, are capable of perceiving light, it does not seem impossible that certain sensitive elements in their sarcode should become aggregated and developed into nerves, endowed with this special sensibility.

In searching for the gradations through which an organ in any species has been perfected, we ought to look exclusively to its lineal progenitors; but this is scarcely ever possible, and we are forced to look to other species and genera of the same group, that is to the collateral descendants from the same parent-form, in order to see what gradations are possible, and for the chance of some gradations having been transmitted in an unaltered or little altered condition. But the state of the same organ in distinct classes may incidentally throw light on the steps by which it has been perfected.

The simplest organ which can be called an eye consists of an optic nerve, surrounded by pigment-cells and covered by translucent skin, but without any lens or other refractive body. We may, however, according to M. Jourdain, descend even a step lower and find aggregates of pigment cells, apparently serving as organs of vision, without any nerves, and resting merely on sarcodic tissue. Eyes of the above simple nature are not capable of distinct vision, and serve only to distinguish light from darkness. In certain starfishes, small depressions in the layer of pigment which surrounds the nerve are filled, as described by the author just quoted, with transparent gelatinous matter, projecting with a convex surface, like the cornea in the higher animals. He suggests that this serves not to form an image, but only to concentrate the luminous rays and render their perception more easy. In this concentration of the rays we gain the first and by far the most important step towards the formation of a true, picture-forming eye; for we have only to place the naked extremity of the optic nerve, which in some of the lower animals lies deeply buried in the body, and in some near the surface, at the right distance from the concentrating apparatus, and an image will be formed on it.

In the great class of the Articulata, we may start from an optic nerve simply coated with pigment, the latter sometimes forming a sort of pupil, but destitute of lens or other optical contrivance. With insects it is now known that the numerous facets on the cornea of their great compound eyes form true lenses, and that the cones include curiously modified nervous filaments. But these organs in the Articulata are so much diversified that Müller formerly made three main classes with seven subdivisions, besides a fourth main class of aggregated simple eyes.

When we reflect on these facts, here given much too briefly, with respect to the wide, diversified, and graduated range of structure in the eyes of the lower animals; and when we bear in mind how small the number of all living forms must be in comparison with those which have become extinct, the difficulty ceases to be very great in believing that natural selection may have converted the simple apparatus of an optic nerve, coated with pigment and invested by transparent membrane, into an optical instrument as perfect as is possessed by any member of the Articulata class.

He who will go thus far, ought not to hesitate to go one step further, if he finds on finishing this volume that large bodies of facts, otherwise inexplicable, can be explained by the theory of modification through natural selection; he ought to admit that a structure even as perfect as an eagle's eye might thus be formed, although in this case he does not know the transitional states. It has been objected that in order to modify the eye and still preserve it as a perfect instrument, many changes would have to be effected simultaneously, which, it is assumed, could not be done through natural selection; but as I have attempted to show in my work on the variation of domestic animals, it is not necessary to suppose that the modifications were all simultaneous, if they were extremely slight and gradual. Different kinds of modification would, also, serve for the same general purpose: as Mr. Wallace has remarked, 'If a lens has too short or too long a focus, it may be amended either by an alteration of curvature, or an alteration of density; if the curvature be irregular, and the rays do not converge to a point,

then any increased regularity of curvature will be an improvement. So the contraction of the iris and the muscular movements of the eye are neither of them essential to vision, but only improvements which might have been added and perfected at any stage of the construction of the instrument.' Within the highest division of the animal kingdom, namely, the Vertebrata, we can start from an eye so simple, that it consists, as in the lancelet, of a little sack of transparent skin, furnished with a nerve and lined with pigment, but destitute of any other apparatus. In fishes and reptiles, as Owen has remarked, 'The range of gradation of dioptric structures is very great.' It is a significant fact that even in man, according to the high authority of Virchow, the beautiful crystalline lens is formed in the embryo by an accumulation of epidermic cells, lying in a sack-like fold of the skin; and the vitreous body is formed from embryonic subcutaneous tissue. To arrive, however, at a just conclusion regarding the formation of the eye, with all its marvellous yet not absolutely perfect characters, it is indispensable that the reason should conquer the imagination; but I have felt the difficulty far too keenly to be surprised at others hesitating to extend the principle of natural selection to so startling a length.

It is scarcely possible to avoid comparing the eye with a telescope. We know that this instrument has been perfected by the long-continued efforts of the highest human intellects; and we naturally infer that the eye has been formed by a somewhat analogous process. But may not this inference be presumptuous? Have we any right to assume that the Creator works by intellectual powers like those of man? If we must compare the eye to an optical instrument, we ought in imagination to take a thick layer of transparent tissue, with spaces filled with fluid, and with a nerve sensitive to light beneath, and then suppose every part of this layer to be continually changing slowly in density, so as to separate into layers of different densities and thicknesses, placed

1. *Gymnodactylus Gaudichaudii*
2. *Naultinus Grayii*

Reptiles observed and drawn during the voyage of H.M.S. *Beagle* at different distances from each other, and with the surfaces of each layer slowly changing in form. Further we must suppose that there is a power, represented by natural selection or the survival of the fittest, always intently watching each slight alteration in the transparent layers; and carefully preserving each which, under varied circumstances, in any way or degree, tends to produce a distincter image. We must suppose each new state of the instrument to be multiplied by the million; each to be preserved until a better is produced, and then the old ones to be all destroyed. In living bodies, variation will cause the slight alteration, generation will multiply them almost infinitely, and natural selection will pick out with unerring skill each improvement. Let this process go on for millions of years; and during each year on millions of individuals of many kinds; and may we not believe that a living optical instrument might thus be formed as superior to one of glass, as the works of the Creator are to those of man?

MODES OF TRANSITION

If it could be demonstrated that any complex organ existed, which could not possibly have been formed by numerous, successive, slight modifications, my theory would absolutely break down. But I can find out no such case. No doubt many organs exist of which we do not know the transitional grades, more especially if we look to much-isolated species, around which, according to the theory, there has been much extinction. Or again, if we take an organ common to all the members of a class, for in this latter case the organ must have been originally formed at a remote period, since which all the many members of the class have been developed; and in order to discover the early transitional grades through which the organ has passed, we should have to look to very ancient ancestral forms, long since become extinct.

We should be extremely cautious in concluding that an organ could not have been

formed by transitional gradations of some kind. Numerous cases could be given amongst the lower animals of the same organ performing at the same time wholly distinct functions; thus in the larva of the dragon-fly and in the fish Cobites the alimentary canal respires, digests, and excretes. In the Hydra, the animal may be turned inside out, and the exterior surface will then digest and the stomach respire. In such cases natural selection might specialise, if any advantage were thus gained, the whole or part of an organ, which had previously performed two functions, for one function alone, and thus by insensible steps greatly change its nature. Many plants are known which regularly produce at the same time differently constructed flowers; and if such plants were to produce one kind alone, a great change would be effected with comparative suddenness in the character of the species. It is, however, probable that the two sorts of flowers borne by the same plant were originally differentiated by finely graduated steps, which may still be followed in some few cases.

Again, two distinct organs, or the same organ under two very different forms, may simultaneously perform in the same individual the same function, and this is an extremely important means of transition: to give one instance – there are fish with gills or branchiae that breathe the air dissolved in the water, at the same time that they breathe free air in their swim-bladders, this latter organ being divided by highly vascular partitions and having a ductus pneumaticus for the supply of air. To give another instance from the vegetable kingdom: plants climb by three distinct means, by spirally twining, by clasping a support with their sensitive tendrils, and by the emission of aerial rootlets; these three means are usually found in distinct groups, but some few species exhibit two of the means, or even all three, combined in the same individual. In all such cases one of the two organs might readily be modified and perfected so as to perform all the work, being aided during the progress of modification by the other organ;

and then this other organ might be modified for some other and quite distinct purpose, or be wholly obliterated.

The illustration of the swim bladder in fishes is a good one, because it shows us clearly the highly important fact that an organ originally constructed for one purpose, namely flotation, may be converted into one for a widely different purpose, namely respiration. The swim bladder has, also, been worked in as an accessory to the auditory organs of certain fishes. All physiologists admit that the swim bladder is homologous, or 'ideally similar' in position and structure with the lungs of the higher vertebrate animals: hence there is no reason to doubt that the swim bladder has actually been converted into lungs, or an organ used exclusively for respiration.

According to this view it may be inferred that all vertebrate animals with true lungs are descended by ordinary generation from an ancient and unknown prototype which was furnished with a floating apparatus or swim-bladder. We can thus, as I infer from Professor Owen's interesting description of these parts, understand the strange fact that every particle of food and drink which we swallow has to pass over the orifice of the trachea, with some risk of falling into the lungs, notwithstanding the beautiful contrivance by which the glottis is closed. In the higher Vertebrata the branchiae have wholly disappeared – but in the embryo the slits on the sides of the neck and the loop-like course of the arteries still mark their former position. But it is conceivable that the now utterly lost branchiae might have been gradually worked in by natural selection for some distinct purpose: for instance, Landois has shown that the wings of insects are developed from the trachea; it is therefore highly probable that in this great class organs which once served for respiration have been actually converted into organs for flight.

In considering transitions of organs, it is so important to bear in mind the probability of conversion from one function to another, that I

will give another instance. Pedunculated cirripedes have two minute folds of skin, called by me the ovigerous frena, which serve, through the means of a sticky secretion, to retain the eggs until they are hatched within the sack. These cirripedes have no branchiae, the whole surface of the body and of the sack, together with the small frena, serving for respiration. The Balanidae or sessile cirripedes, on the other hand, have no ovigerous frena, the eggs lying loose at the bottom of the sack, within the well-enclosed shell; but they have, in the same relative position with the frena, large, much-folded membranes, which freely communicate with the circulatory lacunae of the sack and body, and which have been considered by all naturalists to act as branchiae. Now I think no one will dispute that the ovigerous frena in the one family are strictly homologous with the branchiae of the other family; indeed, they graduate into each other. Therefore it need not be doubted that the two little folds of skin, which originally served as ovigerous frena, but which, likewise, very slightly aided in the act of respiration, have been gradually converted by natural selection into branchiae, simply through an increase in their size and the obliteration of their adhesive glands. If all pedunculated cirripedes had become extinct, and they have suffered far more extinction than have sessile cirripedes, who would ever have imagined that the branchiae in this latter family had originally existed as organs for preventing the ova from being washed out of the sack?

There is another possible mode of transition, namely, through the acceleration or retardation of the period of reproduction. This has lately been insisted on by Professor Cope and others in the United States. It is now known that some animals are capable of reproduction at a very early age, before they have acquired their perfect characters; and if this power became thoroughly well developed in a species, it seems probable that the adult stage of development would sooner or later be lost, and in this case, especially if the larva differed much from the mature form, the character of the species would be greatly changed and degraded. Again, not a few animals, after arriving at maturity, go on changing in character during nearly their whole lives. With mammals, for instance, the form of the skull is often much altered with age, of which Dr. Murie has given some striking instances with seals. Every one knows how the horns of stags become more and more branched, and the plumes of some birds become more finely developed, as they grow older. Professor Cope states that the teeth of certain lizards change much in shape with advancing years. With crustaceans not only many trivial, but some important parts assume a new character, as recorded by Fritz Müller, after maturity. In all such cases, – and many could be given, – if the age

for reproduction were retarded, the character of the species, at least in its adult state, would be modified; nor is it improbable that the previous and earlier stages of development would in some cases be hurried through and finally lost. Whether species have often or ever been modified through this comparatively sudden mode of transition, I can form no opinion; but if this has occurred, it is probable that the differences between the young and the mature, and between the mature and the old, were primordially acquired by graduated steps.

SPECIAL DIFFICULTIES OF THE THEORY OF NATURAL SELECTION

Although we must be extremely cautious in concluding that any organ could not have been

The Leopard Seal of Antarctica

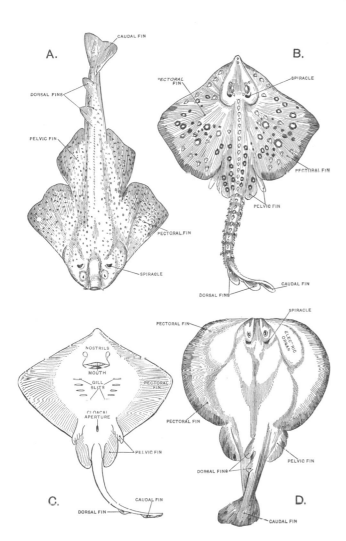

Dorsal view of a Monk-fish (A) and two Rays
(B, C) with the ventral view of a young Skate (D)

yet in the ray, as observed by Matteucci, an analogous organ in the tail manifests but little electricity, even when the animal is greatly irritated; so little that it can hardly be of any use for the above purposes. Moreover, in the ray, besides the organ just referred to, there is, as Dr. R. McDonnell has shown, another organ near the head, not known to be electrical, but which appears to be the real homologue of the electric battery in the torpedo. It is generally admitted that there exists between these organs and ordinary muscle a close analogy, in intimate structure, in the distribution of the nerves, and in the manner in which they are acted on by various reagents. It should, also, be especially observed that muscular contraction is accompanied by an electrical discharge; and, as Dr. Radcliffe insists, 'in the electrical apparatus of the torpedo during rest, there would seem to be a charge in every respect like that which is met with in muscle and nerve during the rest, and the discharge of the torpedo, instead of being peculiar, may be only another form of the discharge which attends upon the action of muscle and motor nerve.' Beyond this we cannot at present go in the way of explanation; but as we know so little about the uses of these organs, and as we know nothing about the habits and structure of the progenitors of the existing electric fishes, it would be extremely bold to maintain that no serviceable transitions are possible by which these organs might have been gradually developed.

These organs appear at first to offer another and far more serious difficulty; for they occur in about a dozen kinds of fish, of which several are widely remote in their affinities. When the same organ is found in several members of the same class, especially if in members having very different habits of life, we may generally attribute its presence to inheritance from a common ancestor; and its absence in some of the members to loss through disuse or natural selection. So that, if the electric organs had been inherited from some one ancient progenitor, we might have

produced by successive, small, transitional gradations, yet undoubtedly serious cases of difficulty occur.

One of the most serious is that of neuter insects, which are often differently constructed from either the males or fertile females; but this case will be treated of in the next chapter. The electric organs of fishes offer another case of special difficulty; for it is impossible to conceive by what steps these wondrous organs have been produced. But this is not surprising, for we do not even know of what use they are. In the Gymnotus and torpedo they no doubt serve as powerful means of defence, and perhaps for securing prey;

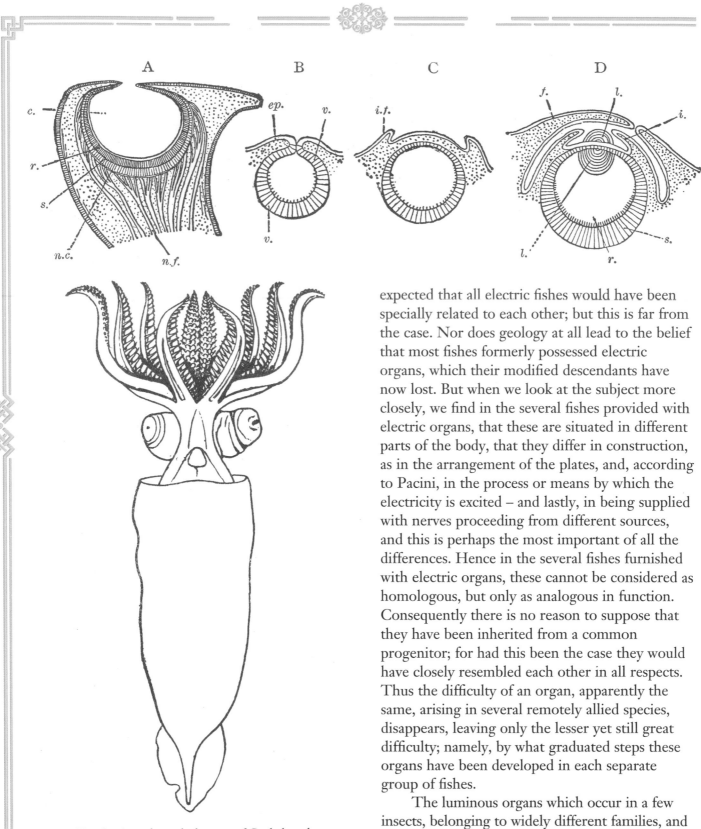

Top: Sections through the eyes of Cephalopods:
A Nautilus
B, C and *D* developmental stages of a Cuttlefish eye
Bottom: Deep-sea Cuttlefish with large eyes

expected that all electric fishes would have been specially related to each other; but this is far from the case. Nor does geology at all lead to the belief that most fishes formerly possessed electric organs, which their modified descendants have now lost. But when we look at the subject more closely, we find in the several fishes provided with electric organs, that these are situated in different parts of the body, that they differ in construction, as in the arrangement of the plates, and, according to Pacini, in the process or means by which the electricity is excited – and lastly, in being supplied with nerves proceeding from different sources, and this is perhaps the most important of all the differences. Hence in the several fishes furnished with electric organs, these cannot be considered as homologous, but only as analogous in function. Consequently there is no reason to suppose that they have been inherited from a common progenitor; for had this been the case they would have closely resembled each other in all respects. Thus the difficulty of an organ, apparently the same, arising in several remotely allied species, disappears, leaving only the lesser yet still great difficulty; namely, by what graduated steps these organs have been developed in each separate group of fishes.

The luminous organs which occur in a few insects, belonging to widely different families, and which are situated in different parts of the body, offer, under our present state of ignorance, a difficulty almost exactly parallel with that of the electric organs. Other similar cases could be given; for instance in plants, the very curious contrivance of a mass of pollen-grains, borne on a foot-stalk with an adhesive gland, is apparently the same in Orchis and Asclepias, genera almost as remote as is possible amongst flowering plants; but here again the parts are not homologous. In all cases of beings, far removed from each other in the scale of organisation, which are furnished with similar and peculiar organs, it will be found that although the general appearance and function of the organs may be the same, yet fundamental differences between them can always be detected. For instance, the eyes of Cephalopods or cuttle-fish and of vertebrate animals appear wonderfully alike; and in such widely sundered groups no part of this resemblance can be due to inheritance from a common progenitor. Mr. Mivart has advanced this case as one of special difficulty, but I am unable to see the force of his argument. An organ for vision must be formed of transparent tissue, and must include some sort of lens for throwing an image at the back of a darkened chamber. Beyond this superficial resemblance, there is hardly any real similarity between the eyes of cuttle-fish and vertebrates, as may be seen by consulting Hensen's admirable memoir on these organs in the Cephalopoda. It is impossible for me here to enter on details, but I may specify a few of the points of difference. The crystalline lens in the higher cuttle-fish consists of two parts, placed one behind the other like two lenses, both having a very different structure and disposition to what occurs in the Vertebrata. The retina is wholly different, with an actual inversion of the elemental parts, and with a large nervous ganglion included within the membranes of the eye. The relations of the muscles are as different as it is possible to conceive, and so in other points. Hence it is not a little difficult to decide how far even the same terms ought to be employed in describing the eyes of the Cephalopoda and Vertebrata. It is, of course, open to any one to deny that the eye in either case could have been developed through the natural selection

of successive slight variations; but if this be admitted in the one case it is clearly possible in the other; and fundamental differences of structure in the visual organs of two groups might have been anticipated, in accordance with this view of their manner of formation. As two men have sometimes independently hit on the same invention, so in the several foregoing cases it appears that natural selection, working for the good of each being, and taking advantage of all favourable variations, has produced similar organs, as far as function is concerned, in distinct organic beings, which owe none of their structure in common to inheritance from a common progenitor.

Fritz Müller, in order to test the conclusions arrived at in this volume, has followed out with much care a nearly similar line of argument. Several families of crustaceans include a few species, possessing an air-breathing apparatus and fitted to live out of the water. In two of these families, which were more especially examined by Müller, and which are nearly related to each other, the species agree most closely in all important characters: namely in their sense organs, circulating systems, in the position of the tufts of hair within their complex stomachs, and lastly in the whole structure of the water-breathing branchiae, even to the microscopical hooks by which they are cleansed. Hence it might have been expected that in the few species belonging to both families which live on the land, the equally important air-breathing apparatus would have been the same; for why should this one apparatus, given for the same purpose, have been made to differ, whilst all the other important organs were closely similar, or rather, identical.

Fritz Müller argues that this close similarity in so many points of structure must, in accordance with the views advanced by me, be accounted for by inheritance from a common progenitor. But as the vast majority of the species in the above two families, as well as most other crustaceans, are aquatic in their habits, it is improbable in the highest degree that their common progenitor should have been adapted for breathing air. Müller was thus led carefully to examine the apparatus in the air-breathing species; and he found it to differ in each in several important points, as in the position of the orifices, in the manner in which they are opened and closed, and in some accessory details. Now such differences are intelligible, and might even have been expected, on the supposition that species belonging to distinct families had slowly become adapted to live more and more out of water, and to breathe the air. For these species, from belonging to distinct families, would have differed to a certain extent, and in accordance with the principle that the nature of each variation depends on two factors, viz., the nature of the organism and that of the surrounding conditions, their variability assuredly would not have been exactly the same. Consequently natural selection would have had different materials or variations to work on, in order to arrive at the same functional result; and the structures thus acquired would almost necessarily have differed. On the hypothesis of separate acts of creation the whole case remains unintelligible. This line of argument seems to have had great weight in leading Fritz Müller to accept the views maintained by me in this volume.

Another distinguished zoologist, the late Professor Claparède, has argued in the same manner, and has arrived at the same result. He shows that there are parasitic mites (Acaridae), belonging to distinct sub-families and families, which are furnished with hair-claspers. These organs must have been independently developed, as they could not have been inherited from a common progenitor; and in the several groups they are formed by the modification of the fore legs, of the hind legs, of the maxillae or lips, and of appendages on the under side of the hind part of the body.

In the foregoing cases, we see the same end gained and the same function performed, in beings not at all or only remotely allied, by organs in appearance, though not in development, closely similar. On the other hand, it is a common rule throughout nature that the same end should be gained, even sometimes in the case of closely related beings, by the most diversified means. How differently constructed is the feathered wing of a bird and the membrane-covered wing of a bat; and still more so the four wings of a butterfly, the two wings of a fly, and the two wings with the elytra of a beetle. Bivalve shells are made to open and shut, but on what a number of patterns is the hinge constructed, – from the long row of neatly interlocking teeth in a Nucula to the simple ligament of a Mussel! Seeds are disseminated by their minuteness, – by their capsule being converted into a light balloon-like envelope, – by being embedded in pulp or flesh, formed of the most diverse parts, and rendered nutritious, as well as conspicuously coloured, so as to attract and be devoured by birds, – by having hooks and grapnels of many kinds and serrated awns, so as to adhere to the fur of quadrupeds, – and by being furnished with wings and plumes, as different in shape as they are elegant in structure, so as to be wafted by every breeze. I will give one other instance: for this subject of the same end being gained by the most diversified means well deserves attention. Some authors maintain that organic beings have been formed in many ways for the sake of mere variety, almost like toys in a shop, but such a view of nature is incredible. With plants having separated sexes, and with those in which, though hermaphrodites, the pollen does not spontaneously fall on the stigma, some aid is necessary for their fertilisation. With several kinds this is effected by the pollen-grains, which are light and incoherent, being blown by the wind through mere chance on to the stigma; and this is the simplest plan which can well be conceived. An almost equally simple, though very different plan occurs in many plants in which a symmetrical flower secretes a few drops of nectar, and is consequently visited by insects; and these carry the pollen from the anthers to the stigma.

From this simple stage we may pass through an inexhaustible number of contrivances, all for the same purpose and effected in essentially the same manner, but entailing changes in every part of the flower. The nectar may be stored in variously shaped receptacles, with the stamens and pistils modified in many ways, sometimes forming trap-like contrivances, and sometimes capable of neatly adapted movements through irritability or elasticity. From such structures we may advance till we come to such a case of extraordinary adaptation as that lately described by Dr. Crüger in the Coryanthes. This orchid has part of its labellum or lower lip hollowed out into a great bucket, into which drops of almost pure water continually fall from two secreting horns which stand above it; and when the bucket is half full, the water overflows by a spout on one side. The basal part of the labellum stands over the bucket, and is itself hollowed out into a sort of chamber with two lateral entrances; within this chamber there are curious fleshy ridges. The most ingenious man, if he had not witnessed what takes place, could never have imagined what purpose all these parts serve. But Dr. Crüger saw crowds of large humble-bees visiting the gigantic flowers of this orchid, not in order to suck nectar, but to gnaw off the ridges within the chamber above the bucket; in doing this they frequently pushed each other into the bucket, and their wings being thus wetted they could not fly away, but were compelled to crawl out through the passage formed by the spout or overflow. Dr. Crüger saw a 'continual procession' of bees thus crawling out of their involuntary bath. The passage is narrow, and is roofed over by the column, so that a bee, in forcing its way out, first rubs its back against the viscid stigma and then against the viscid glands of the pollen-masses. The pollen-masses are thus glued to the back of the bee which first happens to crawl out through the passage of a lately expanded flower, and are thus carried away. Dr. Crüger sent me a flower in spirits of wine, with a bee which he had killed before it had quite crawled out, with a pollen-mass still fastened to its back. When the bee, thus provided, flies to another flower, or to

the same flower a second time, and is pushed by its comrades into the bucket and then crawls out by the passage, the pollen-mass necessarily comes first into contact with the viscid stigma, and adheres to it, and the flower is fertilised. Now at last we see the full use of every part of the flower, of the water-secreting horns, of the bucket half-full of water, which prevents the bees from flying away, and forces them to crawl out through the spout, and rub against the properly placed viscid pollen-masses and the viscid stigma.

The construction of the flower in another closely allied orchid, namely, the Catasetum, is widely different, though serving the same end; and is equally curious. Bees visit these flowers, like those of the Coryanthes, in order to gnaw the labellum; in doing this they inevitably touch a long, tapering, sensitive projection, or, as I have called it, the antenna. This antenna, when touched, transmits a sensation or vibration to a certain membrane which is instantly ruptured; this sets free a spring by which the pollen-mass is shot forth, like an arrow, in the right direction, and adheres by its viscid extremity to the back of the bee. The pollen-mass of the male plant (for the sexes are separate in this orchid) is thus carried to the flower of the female plant, where it is brought into contact with the stigma, which is viscid enough to break certain elastic threads, and retain the pollen, thus effecting fertilisation.

How, it may be asked, in the foregoing and in innumerable other instances, can we understand the graduated scale of complexity and the multifarious means for gaining the same end. The answer no doubt is, as already remarked, that when two forms vary, which already differ from each other in some slight degree, the variability will not be of the same exact nature, and consequently the results obtained through natural selection for the same general purpose will not be the same. We should also bear in mind that every highly developed organism has passed through many changes, and that each modified structure tends to be inherited, so that each modification

Iris Pseudacorus.

The Iris pseudacorus or Yellow Flag, showing the roots (bottom right)
and reproductive organs (bottom left)

will not readily be quite lost, but may be again and again further altered. Hence, the structure of each part of each species, for whatever purpose it may serve, is the sum of many inherited changes, through which the species has passed during its successive adaptations to changed habits and conditions of life.

Finally, then, although in many cases it is most difficult even to conjecture by what transitions organs could have arrived at their present state; yet, considering how small the proportion of living and known forms is to the extinct and unknown, I have been astonished how rarely an organ can be named, towards which no transitional grade is known to lead. It is certainly true, that new organs appearing as if created for some special purpose rarely or never appear in any being; as indeed is shown by that old, but somewhat exaggerated, canon in natural history of 'Natura non facit saltum.' We meet with this admission in the writings of almost every experienced naturalist; or, as Milne Edwards has well expressed it, 'Nature is prodigal in variety, but niggard in innovation.' Why, on the theory of Creation, should there be so much variety and so little real novelty? Why should all the parts and organs of many independent beings, each supposed to have been separately created for its own proper place in nature, be so commonly linked together by graduated steps? Why should not Nature take a sudden leap from structure to structure? On the theory of natural selection, we can clearly understand why she should not; for natural selection acts only by taking advantage of slight successive variations; she can never take a great and sudden leap, but must advance by the short and sure, though slow steps.

ORGANS OF LITTLE APPARENT IMPORTANCE, AS AFFECTED BY NATURAL SELECTION

As natural selection acts by life and death, by the survival of the fittest, by the destruction of the less well-fitted individuals, – I have sometimes felt great difficulty in understanding the origin or formation of parts of little importance; almost as great, though of a very different kind, as in the case of the most perfect and complex organs.

In the first place, we are much too ignorant in regard to the whole economy of any one organic being to say what slight modifications would be of importance or not. In a former chapter I have given instances of very trifling characters, such as the down on fruit and the colour of its flesh, the colour of the skin and hair of quadrupeds, which, from being correlated with constitutional differences, or from determining the attacks of insects, might assuredly be acted on by natural selection. The tail of the giraffe looks like an artificially constructed fly-flapper; and it seems at first incredible that this could have been adapted for its present purpose by successive slight modifications, each better and better fitted, for so trifling an object as to drive away flies; yet we should pause before being too positive even in this case, for we know that the distribution and existence of cattle and other animals in South America absolutely depend on their power of resisting the attacks of insects: so that individuals which could by any means defend themselves from these small enemies, would be able to range into new pastures and thus gain a great advantage. It is not that the larger quadrupeds are actually destroyed (except in some rare cases) by flies, but they are incessantly harassed and their strength reduced, so that they are more subject to disease, or not so well enabled in a coming dearth to search for food, or to escape from beasts of prey.

Organs now of trifling importance have probably in some cases been of high importance to an early progenitor, and, after having been slowly perfected at a former period, have been transmitted to existing species in nearly the same state, although now of very slight use; but any actually injurious deviations in their structure would of course have been checked by natural selection. Seeing how important an organ of locomotion the tail is in most aquatic animals, its general presence and use for many purposes in so many land animals, which in their lungs or modified swim-bladders betray their aquatic origin, may perhaps be thus accounted for. A well-developed tail having been formed in an aquatic animal, it might subsequently come to be worked in for all sorts of purposes, – as a fly-flapper, an organ of prehension, or as an aid in turning, as in the case of the dog, though the aid in this latter respect must be slight, for the hare, with hardly any tail, can double still more quickly.

In the second place, we may easily err in attributing importance to characters, and in believing that they have been developed through natural selection. We must by no means overlook the effects of the definite action of changed conditions of life, of so-called spontaneous variations, which seem to depend in a quite subordinate degree on the nature of the conditions, of the tendency to reversion to long-lost characters, of the complex laws of growth, such as of correlation, comprehension, of the pressure of one part on another, etc., and finally of sexual selection, by which characters of use to one sex are often gained and then transmitted more or less perfectly to the other sex, though of no use to the sex. But structures thus indirectly gained, although at first of no advantage to a species, may subsequently have been taken advantage of by its modified descendants, under new conditions of life and newly acquired habits.

If green woodpeckers alone had existed, and we did not know that there were many black and pied kinds, I dare say that we should have thought that the green colour was a beautiful adaptation to conceal this tree-frequenting bird from its enemies; and consequently that it was a character of importance, and had been acquired through natural selection; as it is, the colour is probably in chief part due to sexual selection. A trailing palm in the Malay Archipelago climbs the loftiest trees by the aid of exquisitely constructed hooks clustered around the ends of the branches, and this contrivance, no doubt, is of the highest service to the plant; but as we see nearly similar hooks on many trees which are not climbers, and which, as there is reason to believe from the distribution of the thorn-bearing species in Africa and South America, serve as a defence against browsing quadrupeds, so the spikes on the palm may at first have been developed for this object, and subsequently have been improved and taken advantage of by the plant, as it underwent further modification and became a climber. The naked skin on the head of a vulture is generally considered as a direct adaptation for wallowing in putridity; and so it may be, or it may possibly be due to the direct action of putrid matter; but we should be very cautious in drawing any such inference, when we see that the skin on the head of the clean-feeding male turkey is likewise naked. The sutures in the skulls of young mammals have been advanced as a beautiful adaptation for aiding parturition, and no doubt they facilitate, or may be indispensable for this act; but as sutures occur in the skulls of young birds and reptiles, which have only to escape from a broken egg, we may infer that this structure has arisen from the laws of growth, and has been taken advantage of in the parturition of the higher animals.

We are profoundly ignorant of the cause of each slight variation or individual difference; and we are immediately made conscious of this by reflecting on the differences between the breeds of our domesticated animals in different countries, more especially in the less civilized countries, where there has been but little methodical selection. Animals kept by savages in different countries often have to struggle for their own subsistence, and are exposed to a certain extent to natural selection, and individuals with slightly different constitutions would succeed best under different climates. With cattle susceptibility to the attacks of flies is correlated with colour, as is the liability to be poisoned by certain plants; so that even colour would be thus subjected to the action of natural selection. Some observers are convinced that a damp climate affects the growth of the hair, and that with the hair the horns are correlated. Mountain breeds always differ from lowland breeds; and a mountainous country would probably affect the hind limbs from exercising them more, and possibly even the form of the pelvis; and then by the law of homologous variation, the front limbs and the head would probably be affected. The shape, also, of the pelvis might affect by pressure the shape of certain parts of the young in the womb. The laborious breathing necessary in high regions tends, as we have good reason to believe, to increase the size of the chest; and again correlation would come into play. The effects of lessened exercise, together with abundant food, on the whole organisation is probably still more important, and this, as H. von Nathusius has lately shown in his excellent treatise, is apparently one chief cause of the great modification which the breeds of swine have undergone. But we are far too ignorant to speculate on the relative importance of the several known and unknown causes of variation; and I have made these remarks only to show that, if we are unable to account for the characteristic differences of our several domestic breeds, which nevertheless are generally admitted to have arisen through ordinary generation from one or a few parent-stocks, we ought not to lay too much stress on our ignorance of the precise cause of the slight analogous differences between true species.

UTILITARIAN DOCTRINE, HOW FAR TRUE: BEAUTY, HOW ACQUIRED

The foregoing remarks lead me to say a few words on the protest lately made by some naturalists against the utilitarian doctrine that every detail of structure has been produced for the good of its possessor. They believe that many structures have been created for the sake of beauty, to delight man or the Creator (but this latter point is beyond the scope of scientific discussion), or for the sake of mere variety, a view already discussed. Such doctrines, if true, would be absolutely fatal to my theory. I fully admit that many structures are now of no direct use to their possessors, and may never have been of any use to their progenitors; but this does not prove that they were formed solely for beauty or variety. No doubt the definite action of changed conditions, and the various causes of modifications, lately specified, have all produced an effect, probably a great effect, independently of any advantage thus gained. But a still more important consideration is that the chief part of the organisation of every living creature is due to inheritance; and consequently, though each being assuredly is well fitted for its place in nature, many structures have now no very close and direct relation to present habits of life. Thus, we can hardly believe that the webbed feet of the upland goose, or of the frigate- bird, are of special use to these birds; we cannot believe that the similar bones in the arm of the monkey, in the fore leg of the horse, in the wing of the bat, and in the flipper of the seal, are of special use to these animals. We may safely attribute these structures to inheritance. But webbed feet no doubt were as useful to the progenitor of the upland goose and of the frigate bird, as they now are to the most aquatic of living birds. So we may believe that the progenitor of the seal did not possess a flipper, but a foot with five toes fitted for walking or grasping; and we may further venture to believe that the several bones in the limbs of the monkey, horse and bat, were originally developed, on the principle of utility, probably through the reduction of more numerous bones in the fin of some ancient fish-like progenitor of the whole class. It is scarcely possible to decide how much allowance ought to be made for such causes of change, as the definite action of external conditions, so-called spontaneous variations, and the complex laws of growth; but with these important exceptions, we may conclude that the structure of every living creature either now is, or was formerly, of some direct or indirect use to its possessor.

With respect to the belief that organic beings have been created beautiful for the delight of man,

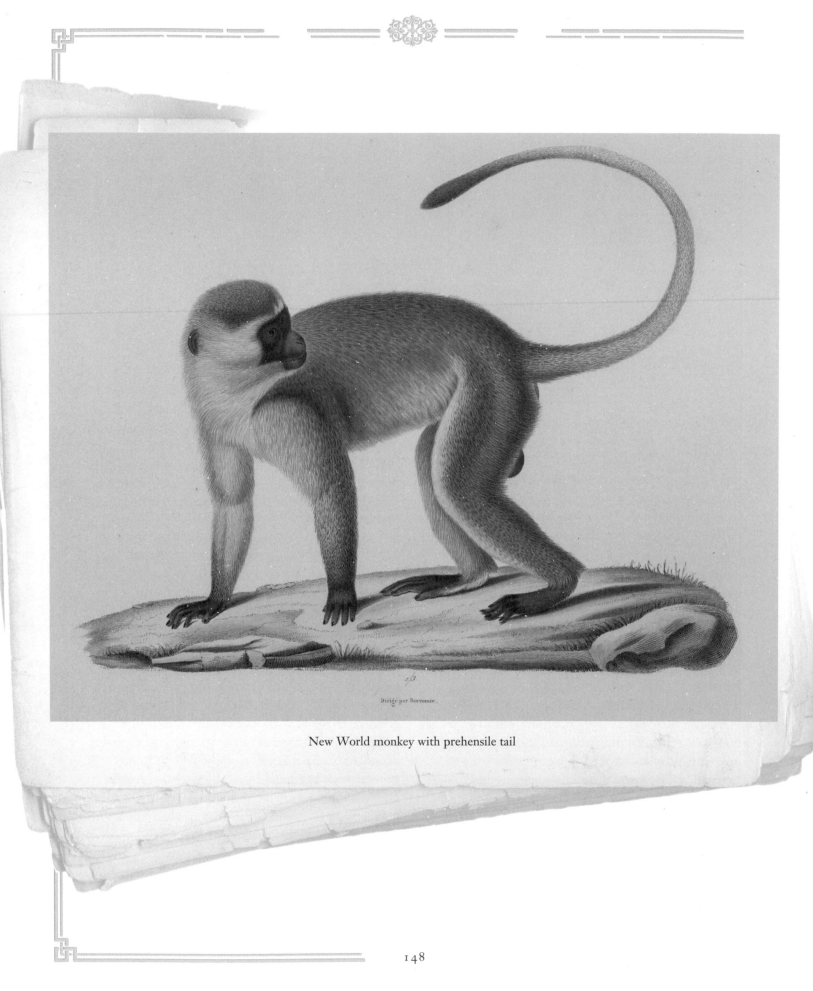

New World monkey with prehensile tail

– a belief which it has been pronounced is subversive of my whole theory, – I may first remark that the sense of beauty obviously depends on the nature of the mind, irrespective of any real quality in the admired object; and that the idea of what is beautiful, is not innate or unalterable. We see this, for instance, in the men of different races admiring an entirely different standard of beauty in their women. If beautiful objects had been created solely for man's gratification, it ought to be shown that before man appeared there was less beauty on the face of the earth than since he came on the stage. Were the beautiful volute and cone shells of the Eocene epoch, and the gracefully sculptured ammonites of the Secondary period, created that man might ages afterwards admire them in his cabinet? Few objects are more beautiful than the minute siliceous cases of the diatomaceae: were these created that they might be examined and admired under the higher powers of the microscope? The beauty in this latter case, and in many others, is apparently wholly due to symmetry of growth. Flowers rank amongst the most beautiful productions of nature; but they have been rendered conspicuous in contrast with the green leaves, and in consequence at the same time beautiful, so that they may be easily observed by insects. I have come to this conclusion from finding it an invariable rule that when a flower is fertilised by the wind it never has a gaily-coloured corolla. Several plants habitually produce two kinds of flowers; one kind open and coloured so as to attract insects; the other closed, not coloured, destitute of nectar, and never visited by insects. Hence, we may conclude that, if insects had not been developed on the face of the earth, our plants would not have been decked with beautiful flowers, but would have produced only such poor flowers as we see on our fir, oak, nut and ash trees, on grasses, spinach, docks, and nettles, which are all fertilised through the agency of the wind. A similar line of argument holds good with fruits; that a ripe strawberry or cherry is as pleasing to the eye as to the palate – that the gaily-coloured fruit of the spindle-wood tree and the scarlet berries of the holly are beautiful objects, – will be admitted by everyone. But this beauty serves merely as a guide to birds and beasts, in order that the fruit may be devoured and the matured seeds disseminated. I infer that this is the case from having as yet found no exception to the rule that seeds are always thus disseminated when embedded within a fruit of any kind (that is within a fleshy or pulpy envelope), if it be coloured of any brilliant tint, or rendered conspicuous by being white or black.

On the other hand, I willingly admit that a great number of male animals, as all our most gorgeous birds, some fishes, reptiles, and mammals, and a host of magnificently coloured butterflies, have been rendered beautiful for beauty's sake. But this has been effected through sexual selection, that is, by the more beautiful males having been continually preferred by the females, and not for the delight of man. So it is with the music of birds. We may infer from all this that a nearly similar taste for beautiful colours and for musical sounds runs through a large part of the animal kingdom. When the female is as beautifully coloured as the male, which is not rarely the case with birds and butterflies, the cause apparently lies in the colours acquired through sexual selection having been transmitted to both sexes, instead of to the males alone. How the sense of beauty in its simplest form – that is, the reception of a peculiar kind of pleasure from certain colours, forms and sounds – was first developed in the mind of man and of the lower animals, is a very obscure subject. The same sort of difficulty is presented if we enquire how it is that certain flavours and odours give pleasure, and others displeasure. Habit in all these cases appears to have come to a certain extent into play; but there must be some fundamental cause in the constitution of the nervous system in each species.

Natural selection cannot possibly produce any modification in a species exclusively for the good of another species; though throughout

Four species of bird catalogued and illustrated in *The Zoology of the Voyage of H.M.S. Beagle*, including *top left* the Galapagos Short-Eared Owl nature one species incessantly takes advantage of, and profits by, the structures of others. But natural selection can and does often produce structures for the direct injury of other animals, as we see in the fang of the adder, and in the ovipositor of the ichneumon, by which its eggs are deposited in the living bodies of other insects. If it could be proved that any part of the structure of any one species had been formed for the exclusive good of another species, it would annihilate my theory, for such could not have been produced through natural selection. Although many statements may be found in works on natural history to this effect, I cannot find even one which seems to me of any weight. It is admitted that the rattlesnake has a poison-fang for its own defence and for the destruction of its prey; but some authors suppose that at the same time it is furnished with a rattle for its own injury, namely, to warn its prey. I would almost as soon believe that the cat curls the end of its tail when preparing to spring, in order to warn the doomed mouse. It is a much more probable view that the rattlesnake uses its rattle, the cobra expands its frill, and the puff-adder swells whilst hissing so loudly and harshly, in order to alarm the many birds and beasts which are known to attack even the most venomous species. Snakes act on the same principle which makes the hen ruffle her feathers and expand her wings when a dog approaches her chickens. But I have not space here to enlarge on the many ways by which animals endeavour to frighten away their enemies.

Natural selection will never produce in a being any structure more injurious than beneficial to that being, for natural selection acts solely by and for the good of each. No organ will be formed, as Paley has remarked, for the purpose of causing pain or for doing an injury to its possessor. If a fair balance be struck between the good and evil caused by each part, each will be found on the whole advantageous. After the lapse of time, under changing conditions of life, if any part comes to be injurious, it will be modified; or

Top: Skull of a Rattlesnake:
a brain-case *b* upper jaw *c* poison fangs *d* lower jaw
Bottom: Rattle of a Rattlesnake:
a cross-section *b* exterior view

if it be not so, the being will become extinct, as myriads have become extinct.

Natural selection tends only to make each organic being as perfect as, or slightly more perfect than, the other inhabitants of the same country with which it comes into competition. And we see that this is the standard of perfection attained under nature. The endemic productions of New Zealand, for instance, are perfect, one compared with another; but they are now rapidly yielding before the advancing legions of plants and animals introduced from Europe. Natural selection will not produce absolute perfection, nor do we always meet, as far as we can judge, with this high standard under nature. The correction for the aberration of light is said by Müller not to be perfect even in that most perfect organ, the human eye. Helmholtz, whose judgment no one will dispute, after describing in the strongest terms the wonderful powers of the human eye, adds these remarkable words: 'That

Sting of a bee

If we look at the sting of the bee, as having existed in a remote progenitor, as a boring and serrated instrument, like that in so many members of the same great order, and that it has since been modified but not perfected for its present purpose, with the poison originally adapted for some other object, such as to produce galls, since intensified, we can perhaps understand how it is that the use of the sting should so often cause the insect's own death: for if on the whole the power of stinging be useful to the social community, it will fulfil all the requirements of natural selection, though it may cause the death of some few members. If we admire the truly wonderful power of scent by which the males of many insects find their females, can we admire the production for this single purpose of thousands of drones, which are utterly useless to the community for any other purpose, and which are ultimately slaughtered by their industrious and sterile sisters? It may be difficult, but we ought to admire the savage instinctive hatred of the queen bee, which urges her to destroy the young queens, her daughters, as soon as they are born, or to perish herself in the combat; for undoubtedly this is for the good of the community; and maternal love or maternal hatred, though the latter fortunately is most rare, is all the same to the inexorable principles of natural selection. If we admire the several ingenious contrivances by which orchids and many other plants are fertilised through insect agency, can we consider as equally perfect the elaboration of dense clouds of pollen by our fir-trees, so that a few granules may be wafted by chance on to the ovules?

which we have discovered in the way of inexactness and imperfection in the optical machine and in the image on the retina, is as nothing in comparison with the incongruities which we have just come across in the domain of the sensations. One might say that nature has taken delight in accumulating contradictions in order to remove all foundation from the theory of a pre-existing harmony between the external and internal worlds.' If our reason leads us to admire with enthusiasm a multitude of inimitable contrivances in nature, this same reason tells us, though we may easily err on both sides, that some other contrivances are less perfect. Can we consider the sting of the bee as perfect, which, when used against many kinds of enemies, cannot be withdrawn, owing to the backward serratures, and thus inevitably causes the death of the insect by tearing out its viscera?

SUMMARY: THE LAW OF UNITY OF TYPE AND OF THE CONDITIONS OF EXISTENCE EMBRACED BY THE THEORY OF NATURAL SELECTION

We have in this chapter discussed some of the difficulties and objections which may be urged against the theory. Many of them are serious; but I think that in the discussion light has been thrown on several facts, which on the belief of independent acts of creation are utterly obscure. We have seen that species at any one period are not indefinitely variable, and are not linked together by a multitude of intermediate gradations, partly because the process of natural selection is always very slow, and at any one time acts only on a few forms; and partly because the very process of natural selection implies the continual supplanting and extinction of preceding and intermediate gradations. Closely allied species, now living on a continuous area, must often have been formed when the area was not continuous, and when the conditions of life did not insensibly graduate away from one part to another. When two varieties are formed in two districts of a continuous area, an intermediate variety will often be formed, fitted for an intermediate zone; but from reasons assigned, the intermediate variety will usually exist in lesser numbers than the two forms which it connects; consequently the two latter, during the course of further modification, from existing in greater numbers, will have a great advantage over the less numerous intermediate variety, and will thus generally succeed in supplanting and exterminating it.

We have seen in this chapter how cautious we should be in concluding that the most different habits of life could not graduate into each other; that a bat, for instance, could not have been formed by natural selection from an animal which at first only glided through the air.

We have seen that a species under new conditions of life may change its habits, or it may have diversified habits, with some very unlike those of its nearest congeners. Hence we can understand, bearing in mind that each organic being is trying to live wherever it can live, how it has arisen that there are upland geese with webbed feet, ground woodpeckers, diving thrushes, and petrels with the habits of auks.

Although the belief that an organ so perfect as the eye could have been formed by natural selection, is enough to stagger any one; yet in the case of any organ, if we know of a long series of gradations in complexity, each good for its possessor, then under changing conditions of life, there is no logical impossibility in the acquirement of any conceivable degree of perfection through natural selection. In the cases in which we know of no intermediate or transitional states, we should be extremely cautious in concluding that none can have existed, for the metamorphoses of many organs show what wonderful changes in function are at least possible. For instance, a swim bladder has apparently been converted into an air-breathing lung. The same organ having performed simultaneously very different functions, and then having been in part or in whole specialised for one function; and two distinct organs having performed at the same time the same function, the one having been perfected whilst aided by the other, must often have largely facilitated transitions.

We have seen that in two beings widely remote from each other in the natural scale, organs serving for the same purpose and in external appearance closely similar may have been separately and independently formed; but when such organs are closely examined, essential differences in their structure can almost always be detected; and this naturally follows from the principle of natural selection. On the other hand, the common rule throughout nature is infinite diversity of structure for gaining the same end; and this again naturally follows from the same great principle.

In many cases we are far too ignorant to be enabled to assert that a part or organ is so unimportant for the welfare of a species, that modifications in its structure could not have been slowly accumulated by means of natural selection. In many other cases, modifications are probably the direct result of the laws of variation or of growth, independently of any good having been thus gained. But even such structures have often, as we may feel assured, been subsequently taken advantage of, and still further modified, for the

Anser melanopterus.

South American Goose

good of species under new conditions of life. We may, also, believe that a part formerly of high importance has frequently been retained (as the tail of an aquatic animal by its terrestrial descendants), though it has become of such small importance that it could not, in its present state, have been acquired by means of natural selection.

Natural selection can produce nothing in one species for the exclusive good or injury of another; though it may well produce parts, organs, and excretions highly useful or even indispensable, or highly injurious to another species, but in all cases at the same time useful to the possessor. In each well-stocked country natural selection acts through the competition of the inhabitants and consequently leads to success in the battle for life, only in accordance with the standard of that particular country. Hence the inhabitants of one country, generally the smaller one, often yield to the inhabitants of another and generally the larger country. For in the larger country there will have existed more individuals, and more diversified forms, and the competition will have been severer, and thus the standard of perfection will have been rendered higher. Natural selection will not necessarily lead to absolute perfection; nor, as far as we can judge by our limited faculties, can absolute perfection be everywhere predicated.

On the theory of natural selection we can clearly understand the full meaning of that old canon in natural history, 'Natura non facit saltum.' This canon, if we look to the present inhabitants alone of the world, is not strictly correct; but if we include all those of past times, whether known or unknown, it must on this theory be strictly true.

It is generally acknowledged that all organic beings have been formed on two great laws – Unity of Type, and the Conditions of Existence. By unity of type is meant that fundamental agreement in structure which we see in organic beings of the same class, and which is quite independent of their habits of life. On my theory, unity of type is explained by unity of descent. The expression of conditions of existence, so often insisted on by the illustrious Cuvier, is fully embraced by the principle of natural selection. For natural selection acts by either now adapting the varying parts of each being to its organic and inorganic conditions of life; or by having adapted them during past periods of time: the adaptations being aided in many cases by the increased use or disuse of parts, being affected by the direct action of external conditions of life, and subjected in all cases to the several laws of growth and variation. Hence, in fact, the law of the Conditions of Existence is the higher law; as it includes, through the inheritance of former variations and adaptations, that of Unity of Type.

Chapter VII

MISCELLANEOUS OBJECTIONS TO THE THEORY OF NATURAL SELECTION

Longevity–Modifications not necessarily simultaneous–Modifications apparently of no direct service–Progressive development–Characters of small functional importance, the most constant–Supposed incompetence of natural selection to account for the incipient stages of useful structures–Causes which interfere with the acquisition through natural selection of useful structures–Gradations of structure with changed functions–Widely different organs in members of the same class, developed from one and the same source–Reasons for disbelieving in great and abrupt modifications.

I will devote this chapter to the consideration of various miscellaneous objections which have been advanced against my views, as some of the previous discussions may thus be made clearer; but it would be useless to discuss all of them, as many have been made by writers who have not taken the trouble to understand the subject. Thus a distinguished German naturalist has asserted that the weakest part of my theory is, that I consider all organic beings as imperfect: what I have really said is, that all are not as perfect as they might have been in relation to their conditions; and this is shown to be the case by so many native forms in many quarters of the world having yielded their places to intruding foreigners. Nor can organic beings, even if they were at any one time perfectly adapted to their conditions of life, have remained so, when their conditions changed, unless they themselves likewise changed; and no one will dispute that the physical conditions of each country, as well as the number and kinds of its inhabitants, have undergone many mutations.

A critic has lately insisted, with some parade of mathematical accuracy, that longevity is a great advantage to all species, so that he who believes in natural selection 'must arrange his genealogical tree' in such a manner that all the descendants have longer lives than their progenitors! Cannot our critic conceive that a biennial plant or one of the lower animals might range into a cold climate and perish there every winter; and yet, owing to advantages gained through natural selection, survive from year to year by means of its seeds or ova? Mr. E. Ray Lankester has recently discussed this subject, and he concludes, as far as its extreme complexity allows him to form a judgment, that longevity is generally related to the standard of each species in the scale of organisation, as well as to the amount of expenditure in reproduction and in general activity. And these conditions have, it is probable, been largely determined through natural selection.

It has been argued that, as none of the animals and plants of Egypt, of which we know anything, have changed during the last three or four thousand years, so probably have none in any part of the world. But, as Mr. G.H. Lewes has remarked, this line of argument proves too much, for the ancient domestic races figured on the Egyptian monuments, or embalmed, are closely similar or even identical with those now living; yet all naturalists admit that such races have been produced through the modification of their original types. The many animals which have remained unchanged since the commencement of the glacial period, would have been an incomparably stronger case, for these have been exposed to great changes of climate and have migrated over great distances; whereas, in Egypt, during the last several thousand years, the conditions of life, as far as we know, have remained absolutely uniform. The fact of little or no modification having been effected since the glacial period would have been of some avail against those who believe in an innate and necessary law of development, but is powerless against the doctrine of natural selection or the survival of the fittest, which implies that when variations or individual differences of a beneficial nature happen to arise, these will be preserved; but this will be effected only under certain favourable circumstances.

The celebrated palaeontologist, Bronn, at the close of his German translation of this work, asks how, on the principle of natural selection, can a variety live side by side with the parent species? If both have become fitted for slightly different habits of life or conditions, they might live together; and if we lay on one side polymorphic species, in which the variability seems to be of a peculiar nature, and all mere temporary variations, such as size, albinism, etc., the more permanent varieties are generally found, as far as I can discover, inhabiting distinct stations, such as high land or low land, dry or moist districts. Moreover, in the case of animals which wander much about and cross freely, their varieties seem to be generally confined to distinct regions.

Bronn also insists that distinct species never differ from each other in single characters, but in many parts; and he asks, how it always comes that many parts of the organisation should have been modified at the same time through variation and natural selection? But there is no necessity for supposing that all the parts of any being have been simultaneously modified. The most striking modifications, excellently adapted for some purpose, might, as was formerly remarked, be acquired by successive variations, if slight, first in one part and then in another; and as they would

be transmitted all together, they would appear to us as if they had been simultaneously developed. The best answer, however, to the above objection is afforded by those domestic races which have been modified, chiefly through man's power of selection, for some special purpose. Look at the race and dray horse, or at the greyhound and mastiff. Their whole frames, and even their mental characteristics, have been modified; but if we could trace each step in the history of their transformation, – and the latter steps can be traced, – we should not see great and simultaneous changes, but first one part and then another slightly modified and improved. Even when selection has been applied by man to some one character alone, – of which our cultivated plants offer the best instances, – it will invariably be found that although this one part, whether it be the flower, fruit, or leaves, has been greatly changed, almost all the other parts have been slightly modified. This may be attributed partly to the principle of correlated growth, and partly to so-called spontaneous variation.

A much more serious objection has been urged by Bronn, and recently by Broca, namely, that many characters appear to be of no service whatever to their possessors, and therefore cannot have been influenced through natural selection. Bronn adduces the length of the ears and tails in the different species of hares and mice, – the complex folds of enamel in the teeth of many animals, and a multitude of analogous cases. With respect to plants, this subject has been discussed by Nägeli in an admirable essay. He admits that natural selection has effected much, but he insists that the families of plants differ chiefly from each other in morphological characters, which appear to be quite unimportant for the welfare of the species. He consequently believes in an innate tendency towards progressive and more perfect development. He specifies the arrangement of the cells in the tissues, and of the leaves on the axis, as cases in which natural selection could not have acted. To these may be added the numerical

Insect galls on an oak leaf

place, we have to allow for the direct and definite action of changed conditions of life, and for so-called spontaneous variations, in which the nature of the conditions apparently plays a quite subordinate part. Bud variations, such as the appearance of a moss-rose on a common rose, or of a nectarine on a peach-tree, offer good instances of spontaneous variations; but even in these cases, if we bear in mind the power of a minute drop of poison in producing complex galls, we ought not to feel too sure that the above variations are not the effect of some local change in the nature of the sap, due to some change in the conditions. There must be some efficient cause for each slight individual difference, as well as for more strongly marked variations which occasionally arise; and if the unknown cause were to act persistently, it is almost certain that all the individuals of the species would be similarly modified.

In the earlier editions of this work I underrated, as it now seems probable, the frequency and importance of modifications due to spontaneous variability. But it is impossible to attribute to this cause the innumerable structures which are so well adapted to the habits of life of each species. I can no more believe in this than that the well-adapted form of a race-horse or greyhound, which before the principle of selection by man was well understood, excited so much surprise in the minds of the older naturalists, can thus be explained.

It may be worth while to illustrate some of the foregoing remarks. With respect to the assumed inutility of various parts and organs, it is hardly necessary to observe that even in the higher and best-known animals many structures exist, which are so highly developed that no one doubts that they are of importance, yet their use has not been, or has only recently been, ascertained. As Bronn gives the length of the ears and tail in the several species of mice as instances, though trifling ones, of differences in structure which can be of no special use, I may mention that, according to Dr. Schöbl, the external ears of the common

divisions in the parts of the flower, the position of the ovules, the shape of the seed, when not of any use for dissemination, etc.

There is much force in the above objection. Nevertheless, we ought, in the first place, to be extremely cautious in pretending to decide what structures now are, or have formerly been, of use to each species. In the second place, it should always be borne in mind that when one part is modified, so will be other parts, through certain dimly seen causes, such as an increased or diminished flow of nutriment to a part, mutual pressure, an early developed part affecting one subsequently developed, and so forth, – as well as through other causes which lead to the many mysterious cases of correlation, which we do not in the least understand. These agencies may be all grouped together, for the sake of brevity, under the expression of the laws of growth. In the third mouse are supplied in an extraordinary manner with nerves, so that they no doubt serve as tactile organs; hence the length of the ears can hardly be quite unimportant. We shall, also, presently see that the tail is a highly useful prehensile organ to some of the species; and its use would be much influenced by its length.

With respect to plants, to which on account of Nägeli's essay I shall confine myself in the following remarks, it will be admitted that the flowers of the orchids present a multitude of curious structures, which a few years ago would have been considered as mere morphological differences without any special function; but they are now known to be of the highest importance for the fertilisation of the species through the aid of insects, and have probably been gained through natural selection. No one until lately would have imagined that in dimorphic and trimorphic plants the different lengths of the stamens and pistils, and their arrangement, could have been of any service, but now we know this to be the case.

In certain whole groups of plants the ovules stand erect, and in others they are suspended; and within the same ovarium of some few plants, one ovule holds the former and a second ovule the latter position. These positions seem at first purely morphological, or of no physiological signification; but Dr. Hooker informs me that within the same ovarium the upper ovules alone in some cases, and in others the lower ones alone are fertilised; and he suggests that this probably depends on the direction in which the pollen-tubes enter the ovarium. If so, the position of the ovules, even when one is erect and the other suspended within the same ovarium, would follow the selection of any slight deviations in position which favoured their fertilisation, and the production of seed.

Several plants belonging to distinct orders habitually produce flowers of two kinds, – the one open, of the ordinary structure, the other closed and imperfect. These two kinds of flowers sometimes differ wonderfully in structure, yet may be seen to graduate into each other on the same plant. The ordinary and open flowers can be intercrossed; and the benefits which certainly are derived from this process are thus secured. The closed and imperfect flowers are, however, manifestly of high importance, as they yield with the utmost safety a large stock of seed, with the expenditure of wonderfully little pollen. The two kinds of flowers often differ much, as just stated, in structure. The petals in the imperfect flowers almost always consist of mere rudiments, and the pollen-grains are reduced in diameter. In Ononis columnae five of the alternate stamens are rudimentary; and in some species of Viola three stamens are in this state, two retaining their proper function, but being of very small size. In six out of thirty of the closed flowers in an Indian violet (name unknown, for the plants have never produced with me perfect flowers), the sepals are reduced from the normal number of five to three. In one section of the Malpighiaceae the closed flowers, according to A. de Jussieu, are still further modified, for the five stamens which stand opposite to the sepals are all aborted, a sixth stamen standing opposite to a petal being alone developed; and this stamen is not present in the ordinary flowers of this species; the style is aborted; and the ovaria are reduced from three to two. Now although natural selection may well have had the power to prevent some of the flowers from expanding, and to reduce the amount of pollen, when rendered by the closure of the flowers superfluous, yet hardly any of the above special modifications can have been thus determined, but must have followed from the laws of growth, including the functional inactivity of parts, during the progress of the reduction of the pollen and the closure of the flowers.

It is so necessary to appreciate the important effects of the laws of growth, that I will give some additional cases of another kind, namely of differences in the same part or organ, due to differences in relative position on the same plant. In the Spanish chestnut, and in certain fir-trees, the angles of divergence of the leaves differ, according to Schacht, in the nearly horizontal and in the upright branches. In the common rue and some other plants, one flower, usually the central or terminal one, opens first, and has five sepals and petals, and five divisions to the ovarium; whilst all the other flowers on the plant are tetramerous. In the British Adoxa the uppermost flower generally has two calyx-lobes with the other organs tetramerous, whilst the surrounding flowers generally have three calyx-lobes with the other organs pentamerous. In many Compositae and Umbelliferae (and in some other plants) the circumferential flowers have their corollas much more developed than those of the centre; and this seems often connected with the abortion of the reproductive organs. It is a more curious fact, previously referred to, that the achenes or seeds of the circumference and centre sometimes differ greatly in form, colour, and other characters. In Carthamus and some other Compositae the central achenes alone are furnished with a pappus; and in Hyoseris the same head yields achenes of three different forms. In certain Umbelliferae the exterior seeds, according to Tausch, are orthospermous, and the central one coelospermous, and this is a character which was considered by De Candolle to be in other species of the highest systematic importance. Professor Braun mentions a Fumariaceous genus, in which the flowers in the lower part of the spike bear oval, ribbed, one-seeded nutlets; and in the upper part of the spike, lanceolate, two-valved and two-seeded siliques. In these several cases, with the exception of that of the well-developed ray-florets, which are of service in making the flowers conspicuous to insects, natural selection cannot, as far as we can judge, have come into play, or only in a quite subordinate manner. All these modifications follow from the relative position and inter-action of the parts; and it can hardly be doubted that if all the flowers and leaves on the same plant had been subjected to the same external and internal condition, as are the flowers and leaves in certain positions, all would have been modified in the same manner.

In numerous other cases we find modifications of structure, which are considered by botanists to be generally of a highly important nature, affecting only some of the flowers on the same plant, or occurring on distinct plants, which grow close together under the same conditions. As these variations seem of no special use to the plants, they cannot have been influenced by natural selection. Of their cause we are quite ignorant; we cannot even attribute them, as in the last class of cases, to any proximate agency, such as relative position. I will give only a few instances. It is so common to observe on the same plant, flowers indifferently tetramerous, pentamerous, etc., that I need not give examples; but as numerical variations are comparatively rare when the parts are few, I may mention that, according to De Candolle, the flowers of Papaver bracteatum offer either two sepals with four petals (which is the common type with poppies), or three sepals with six petals. The manner in which the petals are folded in the bud is in most groups a very constant morphological character; but Professor Asa Gray states that with some species of Mimulus, the aestivation is almost as frequently that of the Rhinanthideae as of the Antirrhinideae, to which latter tribe the genus belongs. Aug. St. Hilaire gives the following cases: the genus Zanthoxylon belongs to a division of the Rutaceae with a single ovary, but in some species flowers may be found on the same plant, and even in the same panicle, with either one or two ovaries. In Helianthemum the capsule has been described as unilocular or tri-locular; and in H. mutabile, 'Une lame, *plus ou moins large*, s'étend entre le pericarpe et le placenta.' In the flowers of Saponaria officinalis, Dr. Masters has observed instances of both marginal and free central placentation. Lastly, St. Hilaire found towards the southern extreme of the range of Gomphia oleaeformis two forms which he did not at first doubt were distinct species, but he subsequently saw them growing on the same bush; and he then adds, 'Voilà donc dans un même individu des loges et un style qui se rattachent tantôt à un axe verticale et tantôt à un gynobase.'

We thus see that with plants many morphological changes may be attributed to the laws of growth and the inter-action of parts, independently of natural selection. But with respect to Nägeli's doctrine of an innate tendency towards perfection or progressive development, can it be said in the case of these strongly pronounced variations, that the plants have been caught in the act of progressing towards a higher state of development? On the contrary, I should infer from the mere fact of the parts in question differing or varying greatly on the same plant, that such modifications were of extremely small importance to the plants themselves, of whatever importance they may generally be to us for our classifications. The acquisition of a useless part can hardly be said to raise an organism in the natural scale; and in the case of the imperfect, closed flowers, above described, if any new principle has to be invoked, it must be one of retrogression rather than of progression; and so it must be with many parasitic and degraded animals. We are ignorant of the exciting cause of the above specified modifications; but if the unknown cause were to act almost uniformly for a length of time, we may infer that the result would be almost uniform; and in this case all the individuals of the species would be modified in the same manner.

From the fact of the above characters being unimportant for the welfare of the species, any slight variations which occurred in them would not have been accumulated and augmented through natural selection. A structure which has been developed through long-continued selection, when it ceases to be of service to a species, generally becomes variable, as we see with rudimentary organs; for it will no longer be regulated by this same power of selection. But when, from the nature of the organism and of the conditions, modifications have been induced which are unimportant for the welfare of the species, they may be, and apparently often have been, transmitted in nearly the same state to numerous, otherwise modified, descendants. It cannot have been of much importance to the greater number of mammals, birds, or reptiles, whether they were clothed with hair, feathers or scales; yet hair has been transmitted to almost all mammals, feathers to all birds, and scales to all true reptiles. A structure, whatever it may be, which is common to many allied forms, is ranked by us as of high systematic importance, and consequently is often assumed to be of high vital importance to the species. Thus, as I am inclined to believe, morphological differences, which we consider as important – such as the arrangement of the leaves, the divisions of the flower or of the ovarium, the position of the ovules, etc., first appeared in many cases as fluctuating variations, which sooner or later became constant through the nature of the organism and of the surrounding conditions, as well as through the intercrossing of distinct individuals, but not through natural selection; for as these morphological characters do not affect the welfare of the species, any slight deviations in them could not have been governed or accumulated through this latter agency. It is a strange result which we thus arrive at, namely, that characters of slight vital importance to the species, are the most important to the systematist; but, as we shall hereafter see when we treat of the genetic principle of classification, this is by no means so paradoxical as it may at first appear.

Although we have no good evidence of the existence in organic beings of an innate tendency towards progressive development, yet this necessarily follows, as I have attempted to show in the fourth chapter, through the continued action of natural selection. For the best definition which has ever been given of a high standard of organisation, is the degree to which the parts have been specialised or differentiated; and natural selection tends towards this end, inasmuch as the

1. Leiocephalus Grayii.
2. Centrura flagellifer.

The Galapagos Iguana

parts are thus enabled to perform their functions more efficiently.

A distinguished zoologist, Mr. St. George Mivart, has recently collected all the objections which have ever been advanced by myself and others against the theory of natural selection, as propounded by Mr. Wallace and myself, and has illustrated them with admirable art and force. When thus marshalled, they make a formidable array; and as it forms no part of Mr. Mivart's plan to give the various facts and considerations opposed to his conclusions, no slight effort of reason and memory is left to the reader, who may wish to weigh the evidence on both sides. When discussing special cases, Mr. Mivart passes over the effects of the increased use and disuse of parts, which I have always maintained to be highly important, and have treated in my 'Variation under Domestication' at greater length than, as I believe, any other writer. He likewise often assumes that I attribute nothing to variation, independently of natural selection, whereas in the work just referred to I have collected a greater number of well-established cases than can be found in any other work known to me. My judgment may not be trustworthy, but after reading with care Mr. Mivart's book, and comparing each section with what I have said on the same head, I never before felt so strongly convinced of the general truth of the conclusions here arrived at, subject, of course, in so intricate a subject, to much partial error.

All Mr. Mivart's objections will be, or have been, considered in the present volume. The one new point which appears to have struck many readers is, 'That natural selection is incompetent to account for the incipient stages of useful structures.' This subject is intimately connected with that of the gradation of the characters, often accompanied by a change of function, for instance, the conversion of a swim bladder into lungs, points which were discussed in the last chapter under two headings. Nevertheless, I will here consider in some detail several of the cases advanced by Mr. Mivart, selecting those which are the most illustrative, as want of space prevents me from considering all.

The giraffe, by its lofty stature, much elongated neck, fore legs, head and tongue, has its whole frame beautifully adapted for browsing on the higher branches of trees. It can thus obtain food beyond the reach of the other Ungulata or hoofed animals inhabiting the same country; and this must be a great advantage to it during dearths. The Niata cattle in South America show us how small a difference in structure may make, during such periods, a great difference in preserving an animal's life. These cattle can browse as well as others on grass, but from the projection of the lower jaw they cannot, during the often recurrent droughts, browse on the twigs of trees, reeds, etc., to which food the common cattle and horses are then driven; so that at these times the Niatas perish, if not fed by their owners. Before coming to Mr. Mivart's objections, it may be well to explain once again how natural selection will act in all ordinary cases. Man has modified some of his animals, without necessarily having attended to special points of structure, by simply preserving and breeding from the fleetest individuals, as with the race-horse and greyhound, or as with the game-cock, by breeding from the victorious birds. So under nature with the nascent giraffe, the individuals which were the highest browsers and were able during dearths to reach even an inch or two above the others, will often have been preserved; for they will have roamed over the whole country in search of food. That the individuals of the same species often differ slightly in the relative lengths of all their parts may be seen in many works of natural history, in which careful measurements are given. These slight proportional differences, due to the laws of growth and variation, are not of the slightest use or importance to most species. But it will have been otherwise with the nascent giraffe, considering its probable habits of life; for those individuals which had some one part or several parts of their bodies rather more elongated than usual, would generally have survived. These will have intercrossed and left offspring, either inheriting the same bodily peculiarities, or with a tendency to vary again in the same manner; whilst the individuals less favoured in the same respects will have been the most liable to perish.

We here see that there is no need to separate single pairs, as man does, when he methodically improves a breed: natural selection will preserve and thus separate all the superior individuals, allowing them freely to intercross, and will destroy all the inferior individuals. By this process long-continued, which exactly corresponds with what I have called unconscious selection by man, combined, no doubt, in a most important manner with the inherited effects of the increased use of parts, it seems to me almost certain that an ordinary hoofed quadruped might be converted into a giraffe.

To this conclusion Mr. Mivart brings forward two objections. One is that the increased size of the body would obviously require an increased supply of food, and he considers it as 'very problematical whether the disadvantages thence arising would not, in times of scarcity, more than counterbalance the advantages.' But as the giraffe does actually exist in large numbers in Africa, and as some of the largest antelopes in the world, taller than an ox, abound there, why should we doubt that, as far as size is concerned, intermediate gradations could formerly have existed there, subjected as now to severe dearths. Assuredly the being able to reach, at each stage of increased size, to a supply of food, left untouched by the other hoofed quadrupeds of the country, would have been of some advantage to the nascent giraffe. Nor must we overlook the fact, that increased bulk would act as a protection against almost all beasts of prey excepting the lion; and against this animal, its tall neck, – and the taller the better, – would, as Mr. Chauncey Wright has remarked, serve as a watch-tower. It is from this cause, as Sir S. Baker remarks, that no animal is

more difficult to stalk than the giraffe. This animal also uses its long neck as a means of offence or defence, by violently swinging its head armed with stump-like horns. The preservation of each species can rarely be determined by any one advantage, but by the union of all, great and small.

Mr. Mivart then asks (and this is his second objection), if natural selection be so potent, and if high browsing be so great an advantage, why has not any other hoofed quadruped acquired a long neck and lofty stature, besides the giraffe, and, in a lesser degree, the camel, guanaco and macrauchenia? Or, again, why has not any member of the group acquired a long proboscis? With respect to South Africa, which was formerly inhabited by numerous herds of the giraffe, the answer is not difficult, and can best be given by an illustration. In every meadow in England, in which trees grow, we see the lower branches trimmed or planed to an exact level by the browsing of the horses or cattle; and what advantage would it be, for instance, to sheep, if kept there, to acquire slightly longer necks? In every district some one kind of animal will almost certainly be able to browse higher than the others; and it is almost equally certain that this one kind alone could have its neck elongated for this purpose, through natural selection and the effects of increased use. In South Africa the competition for browsing on the higher branches of the acacias and other trees must be between giraffe and giraffe, and not with the other ungulate animals.

Why, in other quarters of the world, various animals belonging to this same order have not acquired either an elongated neck or a proboscis, cannot be distinctly answered; but it is as unreasonable to expect a distinct answer to such a question as why some event in the history of mankind did not occur in one country whilst it did in another. We are ignorant with respect to the conditions which determine the numbers and range of each species, and we cannot even conjecture what changes of structure would be favourable to its increase in some new country.

CERVUS.

The Giraffe

We can, however, see in a general manner that various causes might have interfered with the development of a long neck or proboscis. To reach the foliage at a considerable height (without climbing, for which hoofed animals are singularly ill-constructed) implies greatly increased bulk of body; and we know that some areas support singularly few large quadrupeds, for instance South America, though it is so luxuriant, whilst South Africa abounds with them to an unparalleled degree. Why this should be so we do not know; nor why the later tertiary periods should have been much more favourable for their existence than the present time. Whatever the causes may have been, we can see that certain districts and times would have been much more favourable than others for the development of so large a quadruped as the giraffe.

In order that an animal should acquire some structure specially and largely developed, it is almost indispensable that several other parts should be modified and coadapted. Although every part of the body varies slightly, it does not follow that the necessary parts should always vary in the right direction and to the right degree. With the different species of our domesticated animals we know that the parts vary in a different manner and degree, and that some species are much more variable than others. Even if the fitting variations did arise, it does not follow that natural selection would be able to act on them and produce a structure which apparently would be beneficial to the species. For instance, if the number of individuals existing in a country is determined chiefly through destruction by beasts of prey, – by external or internal parasites, etc., – as seems often to be the case, then natural selection will be able to do little, or will be greatly retarded, in modifying any particular structure for obtaining food. Lastly, natural selection is a slow process, and the same favourable conditions must long endure in order that any marked effect should thus be produced. Except by assigning such general and vague reasons, we cannot

explain why, in many quarters of the world, hoofed quadrupeds have not acquired much elongated necks or other means for browsing on the higher branches of trees.

Objections of the same nature as the foregoing have been advanced by many writers. In each case various causes, besides the general ones just indicated, have probably interfered with the acquisition through natural selection of structures, which it is thought would be beneficial to certain species. One writer asks, why has not the ostrich acquired the power of flight? But a moment's reflection will show what an enormous supply of food would be necessary to give to this bird of the desert force to move its huge body through the air. Oceanic islands are inhabited by bats and seals, but by no terrestrial mammals; yet as some of these bats are peculiar species, they must have long inhabited their present homes. Therefore Sir C. Lyell asks, and assigns certain reasons in answer, why have not seals and bats given birth on such islands to forms fitted to live on the land? But seals would necessarily be first converted into terrestrial carnivorous animals of considerable size, and bats into terrestrial insectivorous animals; for the former there would be no prey; for the bats ground-insects would serve as food, but these would already be largely preyed on by the reptiles or birds, which first colonise and abound on most oceanic islands. Gradations of structure, with each stage beneficial to a changing species, will be favoured only under certain peculiar conditions. A strictly terrestrial animal, by occasionally hunting for food in shallow water, then in streams or lakes, might at last be converted into an animal so thoroughly aquatic as to brave the open ocean. But seals would not find on oceanic islands the conditions favourable to their gradual reconversion into a terrestrial form. Bats, as formerly shown, probably acquired their wings by at first gliding through the air from tree to tree, like the so-called flying squirrels, for the sake of escaping from their enemies, or for avoiding falls; but

The Barbary Ape

when the power of true flight had once been acquired, it would never be reconverted back, at least for the above purposes, into the less efficient power of gliding through the air. Bats, might, indeed, like many birds, have had their wings greatly reduced in size, or completely lost, through disuse; but in this case it would be necessary that they should first have acquired the power of running quickly on the ground, by the aid of their hind legs alone, so as to compete with birds or other ground animals; and for such a change a bat seems singularly ill-fitted. These conjectural remarks have been made merely to show that a transition of structure, with each step

beneficial, is a highly complex affair; and that there is nothing strange in a transition not having occurred in any particular case.

Lastly, more than one writer has asked why have some animals had their mental powers more highly developed than others, as such development would be advantageous to all? Why have not apes acquired the intellectual powers of man? Various causes could be assigned; but as they are conjectural, and their relative probability cannot be weighed, it would be useless to give them. A definite answer to the latter question ought not to be expected, seeing that no one can solve the simpler problem, why, of two races of

Caterpillars of the Lobster Moth

savages, one has risen higher in the scale of civilisation than the other; and this apparently implies increased brain power.

We will return to Mr. Mivart's other objections. Insects often resemble for the sake of protection various objects, such as green or decayed leaves, dead twigs, bits of lichen, flowers, spines, excrement of birds, and living insects; but to this latter point I shall hereafter recur. The resemblance is often wonderfully close, and is not confined to colour, but extends to form, and even to the manner in which the insects hold themselves. The caterpillars which project motionless like dead twigs from the bushes on which they feed, offer an excellent instance of a resemblance of this kind. The cases of the imitation of such objects as the excrement of birds, are rare and exceptional. On this head, Mr. Mivart remarks, 'As, according to Mr. Darwin's theory, there is a constant tendency to indefinite variation, and as the minute incipient variations will be in *all directions*, they must tend to neutralize each other, and at first to form such unstable modifications

that it is difficult, if not impossible, to see how such indefinite oscillations of infinitesimal beginnings can ever build up a sufficiently appreciable resemblance to a leaf, bamboo, or other object, for natural selection to seize upon and perpetuate.'

But in all the foregoing cases the insects in their original state no doubt presented some rude and accidental resemblance to an object commonly found in the stations frequented by them. Nor is this at all improbable, considering the almost infinite number of surrounding objects and the diversity in form and colour of the hosts of insects which exist. As some rude resemblance is necessary for the first start, we can understand how it is that the larger and higher animals do not (with the exception, as far as I know, of one fish) resemble for the sake of protection special objects, but only the surface which commonly surrounds them, and this chiefly in colour. Assuming that an insect originally happened to resemble in some degree a dead twig or a decayed leaf, and that it varied slightly in many ways, then all the variations which rendered the insect at all more like any such object, and thus favoured its escape, would be preserved, whilst other variations would be neglected and ultimately lost; or, if they rendered the insect at all less like the imitated object, they would be eliminated. There would indeed be force in Mr. Mivart's objection, if we were to attempt to account for the above resemblances, independently of natural selection, through mere fluctuating variability; but as the case stands there is none.

Nor can I see any force in Mr. Mivart's difficulty with respect to 'the last touches of perfection in the mimicry;' as in the case given by Mr. Wallace, of a walking stick insect (Ceroxylus laceratus), which resembles 'a stick grown over by a creeping moss or jungermannia.' So close was this resemblance, that a native Dyak maintained that the foliaceous excrescences were really moss Insects are preyed on by birds and other enemies whose sight is probably sharper than ours, and every grade in resemblance which aided an insect to escape notice or detection, would tend towards its preservation; and the more perfect the resemblance so much the better for the insect. Considering the nature of the differences between the species in the group which includes the above Ceroxylus, there is nothing improbable in this insect having varied in the irregularities on its surface, and in these having become more or less green-coloured; for in every group the characters which differ in the several species are the most apt to vary, whilst the generic characters, or those common to all the species, are the most constant.

The Greenland whale is one of the most wonderful animals in the world, and the baleen, or whalebone, one of its greatest peculiarities. The baleen consists of a row, on each side of the upper jaw, of about 300 plates or laminae, which stand close together transversely to the longer axis of the mouth. Within the main row there are some subsidiary rows. The extremities and inner margins of all the plates are frayed into stiff bristles, which clothe the whole gigantic palate,

and serve to strain or sift the water, and thus to secure the minute prey on which these great animals subsist. The middle and longest lamina in the Greenland whale is ten, twelve, or even fifteen feet in length; but in the different species of Cetaceans there are gradations in length; the middle lamina being in one species, according to Scoresby, four feet, in another three, in another eighteen inches, and in the Balaenoptera rostrata only about nine inches in length. The quality of the whalebone also differs in the different species.

With respect to the baleen, Mr. Mivart remarks that if it 'had once attained such a size and development as to be at all useful, then its preservation and augmentation within serviceable limits would be promoted by natural selection alone. But how to obtain the beginning of such useful development?' In answer, it may be asked, why should not the early progenitors of the whales with baleen have possessed a mouth constructed something like the lamellated beak of a duck? Ducks, like whales, subsist by sifting the mud and water; and the family has sometimes been called *Criblatores*, or sifters. I hope that I may not be misconstrued into saying that the progenitors of whales did actually possess mouths lamellated like the beak of a duck. I wish only to show that this is not incredible, and that the immense plates of baleen in the Greenland whale might have been developed from such lamellae by finely graduated steps, each of service to its possessor.

The beak of a shoveller duck (Spatula clypeata) is a more beautiful and complex structure than the mouth of a whale. The upper mandible is furnished on each side (in the specimen examined by me) with a row or comb formed of 188 thin, elastic lamellae, obliquely bevelled so as to be pointed, and placed transversely to the longer axis of the mouth. They arise from the palate, and are attached by flexible membrane to the sides of the mandible. Those standing towards the middle are the longest, being about one-third of an inch in length, and they project .14 of an inch beneath the edge. At their bases there is a short subsidiary row of obliquely transverse lamellae. In these several respects they resemble the plates of baleen in the mouth of a whale. But towards the extremity of the beak they differ much, as they project inward, instead of straight downward. The entire head of the shoveller, though incomparably less bulky, is about one-eighteenth of the length of the head of a moderately large Balaenoptera rostrata, in which species the baleen is only nine inches long; so that if we were to make the head of the shoveller as long as that of the Balaenoptera, the lamellae would be six inches in length, that is, two-thirds of the length of the baleen in this species of whale. The lower mandible of the shoveller duck is furnished with lamellae of equal length with these above, but finer; and in being thus furnished it differs conspicuously from the lower jaw of a whale, which is destitute of baleen. On the other hand, the extremities of these lower lamellae are frayed into fine bristly points, so that they thus curiously resemble the plates of baleen. In the genus Prion, a member of the distinct family of the Petrels, the upper mandible alone is furnished with lamellae, which are well developed and project beneath the margin; so that the beak of this bird resembles in this respect the mouth of a whale.

From the highly developed structure of the shoveller's beak we may proceed (as I have learned from information and specimens sent to me by Mr. Salvin), without any great break, as far as fitness for sifting is concerned, through the beak of the Merganetta armata, and in some respects through that of the Aix sponsa, to the beak of the common duck. In this latter species the lamellae are much coarser than in the shoveller, and are firmly attached to the sides of the mandible; they are only about fifty in number on each side, and do not project at all beneath the margin. They are square-topped, and are edged with translucent, hardish tissue, as if for crushing food. The edges of the lower mandible are crossed by numerous fine ridges, which project very little. Although the

beak is thus very inferior as a sifter to that of a shoveller, yet this bird, as every one knows, constantly uses it for this purpose. There are other species, as I hear from Mr. Salvin, in which the lamellae are considerably less developed than in the common duck; but I do not know whether they use their beaks for sifting the water.

Turning to another group of the same family. In the Egyptian goose (Chenalopex) the beak closely resembles that of the common duck; but the lamellae are not so numerous, nor so distinct from each other, nor do they project so much inward; yet this goose, as I am informed by Mr. E. Bartlett, 'uses its bill like a duck by throwing the water out at the corners.' Its chief food, however, is grass, which it crops like the common goose. In this latter bird the lamellae of the upper mandible are much coarser than in the common duck, almost confluent, about twenty-seven in number on each side, and terminating upward in teeth-like knobs. The palate is also covered with hard rounded knobs. The edges of the lower mandible are serrated with teeth much more prominent, coarser, and sharper than in the duck. The common goose does not sift the water, but uses its beak exclusively for tearing or cutting herbage, for which purpose it is so well fitted that it can crop grass closer than almost any other animal. There are other species of geese, as I hear from Mr. Bartlett, in which the lamellae are less developed than in the common goose.

We thus see that a member of the duck family, with a beak constructed like that of a common goose and adapted solely for grazing, or even a member with a beak having less well-developed lamellae, might be converted by small changes into a species like the Egyptian goose, – this into one like the common duck, – and, lastly, into one like the shoveller, provided with a beak almost exclusively adapted for sifting the water; for this bird could hardly use any part of its beak, except the hooked tip, for seizing or tearing solid food. The beak of a goose, as I may add, might also be converted by small changes into one provided with prominent, recurved teeth, like those of the Merganser (a member of the same family), serving for the widely different purpose of securing live fish.

Returning to the whales. The Hyperoodon bidens is destitute of true teeth in an efficient condition, but its palate is roughened, according to Lacepède, with small, unequal, hard points of horn. There is, therefore, nothing improbable in supposing that some early Cetacean form was provided with similar points of horn on the palate, but rather more regularly placed, and which, like the knobs on the beak of the goose, aided it in seizing or tearing its food. If so, it will hardly be denied that the points might have been converted through variation and natural selection into lamellae as well-developed as those of the Egyptian goose, in which case they would have been used both for seizing objects and for sifting the water; then into lamellae like those of the domestic duck; and so onward, until they became as well constructed as those of the shoveller, in which case they would have served exclusively as a sifting apparatus. From this stage, in which the lamellae would be two-thirds of the length of the plates of baleen in the Balaenoptera rostrata, gradations, which may be observed in still-existing Cetaceans, lead us onward to the enormous plates of baleen in the Greenland whale. Nor is there the least reason to doubt that each step in this scale might have been as serviceable to certain ancient Cetaceans, with the functions of the parts slowly changing during the progress of development, as are the gradations in the beaks of the different existing members of the duck-family. We should bear in mind that each species of duck is subjected to a severe struggle for existence, and that the structure of every part of its frame must be well adapted to its conditions of life.

The Pleuronectidae, or flatfish, are remarkable for their asymmetrical bodies. They rest on one side, – in the greater number of species on the left, but in some on the right side; and occasionally reversed adult specimens occur.

The lower, or resting-surface, resembles at first sight the ventral surface of an ordinary fish; it is of a white colour, less developed in many ways than the upper side, with the lateral fins often of smaller size. But the eyes offer the most remarkable peculiarity; for they are both placed on the upper side of the head. During early youth, however, they stand opposite to each other, and the whole body is then symmetrical, with both sides equally coloured. Soon the eye proper to the lower side begins to glide slowly round the head to the upper side; but does not pass right through the skull, as was formerly thought to be the case. It is obvious that unless the lower eye did thus travel round, it could not be used by the fish whilst lying in its habitual position on one side. The lower eye would, also, have been liable to be abraded by the sandy bottom. That the Pleuronectidae are admirably adapted by their flattened and asymmetrical structure for their habits of life, is manifest from several species, such as soles, flounders, etc., being extremely common. The chief advantages thus gained seem to be protection from their enemies, and facility for feeding on the ground. The different members, however, of the family present, as Schiödte remarks, 'a long series of forms exhibiting a gradual transition from Hippoglossus pinguis, which does not in any considerable degree alter the shape in which it leaves the ovum, to the soles, which are entirely thrown to one side.'

Mr. Mivart has taken up this case, and remarks that a sudden spontaneous transformation in the position of the eyes is hardly conceivable, in which I quite agree with him. He then adds: 'If the transit was gradual, then how such transit of one eye a minute fraction of the journey towards the other side of the head could benefit the individual is, indeed, far from clear. It seems, even, that such an incipient transformation must rather have been injurious.' But he might have found an answer to this objection in the excellent observations published in 1867 by Malm. The Pleuronectidae, whilst very young and still symmetrical, with their eyes standing on opposite sides of the head, cannot long retain a vertical position, owing to the excessive depth of their bodies, the small size of their lateral fins, and to their being destitute of a swim-bladder. Hence, soon growing tired, they fall to the bottom on one side. Whilst thus at rest they often twist, as Malm observed, the lower eye upward, to see above them; and they do this so vigorously that the eye is pressed hard against the upper part of the orbit. The forehead between the eyes consequently becomes, as could be plainly seen, temporarily contracted in breadth. On one occasion Malm saw a young fish raise and depress the lower eye through an angular distance of about seventy degrees.

We should remember that the skull at this early age is cartilaginous and flexible, so that it readily yields to muscular action. It is also known with the higher animals, even after early youth, that the skull yields and is altered in shape, if the skin or muscles be permanently contracted through disease or some accident. With long-eared rabbits, if one ear flops forward and downward, its weight drags forward all the bones of the skull on the same side, of which I have given a figure. Malm states that the newly-hatched young of perches, salmon, and several other symmetrical fishes, have the habit of occasionally resting on one side at the bottom; and he has observed that they often then strain their lower eyes so as to look upward; and their skulls are thus rendered rather crooked. These fishes, however, are soon able to hold themselves in a vertical position, and no permanent effect is thus produced. With the Pleuronectidae, on the other hand, the older they grow the more habitually they rest on one side, owing to the increasing flatness of their bodies, and a permanent effect is thus produced on the form of the head, and on the position of the eyes. Judging from analogy, the tendency to distortion would no doubt be increased through the principle of inheritance. Schiödte believes, in opposition to some other

naturalists, that the Pleuronectidae are not quite symmetrical even in the embryo; and if this be so, we could understand how it is that certain species, whilst young, habitually fall over and rest on the left side, and other species on the right side. Malm adds, in confirmation of the above view, that the adult Trachypterus arcticus, which is not a member of the Pleuronectidae, rests on its left side at the bottom, and swims diagonally through the water; and in this fish, the two sides of the head are said to be somewhat dissimilar. Our great authority on Fishes, Dr. Günther, concludes his abstract of Malm's paper, by remarking that 'the author gives a very simple explanation of the abnormal condition of the Pleuronectoids.'

We thus see that the first stages of the transit of the eye from one side of the head to the other, which Mr. Mivart considers would be injurious, may be attributed to the habit, no doubt beneficial to the individual and to the species, of endeavouring to look upward with both eyes, whilst resting on one side at the bottom. We may also attribute to the inherited effects of use the fact of the mouth in several kinds of flatfish being bent towards the lower surface, with the jaw bones stronger and more effective on this, the eyeless side of the head, than on the other, for the sake, as Dr. Traquair supposes, of feeding with ease on the ground. Disuse, on the other hand, will account for the less developed condition of the whole inferior half of the body, including the lateral fins; though Yarrell thinks that the reduced size of these fins is advantageous to the fish, as 'there is so much less room for their action than with the larger fins above.' Perhaps the lesser number of teeth in the proportion of four to seven in the upper halves of the two jaws of the plaice, to twenty-five to thirty in the lower halves, may likewise be accounted for by disuse. From the colourless state of the ventral surface of most fishes and of many other animals, we may reasonably suppose that the absence of colour in flatfish on the side, whether it be the right or left, which is undermost, is due to the exclusion of light. But it cannot be supposed that the peculiar speckled appearance of the upper side of the sole, so like the sandy bed of the sea, or the power in some species, as recently shown by Pouchet, of changing their colour in accordance with the surrounding surface, or the presence of bony tubercles on the upper side of the turbot, are due to the action of the light. Here natural selection has probably come into play, as well as in adapting the general shape of the body of these fishes, and many other peculiarities, to their habits of life. We should keep in mind, as I have before insisted, that the inherited effects of the increased use of parts, and perhaps of their disuse, will be strengthened by natural selection. For all spontaneous variations in the right direction will thus be preserved; as will those individuals which inherit in the highest degree the effects of the increased and beneficial use of any part. How much to attribute in each particular case to the effects of use, and how much to natural selection, it seems impossible to decide.

A Turbot viewed from its upper side, which blends with the sea bed

I may give another instance of a structure which apparently owes its origin exclusively to use or habit. The extremity of the tail in some American monkeys has been converted into a wonderfully perfect prehensile organ, and serves as a fifth hand. A reviewer, who agrees with Mr. Mivart in every detail, remarks on this structure: 'It is impossible to believe that in any number of ages the first slight incipient tendency to grasp could preserve the lives of the individuals possessing it, or favour their chance of having and of rearing offspring.' But there is no necessity for any such belief. Habit, and this almost implies that some benefit great or small is thus derived, would in all probability suffice for the work. Brehm saw the young of an African monkey (Cercopithecus) clinging to the under surface of their mother by their hands, and at the same time they hooked their little tails round that of their mother. Professor Henslow kept in confinement some harvest mice (Mus messorius) which do not possess a structurally prehensile tail; but he frequently observed that they curled their tails round the branches of a bush placed in the cage, and thus aided themselves in climbing. I have received an analogous account from Dr. Günther, who has seen a mouse thus suspend itself. If the harvest mouse had been more strictly arboreal, it would perhaps have had its tail rendered structurally prehensile, as is the case with some members of the same order. Why Cercopithecus, considering its habits whilst young, has not become thus provided, it would be difficult to say. It is, however, possible that the long tail of this monkey may be of more service to it as a balancing organ in making its prodigious leaps, than as a prehensile organ.

The Woolly Spider Monkey, the largest monkey in South America. Asian primates
do not have prehensile tails as their American counterparts do

The mammary glands are common to the whole class of mammals, and are indispensable for their existence; they must, therefore, have been developed at an extremely remote period, and we can know nothing positively about their manner of development. Mr. Mivart asks: 'Is it conceivable that the young of any animal was ever saved from destruction by accidentally sucking a drop of scarcely nutritious fluid from an accidentally hypertrophied cutaneous gland of its mother? And even if one was so, what chance was there of the perpetuation of such a variation?' But the case is not here put fairly. It is admitted by most evolutionists that mammals are descended from a marsupial form; and if so, the mammary glands will have been at first developed within the marsupial sack. In the case of the fish (Hippocampus) the eggs are hatched, and the young are reared for a time, within a sack of this nature; and an American naturalist, Mr. Lockwood, believes from what he has seen of the development of the young, that they are nourished by a secretion from the cutaneous glands of the sack. Now, with the early progenitors of mammals, almost before they deserved to be thus designated, is it not at least possible that the young might have been similarly nourished? And in this case, the individuals which secreted a fluid, in some degree or manner the most nutritious, so as to partake of the nature of milk, would in the long run have reared a larger number of well-nourished offspring, than would the individuals which secreted a poorer fluid; and thus the cutaneous glands, which are the homologues of the mammary glands, would have been improved or rendered more effective. It accords with the widely extended principle of specialisation, that the glands over a certain space of the sack should have become more highly developed than the remainder; and they would then have formed a breast, but at first without a nipple, as we see in the Ornithorhyncus, at the base of the mammalian series. Through what agency the glands over a certain space became more highly specialised than

the others, I will not pretend to decide, whether in part through compensation of growth, the effects of use, or of natural selection.

The development of the mammary glands would have been of no service, and could not have been affected through natural selection, unless the young at the same time were able to partake of the secretion. There is no greater difficulty in understanding how young mammals have instinctively learned to suck the breast, than in understanding how unhatched chickens have learned to break the egg-shell by tapping against it with their specially adapted beaks; or how a few hours after leaving the shell they have learned to pick up grains of food. In such cases the most probable solution seems to be, that the habit was at first acquired by practice at a more advanced age, and afterwards transmitted to the offspring at an earlier age. But the young kangaroo is said not to suck, only to cling to the nipple of its mother, who has the power of injecting milk into the mouth of her helpless, half-formed offspring. On this head Mr. Mivart remarks: 'Did no special provision exist, the young one must infallibly be choked by the intrusion of the milk into the windpipe. But there is a special provision. The larynx is so elongated that it rises up into the posterior end of the nasal passage, and is thus enabled to give free entrance to the air for the lungs, whilst the milk passes harmlessly on each side of this elongated larynx, and so safely attains the gullet behind it.' Mr. Mivart then asks how did natural selection remove in the adult kangaroo (and in most other mammals, on the assumption that they are descended from a marsupial form), 'this at least perfectly innocent and harmless structure?' It may be suggested in answer that the voice, which is certainly of high importance to many animals, could hardly have been used with full force as long as the larynx entered the nasal passage, and Professor Flower has suggested to me that this structure would have greatly interfered with an animal swallowing solid food.

We will now turn for a short space to the lower divisions of the animal kingdom. The Echinodermata (starfishes, sea urchins, etc.) are furnished with remarkable organs, called pedicellariae, which consist, when well developed, of a tridactyle forceps – that is, of one formed of three serrated arms, neatly fitting together and placed on the summit of a flexible stem, moved by muscles. These forceps can seize firmly hold of any object; and Alexander Agassiz has seen an Echinus or sea urchin rapidly passing particles of excrement from forceps to forceps down certain lines of its body, in order that its shell should not be fouled. But there is no doubt that besides removing dirt of all kinds, they subserve other functions; and one of these apparently is defence.

With respect to these organs, Mr. Mivart, as on so many previous occasions, asks: 'What would be the utility of the *first rudimentary beginnings* of such structures, and how could such incipient

Pedicellariae of Sea Urchins, found around the base
of the urchin's spines

buddings have ever preserved the life of a single Echinus?' He adds, 'not even the *sudden* development of the snapping action would have been beneficial without the freely movable stalk, nor could the latter have been efficient without the snapping jaws, yet no minute, nearly indefinite variations could simultaneously evolve these complex co-ordinations of structure; to deny this seems to do no less than to affirm a startling paradox.' Paradoxical as this may appear to Mr. Mivart, tridactyle forcepses, immovably fixed at the base, but capable of a snapping action, certainly exist on some starfishes; and this is intelligible if they serve, at least in part, as a means of defence. Mr. Agassiz, to whose great kindness I am indebted for much information on the subject, informs me that there are other starfishes, in which one of the three arms of the forceps is reduced to a support for the other two; and again, other genera in which the third arm is completely lost. In Echinoneus, the shell is described by M. Perrier as bearing two kinds of pedicellariae, one resembling those of Echinus, and the other those of Spatangus; and such cases are always interesting as affording the means of apparently sudden transitions, through the abortion of one of the two states of an organ.

With respect to the steps by which these curious organs have been evolved, Mr. Agassiz infers from his own researches and those of Mr. Müller, that both in starfishes and sea urchins the pedicellariae must undoubtedly be looked at as modified spines. This may be inferred from their manner of development in the individual, as well as from a long and perfect series of gradations in different species and genera, from simple granules to ordinary spines, to perfect tridactyle pedicellariae. The gradation extends even to the manner in which ordinary spines and the pedicellariae, with their supporting calcareous rods, are articulated to the shell. In certain genera of starfishes, 'the very combinations needed to show that the pedicellariae are only modified branching spines' may be found. Thus we have

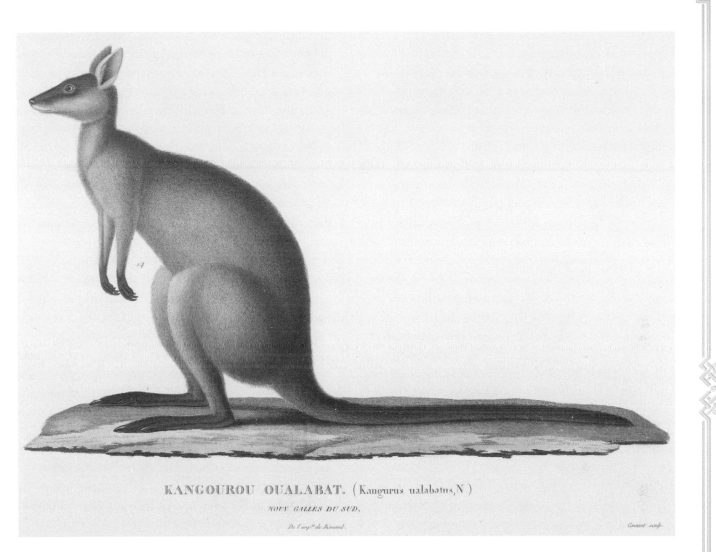

KANGOUROU OUALABAT. (Kangurus ualabatus, N.)

NOUV. GALLES DU SUD.

De l'imp.^{ie} de Rémond.   *Coutant sculp.*

The Kangaroo, the largest species of marsupial, found only on the continent of Australia

fixed spines, with three equi-distant, serrated, movable branches, articulated to near their bases; and higher up, on the same spine, three other movable branches. Now when the latter arise from the summit of a spine they form, in fact, a rude tridactyle pedicellariae, and such may be seen on the same spine together with the three lower branches. In this case the identity in nature between the arms of the pedicellariae and the movable branches of a spine, is unmistakable. It is generally admitted that the ordinary spines serve as a protection; and if so, there can be no reason to doubt that those furnished with serrated and movable branches likewise serve for the same purpose; and they would thus serve still more effectively as soon as by meeting together they acted as a prehensile or snapping apparatus. Thus every gradation, from an ordinary fixed spine to a fixed pedicellariae, would be of service.

In certain genera of starfishes these organs, instead of being fixed or borne on an immovable support, are placed on the summit of a flexible and muscular, though short, stem; and in this case they probably subserve some additional function besides defence. In the sea urchins the steps can be followed by which a fixed spine becomes articulated to the shell, and is thus rendered movable. I wish I had space here to give a fuller abstract of Mr.

Agassiz's interesting observations on the development of the pedicellariae. All possible gradations, as he adds, may likewise be found between the pedicellariae of the starfishes and the hooks of the Ophiurians, another group of the Echinodermata; and again between the pedicellariae of sea urchins and the anchors of the Holothuriae, also belonging to the same great class.

Certain compound animals, or zoophytes, as they have been termed, namely the Polyzoa, are provided with curious organs called avicularia. These differ much in structure in the different species. In their most perfect condition they curiously resemble the head and beak of a vulture in miniature, seated on a neck and capable of movement, as is likewise the lower jaw or mandible. In one species observed by me, all the avicularia on the same branch often moved simultaneously backwards and forwards, with the lower jaw widely open, through an angle of about 90 degrees, in the course of five seconds; and their movement caused the whole polyzoary to tremble. When the jaws are touched with a needle they seize it so firmly that the branch can thus be shaken.

Mr. Mivart adduces this case, chiefly on account of the supposed difficulty of organs, namely the avicularia of the Polyzoa and the pedicellariae of the Echinodermata, which he considers as 'essentially similar,' having been developed through natural selection in widely distinct divisions of the animal kingdom. But, as far as structure is concerned, I can see no similarity between tridactyle pedicellariae and avicularia. The latter resembles somewhat more closely the chelae or pincers of Crustaceans; and Mr. Mivart might have adduced with equal appropriateness this resemblance as a special difficulty, or even their resemblance to the head and beak of a bird. The avicularia are believed by Mr. Busk, Dr. Smitt and Dr. Nitsche – naturalists who have carefully studied this group – to be homologous with the zooids and their cells which compose the zoophyte, the movable lip or lid of the cell corresponding with the lower and movable

mandible of the avicularium. Mr. Busk, however, does not know of any gradations now existing between a zooid and an avicularium. It is therefore impossible to conjecture by what serviceable gradations the one could have been converted into the other, but it by no means follows from this that such gradations have not existed.

As the chelae of Crustaceans resemble in some degree the avicularia of Polyzoa, both serving as pincers, it may be worth while to show that with the former a long series of serviceable gradations still exists. In the first and simplest stage, the terminal segment of a limb shuts down either on the square summit of the broad penultimate segment, or against one whole side, and is thus enabled to catch hold of an object, but the limb still serves as an organ of locomotion. We next find one corner of the broad penultimate segment slightly prominent, sometimes furnished with irregular teeth, and against these the terminal segment shuts down. By an increase in the size of this projection, with its shape, as well as that of the terminal segment, slightly modified and improved, the pincers are rendered more and more perfect, until we have at last an instrument as efficient as the chelae of a lobster. And all these gradations can be actually traced.

Besides the avicularia, the polyzoa possess curious organs called vibracula. These generally consist of long bristles, capable of movement and easily excited. In one species examined by me the vibracula were slightly curved and serrated along the outer margin, and all of them on the same polyzoary often moved simultaneously; so that, acting like long oars, they swept a branch rapidly across the object-glass of my microscope. When a branch was placed on its face, the vibracula became entangled, and they made violent efforts to free themselves. They are supposed to serve as a defence, and may be seen, as Mr. Busk remarks, 'to sweep slowly and carefully over the surface of the polyzoary, removing what might be noxious to the delicate inhabitants of the cells when their tentacula are protruded.' The avicularia, like the

The Rock Lobster

vibracula, probably serve for defence, but they also catch and kill small living animals, which, it is believed, are afterwards swept by the currents within reach of the tentacula of the zooids. Some species are provided with avicularia and vibracula, some with avicularia alone and a few with vibracula alone.

It is not easy to imagine two objects more widely different in appearance than a bristle or vibraculum, and an avicularium like the head of a bird; yet they are almost certainly homologous and have been developed from the same common source, namely a zooid with its cell. Hence, we can understand how it is that these organs graduate in some cases, as I am informed by Mr. Busk, into each other. Thus, with the avicularia of several species of Lepralia, the movable mandible is so much produced and is so like a bristle that the presence of the upper or fixed beak alone serves to determine its avicularian nature. The vibracula may have been directly developed from the lips of the cells, without having passed through the avicularian stage; but it seems more probable that they have passed through this stage, as during the early stages of the transformation, the other parts of the cell, with the included zooid, could hardly have disappeared at once. In many cases the vibracula have a grooved support at the base, which seems to represent the fixed beak; though

this support in some species is quite absent. This view of the development of the vibracula, if trustworthy, is interesting; for supposing that all the species provided with avicularia had become extinct, no one with the most vivid imagination would ever have thought that the vibracula had originally existed as part of an organ, resembling a bird's head or an irregular box or hood. It is interesting to see two such widely different organs developed from a common origin; and as the movable lip of the cell serves as a protection to the zooid, there is no difficulty in believing that all the gradations, by which the lip became converted first into the lower mandible of an avicularium, and then into an elongated bristle, likewise served as a protection in different ways and under different circumstances.

In the vegetable kingdom Mr. Mivart only alludes to two cases, namely the structure of the flowers of orchids, and the movements of climbing plants. With respect to the former, he says: 'The explanation of their *origin* is deemed thoroughly unsatisfactory – utterly insufficient to explain the incipient, infinitesimal beginnings of structures which are of utility only when they are considerably developed.' As I have fully treated this subject in another work, I will here give only a few details on one alone of the most striking peculiarities of the flowers of orchids, namely, their pollinia. A pollinium, when highly developed, consists of a mass of pollen-grains, affixed to an elastic foot-stalk or caudicle, and this to a little mass of extremely viscid matter. The pollinia are by this means transported by insects from one flower to the stigma of another. In some orchids there is no caudicle to the pollen-masses, and the grains are merely tied together by fine threads; but as these are not confined to orchids, they need not here be considered; yet I may mention that at the base of the orchidaceous series, in Cypripedium, we can see how the threads were probably first developed. In other orchids the threads cohere at one end of the pollen-masses; and this forms the first or nascent trace of a caudicle.

That this is the origin of the caudicle, even when of considerable length and highly developed, we have good evidence in the aborted pollen-grains which can sometimes be detected embedded within the central and solid parts.

With respect to the second chief peculiarity, namely, the little mass of viscid matter attached to the end of the caudicle, a long series of gradations can be specified, each of plain service to the plant. In most flowers belonging to other orders the stigma secretes a little viscid matter. Now, in certain orchids similar viscid matter is secreted, but in much larger quantities by one alone of the three stigmas; and this stigma, perhaps in consequence of the copious secretion, is rendered sterile. When an insect visits a flower of this kind, it rubs off some of the viscid matter, and thus at the same time drags away some of the pollen-grains. From this simple condition, which differs but little from that of a multitude of common flowers, there are endless gradations, – to species in which the pollen-mass terminates in a very short, free caudicle, – to others in which the caudicle becomes firmly attached to the viscid matter, with the sterile stigma itself much modified. In this latter case we have a pollinium in its most highly developed and perfect condition. He who will carefully examine the flowers of orchids for himself will not deny the existence of the above series of gradations – from a mass of pollen-grains merely tied together by threads, with the stigma differing but little from that of the ordinary flowers, to a highly complex pollinium, admirably adapted for transportal by insects; nor will he deny that all the gradations in the several species are admirably adapted in relation to the general structure of each flower for its fertilisation by different insects. In this, and in almost every other case, the enquiry may be pushed further backwards; and it may be asked how did the stigma of an ordinary flower become viscid, but as we do not know the full history of any one group of beings, it is as useless to ask, as it is hopeless to attempt answering, such questions.

We will now turn to climbing plants. These can be arranged in a long series, from those which simply twine round a support, to those which I have called leaf-climbers, and to those provided with tendrils. In these two latter classes the stems have generally, but not always, lost the power of twining, though they retain the power of revolving, which the tendrils likewise possess. The gradations from leaf-climbers to tendril bearers are wonderfully close, and certain plants may be differently placed in either class. But in ascending the series from simple twiners to leaf-climbers, an important quality is added, namely sensitiveness to a touch, by which means the foot-stalks of the leaves or flowers, or these modified and converted into tendrils, are excited to bend round and clasp the touching object. He who will read my memoir on these plants will, I think, admit that all the many gradations in function and structure between simple twiners and tendril-bearers are in each case beneficial in a high degree to the species. For instance, it is clearly a great advantage to a twining plant to become a leaf-climber; and it is probable that every twiner which possessed leaves with long foot-stalks would have been developed into a leaf-climber, if the foot-stalks had possessed in any slight degree the requisite sensitiveness to a touch.

As twining is the simplest means of ascending a support, and forms the basis of our series, it may naturally be asked how did plants acquire this power in an incipient degree, afterwards to be improved and increased through natural selection. The power of twining depends, firstly, on the stems whilst young being extremely flexible (but this is a character common to many plants which are not climbers); and, secondly, on their continually bending to all points of the compass, one after the other in succession, in the same order. By this movement the stems are inclined to all sides, and are made to move round and round. As soon as the lower part of a stem strikes against any object and is stopped, the upper part still goes on bending and revolving, and thus necessarily twines round and up the support. The revolving movement ceases after the early growth of each shoot. As in many widely separated families of plants, single species and single genera possess the power of revolving, and have thus become twiners, they must have independently acquired it, and cannot have inherited it from a common progenitor. Hence, I was led to predict that some slight tendency to a movement of this kind would be found to be far from uncommon with plants which did not climb; and that this had afforded the basis for natural selection to work on and improve. When I made this prediction, I knew of only one imperfect case, namely, of the young flower-peduncles of a Maurandia which revolved slightly and irregularly, like the stems of twining plants, but without making any use of this habit. Soon afterwards Fritz Müller discovered that the young stems of an Alisma and of a Linum, – plants which do not climb and are widely separated in the natural system, – revolved plainly, though irregularly, and he states that he has reason to suspect that this occurs with some other plants. These slight movements appear to be of no service to the plants in question; anyhow, they are not of the least use in the way of climbing, which is the point that concerns us. Nevertheless we can see that if the stems of these plants had been flexible, and if under the conditions to which they are exposed it had profited them to ascend to a height, then the habit of slightly and irregularly revolving might have been increased and utilised through natural selection, until they had become converted into well-developed twining species.

With respect to the sensitiveness of the foot-stalks of the leaves and flowers, and of tendrils, nearly the same remarks are applicable as in the case of the revolving movements of twining plants. As a vast number of species, belonging to widely distinct groups, are endowed with this kind of sensitiveness, it ought to be found in a nascent condition in many plants which have not become climbers. This is the case: I observed that the young flower-peduncles of the above Maurandia

curved themselves a little towards the side which was touched. Morren found in several species of Oxalis that the leaves and their foot-stalks moved, especially after exposure to a hot sun, when they were gently and repeatedly touched, or when the plant was shaken. I repeated these observations on some other species of Oxalis with the same result; in some of them the movement was distinct, but was best seen in the young leaves; in others it was extremely slight. It is a more important fact that according to the high authority of Hofmeister, the young shoots and leaves of all plants move after being shaken; and with climbing plants it is, as we know, only during the early stages of growth that the foot-stalks and tendrils are sensitive.

It is scarcely possible that the above slight movements, due to a touch or shake, in the young and growing organs of plants, can be of any functional importance to them. But plants possess, in obedience to various stimuli, powers of movement, which are of manifest importance to them; for instance, towards and more rarely from the light – in opposition to, and more rarely in the direction of, the attraction of gravity. When the nerves and muscles of an animal are excited by galvanism or by the absorption of strychnine, the consequent movements may be called an incidental result, for the nerves and muscles have not been rendered specially sensitive to these stimuli. So with plants it appears that, from having the power of movement in obedience to certain stimuli, they are excited in an incidental manner by a touch, or by being shaken. Hence there is no great difficulty in admitting that in the case of leaf-climbers and tendril-bearers, it is this tendency which has been taken advantage of and increased through natural selection. It is, however, probable, from reasons which I have assigned in my memoir, that this will have occurred only with plants which had already acquired the power of revolving, and had thus become twiners.

I have already endeavoured to explain how plants became twiners, namely, by the increase of a tendency to slight and irregular revolving movements, which were at first of no use to them; this movement, as well as that due to a touch or shake, being the incidental result of the power of moving, gained for other and beneficial purposes. Whether, during the gradual development of climbing plants, natural selection has been aided by the inherited effects of use, I will not pretend to decide; but we know that certain periodical movements, for instance the so-called sleep of plants, are governed by habit.

I have now considered enough, perhaps more than enough, of the cases, selected with care by a skilful naturalist, to prove that natural selection is incompetent to account for the incipient stages of useful structures; and I have shown, as I hope, that there is no great difficulty on this head. A good opportunity has thus been afforded for enlarging a little on gradations of structure, often associated with strange functions, – an important subject, which was not treated at sufficient length in the former editions of this work. I will now briefly recapitulate the foregoing cases.

With the giraffe, the continued preservation of the individuals of some extinct high-reaching ruminant, which had the longest necks, legs, etc., and could browse a little above the average height, and the continued destruction of those which could not browse so high, would have sufficed for the production of this remarkable quadruped; but the prolonged use of all the parts, together with inheritance, will have aided in an important manner in their co-ordination. With the many insects which imitate various objects, there is no improbability in the belief that an accidental resemblance to some common object was in each case the foundation for the work of natural selection, since perfected through the occasional preservation of slight variations which made the resemblance at all closer; and this will have been carried on as long as the insect continued to vary, and as long as a more and more perfect resemblance led to its escape from sharp-sighted enemies. In certain species of whales there is a tendency to the formation of irregular little points

of horn on the palate; and it seems to be quite within the scope of natural selection to preserve all favourable variations, until the points were converted, first into lamellated knobs or teeth, like those on the beak of a goose, – then into short lamellae, like those of the domestic ducks – and then into lamellae, as perfect as those of the shoveller duck, – and finally into the gigantic plates of baleen, as in the mouth of the Greenland whale. In the family of the ducks, the lamellae are first used as teeth, then partly as teeth and partly as a sifting apparatus, and at last almost exclusively for this latter purpose.

With such structures as the above lamellae of horn or whalebone, habit or use can have done little or nothing, as far as we can judge, towards their development. On the other hand, the transportal of the lower eye of a flatfish to the upper side of the head, and the formation of a prehensile tail, may be attributed almost wholly to continued use, together with inheritance. With respect to the mammae of the higher animals, the most probable conjecture is that primordially the cutaneous glands over the whole surface of a marsupial sack secreted a nutritious fluid; and that these glands were improved in function through natural selection, and concentrated into a confined area, in which case they would have formed a mamma. There is no more difficulty in understanding how the branched spines of some ancient Echinoderm, which served as a defence, became developed through natural selection into tridactyle pedicellariae, than in understanding the development of the pincers of crustaceans, through slight, serviceable modifications in the ultimate and penultimate segments of a limb, which was at first used solely for locomotion. In the avicularia and vibracula of the Polyzoa we have organs widely different in appearance developed from the same source; and with the vibracula we can understand how the successive gradations might have been of service. With the pollinia of orchids, the threads which originally served to tie together the pollen-grains, can be traced cohering into caudicles; and the steps can likewise be followed by which viscid matter, such as that secreted by the stigmas of ordinary flowers, and still subserving nearly but not quite the same purpose, became attached to the free ends of the caudicles; – all these gradations being of manifest benefit to the plants in question. With respect to climbing plants, I need not repeat what has been so lately said.

It has often been asked, if natural selection be so potent, why has not this or that structure been gained by certain species, to which it would apparently have been advantageous? But it is unreasonable to expect a precise answer to such questions, considering our ignorance of the past history of each species, and of the conditions which at the present day determine its numbers and range. In most cases only general reasons, but in some few cases special reasons, can be assigned. Thus to adapt a species to new habits of life, many co-ordinated modifications are almost indispensable, and it may often have happened that the requisite parts did not vary in the right manner or to the right degree. Many species must have been prevented from increasing in numbers through destructive agencies, which stood in no relation to certain structures, which we imagine would have been gained through natural selection from appearing to us advantageous to the species. In this case, as the struggle for life did not depend on such structures, they could not have been acquired through natural selection. In many cases complex and long-enduring conditions, often of a peculiar nature, are necessary for the development of a structure; and the requisite conditions may seldom have concurred. The belief that any given structure, which we think, often erroneously, would have been beneficial to a species, would have been gained under all circumstances through natural selection, is opposed to what we can understand of its manner of action. Mr. Mivart does not deny that natural selection has effected something; but he considers it as 'demonstrably insufficient' to account for the phenomena which I

explain by its agency. His chief arguments have now been considered, and the others will hereafter be considered. They seem to me to partake little of the character of demonstration, and to have little weight in comparison with those in favour of the power of natural selection, aided by the other agencies often specified. I am bound to add, that some of the facts and arguments here used by me, have been advanced for the same purpose in an able article lately published in the 'Medico-Chirurgical Review.'

At the present day almost all naturalists admit evolution under some form. Mr. Mivart believes that species change through 'an internal force or tendency,' about which it is not pretended that anything is known. That species have a capacity for change will be admitted by all evolutionists; but there is no need, as it seems to me, to invoke any internal force beyond the tendency to ordinary variability, which through the aid of selection, by man has given rise to many well-adapted domestic races, and which, through the aid of natural selection, would equally well give rise by graduated steps to natural races or species. The final result will generally have been, as already explained, an advance, but in some few cases a retrogression, in organisation.

Mr. Mivart is further inclined to believe, and some naturalists agree with him, that new species manifest themselves 'with suddenness and by modifications appearing at once.' For instance, he supposes that the differences between the extinct three-toed Hipparion and the horse arose suddenly. He thinks it difficult to believe that the wing of a bird 'was developed in any other way than by a comparatively sudden modification of a marked and important kind;' and apparently he would extend the same view to the wings of bats and pterodactyles. This conclusion, which implies great breaks or discontinuity in the series, appears to me improbable in the highest degree.

Everyone who believes in slow and gradual evolution, will of course admit that specific changes may have been as abrupt and as great as any single variation which we meet with under nature, or even under domestication. But as species are more variable when domesticated or cultivated than under their natural conditions, it is not probable that such great and abrupt variations have often occurred under nature, as are known occasionally to arise under domestication. Of these latter variations several may be attributed to reversion; and the characters which thus reappear were, it is probable, in many cases at first gained in a gradual manner. A still greater number must be called monstrosities, such as six-fingered men, porcupine men, Ancon sheep, Niata cattle, etc.; and as they are widely different in character from natural species, they throw very little light on our subject. Excluding such cases of abrupt variations, the few which remain would at best constitute, if found in a state of nature, doubtful species, closely related to their parental types.

My reasons for doubting whether natural species have changed as abruptly as have occasionally domestic races, and for entirely disbelieving that they have changed in the wonderful manner indicated by Mr. Mivart, are as follows. According to our experience, abrupt and strongly marked variations occur in our domesticated productions, singly and at rather long intervals of time. If such occurred under nature, they would be liable, as formerly explained, to be lost by accidental causes of destruction and by subsequent intercrossing; and so it is known to be under domestication, unless abrupt variations of this kind are specially preserved and separated by the care of man. Hence, in order that a new species should suddenly appear in the manner supposed by Mr. Mivart, it is almost necessary to believe, in opposition to all analogy, that several wonderfully changed individuals appeared simultaneously within the same district. This difficulty, as in the case of unconscious selection by man, is avoided on the theory of gradual evolution, through the preservation of a large number of individuals, which varied more or less in any favourable

direction, and of the destruction of a large number which varied in an opposite manner.

That many species have been evolved in an extremely gradual manner, there can hardly be a doubt. The species and even the genera of many large natural families are so closely allied together that it is difficult to distinguish not a few of them. On every continent, in proceeding from north to south, from lowland to upland, etc., we meet with a host of closely related or representative species; as we likewise do on certain distinct continents, which we have reason to believe were formerly connected. But in making these and the following remarks, I am compelled to allude to subjects hereafter to be discussed. Look at the many outlying islands round a continent, and see how many of their inhabitants can be raised only to the rank of doubtful species. So it is if we look to past times, and compare the species which have just passed away with those still living within the same areas; or if we compare the fossil species embedded in the sub-stages of the same geological formation. It is indeed manifest that multitudes of species are related in the closest manner to other species that still exist, or have lately existed; and it will hardly be maintained that such species have been developed in an abrupt or sudden manner. Nor should it be forgotten, when we look to the special parts of allied species, instead of to distinct species, that numerous and wonderfully fine gradations can be traced, connecting together widely different structures.

Many large groups of facts are intelligible only on the principle that species have been evolved by very small steps. For instance, the fact that the species included in the larger genera are more closely related to each other, and present a greater number of varieties than do the species in the smaller genera. The former are also grouped in little clusters, like varieties round species; and they present other analogies with varieties, as was shown in our second chapter. On this same principle we can understand how it is that specific characters are more variable than generic characters; and how the parts which are developed in an extraordinary degree or manner are more variable than other parts of the same species. Many analogous facts, all pointing in the same direction, could be added.

Although very many species have almost certainly been produced by steps not greater than those separating fine varieties, yet it may be maintained that some have been developed in a different and abrupt manner. Such an admission, however, ought not to be made without strong evidence being assigned. The vague and in some respects false analogies, as they have been shown to be by Mr. Chauncey Wright, which have been advanced in favour of this view, such as the sudden crystallisation of inorganic substances, or the falling of a facetted spheroid from one facet to another, hardly deserve consideration. One class of facts, however, namely, the sudden appearance of new and distinct forms of life in our geological formations supports at first sight the belief in abrupt development. But the value of this evidence depends entirely on the perfection of the geological record, in relation to periods remote in the history of the world. If the record is as fragmentary as many geologists strenuously assert, there is nothing strange in new forms appearing as if suddenly developed.

Unless we admit transformations as prodigious as those advocated by Mr. Mivart, such as the sudden development of the wings of birds or bats, or the sudden conversion of a Hipparion into a horse, hardly any light is thrown by the belief in abrupt modifications on the deficiency of connecting links in our geological formations. But against the belief in such abrupt changes, embryology enters a strong protest. It is notorious that the wings of birds and bats, and the legs of horses or other quadrupeds, are undistinguishable at an early embryonic period, and that they become differentiated by insensibly fine steps. Embryological resemblances of all kinds can be accounted for, as we shall hereafter see, by the progenitors of our existing species having varied after early youth, and having transmitted their newly acquired characters to their offspring, at a corresponding age. The embryo is thus left almost unaffected, and serves as a record of the past condition of the species. Hence it is that existing species during the early stages of their development so often resemble ancient and extinct forms belonging to the same class. On this view of the meaning of embryological resemblances, and indeed on any view, it is incredible that an animal should have undergone such momentous and abrupt transformations as those above indicated, and yet should not bear even a trace in its embryonic condition of any sudden modification, every detail in its structure being developed by insensibly fine steps.

He who believes that some ancient form was transformed suddenly through an internal force or tendency into, for instance, one furnished with wings, will be almost compelled to assume, in opposition to all analogy, that many individuals varied simultaneously. It cannot be denied that such abrupt and great changes of structure are widely different from those which most species apparently have undergone. He will further be compelled to believe that many structures beautifully adapted to all the other parts of the same creature and to the surrounding conditions, have been suddenly produced; and of such complex and wonderful co-adaptations, he will not be able to assign a shadow of an explanation. He will be forced to admit that these great and sudden transformations have left no trace of their action on the embryo. To admit all this is, as it seems to me, to enter into the realms of miracle, and to leave those of science.

Chapter VIII

INSTINCT

Instincts comparable with habits, but different in their origin–Instincts graduated–Aphides and ants–Instincts variable–Domestic instincts, their origin–Natural instincts of the cuckoo, molothrus, ostrich, and parasitic bees–Slave-making ants–Hive bee, its cell-making instinct–Changes of instinct and structure not necessarily simultaneous–Difficulties of the theory of the Natural Selection of instincts–Neuter or sterile insects–Summary.

Many instincts are so wonderful that their development will probably appear to the reader a difficulty sufficient to overthrow my whole theory. I may here premise, that I have nothing to do with the origin of the mental powers, any more than I have with that of life itself. We are concerned only with the diversities of instinct and of the other mental faculties in animals of the same class.

I will not attempt any definition of instinct. It would be easy to show that several distinct mental actions are commonly embraced by this term; but every one understands what is meant, when it is said that instinct impels the cuckoo to migrate and to lay her eggs in other birds' nests. An action, which we ourselves require experience to enable us to perform, when performed by an animal, more especially by a very young one, without experience, and when performed by many individuals in the same way, without their knowing for what purpose it is performed, is usually said to be instinctive. But I could show that none of these characters are universal. A little dose of judgment or reason, as Pierre Huber expresses it, often comes into play, even with animals low in the scale of nature.

Frederick Cuvier and several of the older metaphysicians have compared instinct with habit. This comparison gives, I think, an accurate notion of the frame of mind under which an instinctive action is performed, but not necessarily of its origin. How unconsciously many habitual actions are performed, indeed not rarely in direct opposition to our conscious will! yet they may be modified by the will or reason. Habits easily become associated with other habits, with certain periods of time and states of the body. When once acquired, they often remain constant throughout life. Several other points of resemblance between instincts and habits could be pointed out. As in repeating a well-known song, so in instincts, one action follows another by a sort of rhythm; if a person be interrupted in a song, or in repeating anything by rote, he is generally forced to go back to recover the habitual train of thought; so P. Huber found it was with a caterpillar, which makes a very complicated hammock; for if he took a caterpillar which had completed its hammock up to, say, the sixth stage of construction, and put it into a hammock completed up only to the third stage, the caterpillar simply re-performed the fourth, fifth, and sixth stages of construction. If, however, a caterpillar were taken out of a hammock made up, for instance, to the third stage, and were put into one finished up to the sixth stage, so that much of its work was already done for it, far from deriving any benefit from this, it was much embarrassed, and, in order to complete its hammock, seemed forced to start from the third stage, where it had left off, and thus tried to complete the already finished work.

If we suppose any habitual action to become inherited – and it can be shown that this does sometimes happen – then the resemblance between what originally was a habit and an instinct becomes so close as not to be distinguished. If Mozart, instead of playing the pianoforte at three years old with wonderfully little practice, had played a tune with no practice at all, be might truly be said to have done so instinctively. But it would be a serious error to suppose that the greater number of instincts have been acquired by habit in one generation, and then transmitted by inheritance to succeeding generations. It can be clearly shown that the most wonderful instincts with which we are acquainted, namely, those of the hive bee and of many ants, could not possibly have been acquired by habit.

It will be universally admitted that instincts are as important as corporeal structures for the welfare of each species, under its present conditions of life. Under changed conditions of life, it is at least possible that slight modifications of instinct might be profitable to a species; and if it can be shown that instincts do vary ever so little, then I can see no difficulty in natural selection preserving and continually accumulating variations of instinct to any extent that was profitable. It is thus, as I believe, that all the most complex and wonderful instincts have originated. As modifications of corporeal structure arise from, and are increased by, use or habit, and are diminished or lost by disuse, so I do not doubt it has been with instincts. But I believe that the effects of habit are in many cases of subordinate importance to the effects of the natural selection of what may be called spontaneous variations of instincts; – that is of variations produced by the same unknown causes which produce slight deviations of bodily structure.

No complex instinct can possibly be produced through natural selection, except by the slow and gradual accumulation of numerous, slight, yet profitable, variations. Hence, as in the case of corporeal structures, we ought to find in nature, not the actual transitional gradations by which each complex instinct has been acquired – for these could be found only in the lineal ancestors of each species – but we ought to find in the collateral lines of descent some evidence of such gradations; or we ought at least to be able to show that gradations of some kind are possible; and this we certainly can do. I have been surprised to find, making allowance for the instincts of animals having been but little observed except in Europe and North America, and for no instinct being known amongst extinct species, how very generally gradations, leading to the most complex instincts, can be discovered. Changes of instinct may sometimes be facilitated by the same species having different instincts at different periods of life, or at different seasons of the year, or when placed under different circumstances, etc.; in which case either the one or the other instinct might be preserved by natural selection. And such instances of diversity of instinct in the same species can be shown to occur in nature.

Again, as in the case of corporeal structure, and conformably to my theory, the instinct of each species is good for itself, but has never, as far as we can judge, been produced for the exclusive good of others. One of the strongest instances of an animal apparently performing an action for the sole good of another, with which I am acquainted, is that of aphides voluntarily yielding, as was first observed by Huber, their sweet excretion to ants: that they do so voluntarily, the following facts show. I removed all the ants from a group of about a dozen aphides on a dock-plant, and prevented their attendance during several hours. After this interval, I felt sure that the aphides would want to excrete. I watched them for some time through a lens, but not one excreted; I then tickled and stroked them with a hair in the same manner, as well as I could, as the ants do with their antennae; but not one excreted. Afterwards, I

An Ant in the act of 'milking' an Aphid

allowed an ant to visit them, and it immediately seemed, by its eager way of running about to be well aware what a rich flock it had discovered; it then began to play with its antennae on the abdomen first of one aphis and then of another; and each, as soon as it felt the antennae, immediately lifted up its abdomen and excreted a limpid drop of sweet juice, which was eagerly devoured by the ant. Even the quite young aphides behaved in this manner, showing that the action was instinctive, and not the result of experience. It is certain, from the observations of Huber, that the aphides show no dislike to the ants: if the latter be not present they are at last compelled to eject their excretion. But as the excretion is extremely viscid, it is no doubt a convenience to the aphides to have it removed; therefore probably they do not excrete solely for the good of the ants. Although there is no evidence that any animal performs an action for the exclusive good of another species, yet each tries to take advantage of the instincts of others, as each takes advantage of the weaker bodily structure of other species. So again certain instincts cannot be considered as absolutely perfect; but as details on this and other such

points are not indispensable, they may be here passed over.

As some degree of variation in instincts under a state of nature, and the inheritance of such variations, are indispensable for the action of natural selection, as many instances as possible ought to be given; but want of space prevents me. I can only assert that instincts certainly do vary – for instance, the migratory instinct, both in extent and direction, and in its total loss. So it is with the nests of birds, which vary partly in dependence on the situations chosen, and on the nature and temperature of the country inhabited, but often from causes wholly unknown to us. Audubon has given several remarkable cases of differences in the nests of the same species in the northern and southern United States. Why, it has been asked, if instinct be variable, has it not granted to the bee 'the ability to use some other material when wax was deficient?' But what other natural material could bees use? They will work, as I have seen, with wax hardened with vermilion or softened with lard. Andrew Knight observed that his bees, instead of laboriously collecting propolis, used a cement of wax and turpentine, with which he had covered decorticated trees. It has lately been shown that bees, instead of searching for pollen, will gladly use a very different substance, namely, oatmeal. Fear of any particular enemy is certainly an instinctive quality, as may be seen in nestling birds, though it is strengthened by experience, and by the sight of fear of the same enemy in other animals. The fear of man is slowly acquired, as I have elsewhere shown, by the various animals which inhabit desert islands; and we see an instance of this, even in England, in the greater wildness of all our large birds in comparison with our small birds; for the large birds have been most persecuted by man. We may safely attribute the greater wildness of our large birds to this cause; for in uninhabited islands large birds are not more fearful than small; and the magpie, so wary in England, is tame in Norway, as is the hooded crow in Egypt.

That the mental qualities of animals of the same kind, born in a state of nature, vary much, could be shown by many facts. Several cases could also be adduced of occasional and strange habits in

The Cabbage White Butterfly:
1 adult female *2* eggs laid on underside of leaf
3 caterpillar *4* chrysalis

wild animals, which, if advantageous to the species, might have given rise, through natural selection, to new instincts. But I am well aware that these general statements, without the facts in detail, can produce but a feeble effect on the reader's mind. I can only repeat my assurance, that I do not speak without good evidence.

INHERITED CHANGES OF HABIT OR INSTINCT IN DOMESTICATED ANIMALS

The possibility, or even probability, of inherited variations of instinct in a state of nature will be strengthened by briefly considering a few cases under domestication. We shall thus be enabled to see the part which habit and the selection of so-called spontaneous variations have played in modifying the mental qualities of our domestic animals. It is notorious how much domestic animals vary in their mental qualities. With cats, for instance, one naturally takes to catching rats, and another mice, and these tendencies are known to be inherited. One cat, according to Mr. St. John, always brought home game birds, another hares or rabbits, and another hunted on marshy ground and almost nightly caught woodcocks or snipes. A number of curious and authentic instances could be given of various shades of disposition and taste, and likewise of the oddest tricks, associated with certain frames of mind or periods of time. But let us look to the familiar case of the breeds of dogs: it cannot be doubted that young pointers (I have myself seen striking instances) will sometimes point and even back other dogs the very first time that they are taken out; retrieving is certainly in some degree inherited by retrievers; and a tendency to run round, instead of at, a flock of sheep, by shepherd dogs. I cannot see that these actions, performed without experience by the young, and in nearly the same manner by each individual, performed with eager delight by each breed, and without the end being known – for the young pointer can no more know that he points to aid his master, than the white butterfly knows why she lays her eggs on the leaf of the cabbage – I cannot see that these actions differ essentially from true instincts. If we were to behold one kind of wolf, when young and without any training, as soon as it scented its prey, stand motionless like a statue, and then slowly crawl forward with a peculiar gait; and another kind of wolf rushing round, instead of at, a herd of deer, and driving them to a distant point, we should assuredly call these actions instinctive. Domestic instincts, as they may be called, are certainly far less fixed than natural instincts; but they have been acted on by far less rigorous selection, and have been transmitted for an incomparably shorter period, under less fixed conditions of life.

How strongly these domestic instincts, habits, and dispositions are inherited, and how curiously they become mingled, is well shown when different breeds of dogs are crossed. Thus it is known that a cross with a bull-dog has affected for many generations the courage and obstinacy of greyhounds; and a cross with a greyhound has given to a whole family of shepherd dogs a tendency to hunt hares. These domestic instincts, when thus tested by crossing, resemble natural instincts, which in a like manner become curiously blended together, and for a long period exhibit traces of the instincts of either parent: for example, Le Roy describes a dog, whose great-grandfather was a wolf, and this dog showed a trace of its wild parentage only in one way, by not coming in a straight line to his master, when called.

Domestic instincts are sometimes spoken of as actions which have become inherited solely from long-continued and compulsory habit, but this is not true. No one would ever have thought of teaching, or probably could have taught, the tumbler-pigeon to tumble, – an action which, as I have witnessed, is performed by young birds, that have never seen a pigeon tumble. We may believe that some one pigeon showed a slight tendency to this strange habit, and that the long-continued selection of the best individuals in successive generations made tumblers what they now are;

and near Glasgow there are house-tumblers, as I hear from Mr. Brent, which cannot fly eighteen inches high without going head over heels. It may be doubted whether any one would have thought of training a dog to point, had not some one dog naturally shown a tendency in this line; and this is known occasionally to happen, as I once saw, in a pure terrier: the act of pointing is probably, as many have thought, only the exaggerated pause of an animal preparing to spring on its prey. When the first tendency to point was once displayed, methodical selection and the inherited effects of compulsory training in each successive generation would soon complete the work; and unconscious selection is still in progress, as each man tries to procure, without intending to improve the breed, dogs which stand and hunt best. On the other hand, habit alone in some cases has sufficed; hardly any animal is more difficult to tame than the young of the wild rabbit; scarcely any animal is tamer than the young of the tame rabbit; but I can hardly suppose that domestic rabbits have often been selected for tameness alone; so that we must attribute at least the greater part of the inherited change from extreme wildness to extreme tameness, to habit and long-continued close confinement.

Natural instincts are lost under domestication: a remarkable instance of this is seen in those breeds of fowls which very rarely or never become 'broody,' that is, never wish to sit on their eggs. Familiarity alone prevents our seeing how largely and how permanently the minds of our domestic animals have been modified. It is scarcely possible to doubt that the love of man has become instinctive in the dog. All wolves, foxes, jackals and species of the cat genus, when kept tame, are most eager to attack poultry, sheep, and pigs; and this tendency has been found incurable in dogs which have been brought home as puppies from countries such as Tierra del Fuego and Australia, where the savages do not keep these domestic animals. How rarely, on the other hand, do our civilised dogs, even when quite

Canis antarcticus.

The Falkland Islands Fox

young, require to be taught not to attack poultry, sheep, and pigs! No doubt they occasionally do make an attack, and are then beaten; and if not cured, they are destroyed; so that habit and some degree of selection have probably concurred in civilising by inheritance our dogs. On the other hand, young chickens have lost wholly by habit, that fear of the dog and cat which no doubt was originally instinctive in them, for I am informed by Captain Hutton that the young chickens of the parent stock, the Gallus bankiva, when reared in India under a hen, are at first excessively wild. So it is with young pheasants reared in England under a hen. It is not that chickens have lost all fear, but fear only of dogs and cats, for if the hen gives the danger chuckle they will run (more especially young turkeys) from under her and conceal themselves in the surrounding grass or thickets; and this is evidently done for the instinctive purpose of allowing, as we see in wild ground-birds, their mother to fly away. But this instinct retained by our chickens has become useless under domestication, for the mother-hen has almost lost by disuse the power of flight.

Hence, we may conclude that under domestication instincts have been acquired and natural instincts have been lost, partly by habit and partly by man selecting and accumulating, during successive generations, peculiar mental habits and actions, which at first appeared from what we must in our ignorance call an accident. In some cases compulsory habit alone has sufficed to produce inherited mental changes; in other cases compulsory habit has done nothing, and all has been the result of selection, pursued both methodically and unconsciously; but in most cases habit and selection have probably concurred.

SPECIAL INSTINCTS
We shall, perhaps, best understand how instincts in a state of nature have become modified by selection by considering a few cases. I will select only three, namely, the instinct which leads the cuckoo to lay her eggs in other birds' nests; the

slave-making instinct of certain ants; and the cell-making power of the hive bee: these two latter instincts have generally and justly been ranked by naturalists as the most wonderful of all known instincts.

INSTINCTS OF THE CUCKOO It is supposed by some naturalists that the more immediate cause of the instinct of the cuckoo is that she lays her eggs, not daily, but at intervals of two or three days; so that, if she were to make her own nest and sit on her own eggs, those first laid would have to be left for some time unincubated or there would be eggs and young birds of different ages in the same nest. If this were the case, the process of laying and hatching might be inconveniently long, more especially as she migrates at a very early period; and the first hatched young would probably have to be fed by the male alone. But the American cuckoo is in this predicament; for she makes her own nest, and has eggs and young successively hatched, all at the same time. It has been both asserted and denied that the American cuckoo occasionally lays her eggs in other birds' nests; but I have lately heard from Dr. Merrell, of Iowa, that he once found in Illinois a young cuckoo, together with a young jay in the nest of a blue jay (Garrulus cristatus); and as both were nearly full feathered, there could be no mistake in their identification. I could also give several instances of various birds which have been known occasionally to lay their eggs in other birds' nests. Now let us suppose that the ancient progenitor of our European cuckoo had the habits of the American cuckoo, and that she occasionally laid an egg in another bird's nest. If the old bird profited by this occasional habit through being enabled to migrate earlier or through any other cause; or if the young were made more vigorous by advantage being taken of the mistaken instinct of another species than when reared by their own mother, encumbered as she could hardly fail to be by having eggs and young of different ages at the same time, then the old birds or the fostered

The Fantail Cuckoo

young would gain an advantage. And analogy would lead us to believe that the young thus reared would be apt to follow by inheritance the occasional and aberrant habit of their mother, and in their turn would be apt to lay their eggs in other birds' nests, and thus be more successful in rearing their young. By a continued process of this nature, I believe that the strange instinct of our cuckoo has been generated. It has, also, recently been ascertained on sufficient evidence, by Adolf Müller, that the cuckoo occasionally lays her eggs on the bare ground, sits on them, and feeds her young. This rare event is probably a case of reversion to the long-lost, aboriginal instinct of nidification.

It has been objected that I have not noticed other related instincts and adaptations of structure in the cuckoo, which are spoken of as necessarily co-ordinated. But in all cases, speculation on an instinct known to us only in a single species, is useless, for we have hitherto had no facts to guide us. Until recently the instincts of the European and of the non-parasitic American cuckoo alone were known; now, owing to Mr. Ramsay's observations, we have learned something about three Australian species, which lay their eggs in other birds' nests. The chief points to be referred to are three: first, that the common cuckoo, with rare exceptions, lays only one egg in a nest, so that the large and voracious young bird receives ample food. Secondly, that the eggs are remarkably small, not exceeding those of the skylark, – a bird about one-fourth as large as the cuckoo. That the small size of the egg is a real case of adaptation we may infer from the fact of the non-parasitic American cuckoo laying full-sized eggs. Thirdly, that the young cuckoo, soon after birth, has the instinct, the strength, and a properly shaped back for ejecting its foster-brothers, which then perish from cold and hunger. This has been boldly called a beneficial arrangement, in order that the young cuckoo may get sufficient food, and that its foster-brothers may perish before they had acquired much feeling!

Turning now to the Australian species: though these birds generally lay only one egg in a nest, it is not rare to find two and even three eggs in the same nest. In the Bronze cuckoo the eggs vary greatly in size, from eight to ten lines in length. Now, if it had been of an advantage to this species to have laid eggs even smaller than those now laid, so as to have deceived certain foster-parents, or, as is more probable, to have been hatched within a shorter period (for it is asserted that there is a relation between the size of eggs and the period of their incubation), then there is no difficulty in believing that a race or species might have been formed which would have laid smaller and smaller eggs; for these would have been more safely hatched and reared. Mr. Ramsay remarks that two of the Australian cuckoos, when they lay their eggs in an open nest, manifest a decided preference for nests containing eggs similar in colour to their own. The European species apparently manifests some tendency towards a similar instinct, but not rarely departs from it, as is shown by her laying her dull and pale-coloured eggs in the nest of the Hedge-warbler with bright greenish-blue eggs. Had our cuckoo invariably displayed the above instinct, it would assuredly have been added to those which it is assumed must all have been acquired together. The eggs of the Australian Bronze cuckoo vary, according to Mr. Ramsay, to an extraordinary degree in colour; so that in this respect, as well as in size, natural selection might have secured and fixed any advantageous variation.

In the case of the European cuckoo, the offspring of the foster-parents are commonly ejected from the nest within three days after the cuckoo is hatched; and as the latter at this age is in a most helpless condition, Mr. Gould was formerly inclined to believe that the act of ejection was performed by the foster-parents themselves. But he has now received a trustworthy account of a young cuckoo which was actually seen, whilst still blind and not able even to hold up its own head, in the act of ejecting its foster-brothers. One of these was replaced in the nest by the observer, and was again thrown out. With respect to the means by which this strange and odious instinct was acquired, if it were of great importance for the young cuckoo, as is probably the case, to receive as much food as possible soon after birth, I can see no special difficulty in its having gradually acquired, during successive generations, the blind desire, the strength, and structure necessary for the work of ejection; for those cuckoos which had such habits and structure best developed would be the most securely reared. The first step towards the acquisition of the proper instinct might have been mere unintentional restlessness on the part of the young bird, when somewhat advanced in age and strength; the habit having been afterwards improved, and transmitted to an earlier age. I can see no more difficulty in this than in the unhatched young of other birds acquiring the instinct to break through their own shells; or than in young snakes acquiring in their upper jaws, as Owen has remarked, a transitory sharp tooth for cutting through the tough egg-shell. For if each part is liable to individual variations at all ages, and the variations tend to be inherited at a corresponding or earlier age, – propositions which cannot be disputed, – then the instincts and structure of the young could be slowly modified as surely as those of the adult; and both cases must stand or fall together with the whole theory of natural selection.

Some species of Molothrus, a widely distinct genus of American birds, allied to our starlings, have parasitic habits like those of the cuckoo; and the species present an interesting gradation in the perfection of their instincts. The sexes of Molothrus badius are stated by an excellent observer, Mr. Hudson, sometimes to live promiscuously together in flocks, and sometimes to pair. They either build a nest of their own, or seize on one belonging to some other bird, occasionally throwing out the nestlings of the stranger. They either lay their eggs in the nest thus appropriated, or oddly enough build one for themselves on the top of it. They usually sit on their own eggs and rear their own young; but Mr. Hudson says it is probable that they are occasionally parasitic, for he has seen the young of this species following old birds of a distinct kind and clamouring to be fed by them. The parasitic habits of another species of Molothrus, the M. bonariensis, are much more highly developed than those of the last, but are still far from perfect. This bird, as far as it is known, invariably lays its eggs in the nests of strangers; but it is remarkable that several together sometimes commence to build an irregular untidy nest of their own, placed in singularly ill-adapted situations, as on the leaves of a large thistle. They never, however, as far as Mr. Hudson has ascertained, complete a nest for themselves. They often lay so many eggs – from fifteen to twenty – in the same foster-nest, that few or none can possibly be hatched. They have, moreover, the extraordinary habit of pecking holes in the eggs, whether of their own species or of their foster-parents, which they find in the appropriated nests. They drop also many eggs on the bare ground, which are thus wasted. A third species, the M. pecoris of North America, has acquired instincts as perfect as those of the cuckoo, for it never lays more than one egg in a foster-nest, so that the young bird is securely reared. Mr. Hudson is a strong disbeliever in evolution, but he appears to have been so much struck by the imperfect instincts of the Molothrus bonariensis that he quotes my words, and asks, 'Must we consider these habits, not as especially endowed or created instincts, but as small consequences of one general law, namely, transition?'

Various birds, as has already been remarked, occasionally lay their eggs in the nests of other birds. This habit is not very uncommon with the Gallinaceae, and throws some light on the singular instinct of the ostrich. In this family several hen birds unite and lay first a few eggs in one nest and then in another, and these are hatched by the males. This instinct may probably be accounted for by the fact of the hens laying a large number of eggs, but, as with the cuckoo, at intervals of two or three days. The instinct, however, of the American ostrich, as in the case of the Molothrus bonariensis, has not as yet been perfected; for a surprising number of eggs lie strewed over the plains, so that in one day's hunting I picked up no less than twenty lost and wasted eggs.

Many bees are parasitic, and regularly lay their eggs in the nests of other kinds of bees. This case is more remarkable than that of the cuckoo; for these bees have not only had their instincts but their structure modified in accordance with their parasitic habits; for they do not possess the pollen-collecting apparatus which would have been indispensable if they had stored up food for their own young. Some species of Sphegidae (wasp-like insects) are likewise parasitic; and M. Fabre has lately shown good reason for believing that, although the Tachytes nigra generally makes its own burrow and stores it with paralysed prey for its own larvae, yet that, when this insect finds a burrow already made and stored by another sphex, it takes advantage of the prize, and becomes for the occasion parasitic. In this case, as with that of the Molothrus or cuckoo, I can see no difficulty in natural selection making an occasional habit permanent, if of advantage to the species, and if the insect whose nest and stored food are feloniously appropriated, be not thus exterminated.

SLAVE-MAKING INSTINCT This remarkable instinct was first discovered in the Formica (Polyerges) rufescens by Pierre Huber, a better observer even than his celebrated father. This ant is absolutely dependent on its slaves; without their aid, the species would certainly become extinct in a single year. The males and fertile females do no work of any kind, and the workers or sterile females, though most energetic and courageous in capturing slaves, do no other work. They are incapable of making their own nests, or of feeding their own larvae. When the old nest is found inconvenient, and they have to migrate, it is the slaves which determine the migration, and actually carry their masters in their jaws. So utterly helpless are the masters, that when Huber shut up thirty of them without a slave, but with plenty of the food which they like best, and with their larvae and pupae to stimulate them to work, they did nothing; they could not even feed themselves, and many perished of hunger. Huber then introduced a single slave (F. fusca), and she instantly set to work, fed and saved the survivors; made some cells and tended the larvae, and put all to rights. What can be more extraordinary than these well-ascertained facts? If we had not known of any other slave-making ant, it would have been hopeless to speculate how so wonderful an instinct could have been perfected.

Another species, Formica sanguinea, was likewise first discovered by P. Huber to be a slave-making ant. This species is found in the southern parts of England, and its habits have been attended to by Mr. F. Smith, of the British Museum, to whom I am much indebted for information on this and other subjects. Although fully trusting to the statements of Huber and Mr. Smith, I tried to approach the subject in a sceptical frame of mind, as any one may well be excused for doubting the existence of so extraordinary an instinct as that of making slaves. Hence, I will give the observations which I made in some little detail. I opened fourteen nests of F. sanguinea, and found a few slaves in all. Males and fertile females of the slave species (F. fusca) are found only in their own proper communities, and have never been observed in the nests of F. sanguinea. The slaves are black and not above half the size of their red masters, so that the contrast in their appearance is great. When the nest is slightly disturbed, the slaves occasionally come out, and like their masters are much agitated and defend the nest: when the nest is much disturbed, and the larvae and pupae are exposed, the slaves work energetically together with their masters in carrying them away to a place of safety. Hence, it is clear that the slaves feel quite at home. During the months of June and

July, on three successive years, I watched for many hours several nests in Surrey and Sussex, and never saw a slave either leave or enter a nest. As, during these months, the slaves are very few in number, I thought that they might behave differently when more numerous; but Mr. Smith informs me that he has watched the nests at various hours during May, June and August, both in Surrey and Hampshire, and has never seen the slaves, though present in large numbers in August, either leave or enter the nest. Hence, he considers them as strictly household slaves. The masters, on the other hand, may be constantly seen bringing in materials for the nest, and food of all kinds. During the year 1860, however, in the month of July, I came across a community with an unusually large stock of slaves, and I observed a few slaves mingled with their masters leaving the nest, and marching along the same road to a tall Scotch-fir tree, twenty-five yards distant, which they ascended together, probably in search of aphides or cocci. According to Huber, who had ample opportunities for observation, the slaves in Switzerland habitually work with their masters in making the nest, and they alone open and close the doors in the morning and evening; and, as Huber expressly states, their principal office is to search for aphides. This difference in the usual habits of the masters and slaves in the two countries, probably depends merely on the slaves being captured in greater numbers in Switzerland than in England.

One day I fortunately witnessed a migration of F. sanguinea from one nest to another, and it was a most interesting spectacle to behold the masters carefully carrying their slaves in their jaws instead of being carried by them, as in the case of F. rufescens. Another day my attention was struck by about a score of the slave-makers haunting the same spot, and evidently not in search of food; they approached and were vigorously repulsed by an independent community of the slave species (F. fusca); sometimes as many as three of these ants clinging to the legs of the slave-making

F. sanguinea. The latter ruthlessly killed their small opponents and carried their dead bodies as food to their nest, twenty-nine yards distant; but they were prevented from getting any pupae to rear as slaves. I then dug up a small parcel of the pupae of F. fusca from another nest, and put them down on a bare spot near the place of combat; they were eagerly seized and carried off by the tyrants, who perhaps fancied that, after all, they had been victorious in their late combat.

At the same time I laid on the same place a small parcel of the pupae of another species, F. flava, with a few of these little yellow ants still clinging to the fragments of their nest. This species is sometimes, though rarely, made into slaves, as has been described by Mr. Smith. Although so small a species, it is very courageous, and I have seen it ferociously attack other ants. In one instance I found to my surprise an independent community of F. flava under a stone beneath a nest of the slave-making F. sanguinea; and when I had accidentally disturbed both nests, the little ants attacked their big neighbours with surprising courage. Now I was curious to ascertain whether F. sanguinea could distinguish the pupae of F. fusca, which they habitually make into slaves, from those of the little and furious F. flava, which they rarely capture, and it was evident that they did at once distinguish them; for we have seen that they eagerly and instantly seized the pupae of F. fusca, whereas they were much terrified when they came across the pupae, or even the earth from the nest, of F. flava, and quickly ran away; but in about a quarter of an hour, shortly after all the little yellow ants had crawled away, they took heart and carried off the pupae.

One evening I visited another community of F. sanguinea, and found a number of these ants returning home and entering their nests, carrying the dead bodies of F. fusca (showing that it was not a migration) and numerous pupae. I traced a long file of ants burdened with booty, for about forty yards back, to a very thick clump of heath, whence I saw the last individual of F. sanguinea emerge, carrying a pupa; but I was not able to find the desolated nest in the thick heath. The nest, however, must have been close at hand, for two or three individuals of F. fusca were rushing about in the greatest agitation, and one was perched motionless with its own pupa in its mouth on the top of a spray of heath, an image of despair over its ravaged home.

Such are the facts, though they did not need confirmation by me, in regard to the wonderful instinct of making slaves. Let it be observed what a contrast the instinctive habits of F. sanguinea present with those of the continental F. rufescens. The latter does not build its own nest, does not determine its own migrations, does not collect food for itself or its young, and cannot even feed itself: it is absolutely dependent on its numerous slaves. Formica sanguinea, on the other hand, possesses much fewer slaves, and in the early part of the summer extremely few. The masters determine when and where a new nest shall be formed, and when they migrate, the masters carry the slaves. Both in Switzerland and England the slaves seem to have the exclusive care of the larvae, and the masters alone go on slave-making expeditions. In Switzerland the slaves and masters work together, making and bringing materials for the nest: both, but chiefly the slaves, tend, and milk, as it may be called, their aphides; and thus both collect food for the community. In England the masters alone usually leave the nest to collect building materials and food for themselves, their slaves and larvae. So that the masters in this country receive much less service from their slaves than they do in Switzerland.

By what steps the instinct of F. sanguinea originated I will not pretend to conjecture. But as ants which are not slave-makers, will, as I have seen, carry off pupae of other species, if scattered near their nests, it is possible that such pupae originally stored as food might become developed; and the foreign ants thus unintentionally reared would then follow their proper instincts, and do what work they could.

If their presence proved useful to the species which had seized them – if it were more advantageous to this species, to capture workers than to procreate them – the habit of collecting pupae, originally for food, might by natural selection be strengthened and rendered permanent for the very different purpose of raising slaves. When the instinct was once acquired, if carried out to a much less extent even than in our British F. sanguinea, which, as we have seen, is less aided by its slaves than the same species in Switzerland, natural selection might increase and modify the instinct – always supposing each modification to be of use to the species – until an ant was formed as abjectly dependent on its slaves as is the Formica rufescens.

CELL-MAKING INSTINCT OF THE HIVE BEE

I will not here enter on minute details on this subject, but will merely give an outline of the conclusions at which I have arrived. He must be a dull man who can examine the exquisite structure of a comb, so beautifully adapted to its end, without enthusiastic admiration. We hear from mathematicians that bees have practically solved a recondite problem, and have made their cells of the proper shape to hold the greatest possible amount of honey, with the least possible consumption of precious wax in their construction. It has been remarked that a skilful workman, with fitting tools and measures, would find it very difficult to make cells of wax of the true form, though this is effected by a crowd of bees working in a dark hive. Granting whatever instincts you please, it seems at first quite inconceivable how they can make all the necessary angles and planes, or even perceive when they are correctly made. But the difficulty is not nearly so great as at first appears: all this beautiful work can be shown, I think, to follow from a few simple instincts.

I was led to investigate this subject by Mr. Waterhouse, who has shown that the form of the cell stands in close relation to the presence of adjoining cells; and the following view may, perhaps, be considered only as a modification of his theory. Let us look to the great principle of gradation, and see whether Nature does not reveal to us her method of work. At one end of a short series we have humble-bees, which use their old cocoons to hold honey, sometimes adding to them short tubes of wax, and likewise making separate and very irregular rounded cells of wax. At the other end of the series we have the cells of the hive bee, placed in a double layer: each cell, as is well known, is an hexagonal prism, with the basal edges of its six sides bevelled so as to join an inverted pyramid, of three rhombs. These rhombs have certain angles, and the three which form the pyramidal base of a single cell on one side of the comb, enter into the composition of the bases of three adjoining cells on the opposite side. In the

series between the extreme perfection of the cells of the hive bee and the simplicity of those of the humble-bee, we have the cells of the Mexican Melipona domestica, carefully described and figured by Pierre Huber. The Melipona itself is intermediate in structure between the hive and humble-bee, but more nearly related to the latter: it forms a nearly regular waxen comb of cylindrical cells, in which the young are hatched, and, in addition, some large cells of wax for holding honey. These latter cells are nearly spherical and of nearly equal sizes, and are aggregated into an irregular mass. But the important point to notice is, that these cells are always made at that degree of nearness to each other that they would have intersected or broken into each other if the spheres had been completed; but this is never permitted, the bees building perfectly flat walls of wax between the spheres which thus tend to intersect. Hence, each cell consists of an outer spherical portion, and of two, three, or more flat surfaces, according as the cell adjoins two, three or more other cells. When one cell rests on three other cells, which, from the spheres being nearly of the same size, is very frequently and necessarily the case, the three flat surfaces are united into a pyramid; and this pyramid, as Huber has remarked, is manifestly a gross imitation of the three-sided pyramidal base of the cell of the hive bee. As in the cells of the hive bee, so here, the

Ventral view of a Worker Bee, showing plates of wax

three plane surfaces in any one cell necessarily enter into the construction of three adjoining cells. It is obvious that the Melipona saves wax, and what is more important, labour, by this manner of building; for the flat walls between the adjoining cells are not double, but are of the same thickness as the outer spherical portions, and yet each flat portion forms a part of two cells.

Reflecting on this case, it occurred to me that if the Melipona had made its spheres at some given distance from each other, and had made them of equal sizes and had arranged them symmetrically in a double layer, the resulting structure would have been as perfect as the comb of the hive bee. Accordingly I wrote to Professor Miller, of Cambridge, and this geometer has kindly read over the following statement, drawn up from his information, and tells me that it is strictly correct:–

If a number of equal spheres be described with their centres placed in two parallel layers; with the centre of each sphere at the distance of radius x $\sqrt{2}$, or radius $\times$ 1.41421 (or at some lesser distance), from the centres of the six surrounding spheres in the same layer; and at the same distance from the centres of the adjoining spheres in the other and parallel layer; then, if planes of intersection between the several spheres in both layers be formed, there will result a double layer of hexagonal prisms united together by pyramidal bases formed of three rhombs; and the rhombs and the sides of the hexagonal prisms will have every angle identically the same with the best measurements which have been made of the cells of the hive bee. But I hear from Professor Wyman, who has made numerous careful measurements, that the accuracy of the workmanship of the bee has been greatly exaggerated; so much so, that whatever the typical form of the cell may be, it is rarely, if ever, realised.

Hence we may safely conclude that, if we could slightly modify the instincts already possessed by the Melipona, and in themselves not very wonderful, this bee would make a structure as wonderfully perfect as that of the hive bee. We must suppose the Melipona to have the power of forming her cells truly spherical, and of equal sizes; and this would not be very surprising, seeing that she already does so to a certain extent, and seeing what perfectly cylindrical burrows many insects make in wood, apparently by turning round on a fixed point. We must suppose the Melipona to arrange her cells in level layers, as she already does her cylindrical cells; and we must further suppose, and this is the greatest difficulty, that she can somehow judge accurately at what distance to stand from her fellow-labourers when several are making their spheres; but she is already so far enabled to judge of distance, that she always describes her spheres so as to intersect to a certain extent; and then she unites the points of intersection by perfectly flat surfaces. By such modifications of instincts which in themselves are not very wonderful – hardly more wonderful than those which guide a bird to make its nest – I believe that the hive bee has acquired, through natural selection, her inimitable architectural powers.

But this theory can be tested by experiment. Following the example of Mr. Tegetmeier, I separated two combs, and put between them a long, thick, rectangular strip of wax: the bees instantly began to excavate minute circular pits in it; and as they deepened these little pits, they made them wider and wider until they were converted into shallow basins, appearing to the eye perfectly true or parts of a sphere, and of about the diameter of a cell. It was most interesting to observe that, wherever several bees had begun to excavate these basins near together, they had begun their work at such a distance from each other that by the time the basins had acquired the above stated width (i.e. about the width of an ordinary cell), and were in depth about one sixth of the diameter of the sphere of which they formed a part, the rims of the basins intersected or broke into each other. As soon as this occurred, the bees ceased to excavate, and began to build up flat walls of wax on the lines of intersection between the basins, so that each hexagonal prism was built upon the scalloped edge of a smooth basin, instead of on the straight edges of a three-sided pyramid as in the case of ordinary cells.

I then put into the hive, instead of a thick, rectangular piece of wax, a thin and narrow, knife edged ridge, coloured with vermilion. The bees instantly began on both sides to excavate little basins near to each other, in the same way as before; but the ridge of wax was so thin, that the bottoms of the basins, if they had been excavated to the same depth as in the former experiment, would have broken into each other from the opposite sides. The bees, however, did not suffer this to happen, and they stopped their excavations in due time; so that the basins, as soon as they had been a little deepened, came to have flat bases, and these flat bases, formed by thin little plates of the vermilion wax left ungnawed, were situated, as far as the eye could judge, exactly along the planes of imaginary intersection between the basins on the opposite side of the ridge of wax. In some parts, only small portions, in other parts, large portions of a rhombic plate were thus left between the opposed basins, but the work, from the unnatural state of things, had not been neatly performed. The bees must have worked at very nearly the same rate in circularly gnawing away and deepening the basins on both sides of the ridge of vermilion wax, in order to have thus succeeded in leaving flat plates between the basins, by stopping work at the planes of intersection.

Considering how flexible thin wax is, I do not see that there is any difficulty in the bees, whilst at work on the two sides of a strip of wax, perceiving when they have gnawed the wax away to the proper thinness, and then stopping their work. In ordinary combs it has appeared to me that the bees do not always succeed in working at exactly the same rate from the opposite sides; for I have noticed half-completed rhombs at the base of a just-commenced cell, which were slightly concave on one side, where I suppose that the bees had excavated too quickly, and convex on the opposed side where the bees had worked less quickly. In one well-marked instance, I put the comb back into the hive, and allowed the bees to go on working for a short time, and again examined the cell, and I found that the rhombic plate had been completed, and had become *perfectly flat*: it was absolutely impossible, from the extreme thinness of the little plate, that they could have effected this by gnawing away the convex side; and I suspect that the bees in such cases stand in the opposed cells and push and bend the ductile and warm wax (which as I have tried is easily done) into its proper intermediate plane, and thus flatten it.

From the experiment of the ridge of vermilion wax we can see that, if the bees were to build for themselves a thin wall of wax, they could make their cells of the proper shape, by standing at the proper distance from each other, by excavating at the same rate, and by endeavouring to make equal spherical hollows, but never allowing the spheres to break into each other. Now bees, as may be clearly seen by examining the edge of a growing comb, do make a rough, circumferential wall or rim all round the comb; and they gnaw this away from the opposite sides, always working circularly as they deepen each cell. They do not make the whole three-sided pyramidal base of any one cell at the same time, but only that one rhombic plate which stands on the extreme growing margin, or the two plates, as the case may be; and they never complete the upper edges of the rhombic plates, until the hexagonal walls are commenced. Some of these statements differ from those made by the justly celebrated elder Huber, but I am convinced of their accuracy; and if I had space, I could show that they are conformable with my theory.

Huber's statement, that the very first cell is excavated out of a little parallel-sided wall of wax, is not, as far as I have seen, strictly correct; the first commencement having always been a little hood of wax; but I will not here enter on details.

We see how important a part excavation plays in the construction of the cells; but it would be a great error to suppose that the bees cannot build up a rough wall of wax in the proper position – that is, along the plane of intersection between two adjoining spheres. I have several specimens showing clearly that they can do this. Even in the rude circumferential rim or wall of wax round a growing comb, flexures may sometimes be observed, corresponding in position to the planes of the rhombic basal plates of future cells. But the rough wall of wax has in every case to be finished off, by being largely gnawed away on both sides. The manner in which the bees build is curious; they always make the first rough wall from ten to twenty times thicker than the excessively thin finished wall of the cell, which will ultimately be left. We shall understand how they work, by supposing masons first to pile up a broad ridge of cement, and then to begin cutting it away equally on both sides near the ground, till a smooth, very thin wall is left in the middle; the masons always piling up the cut-away cement, and adding fresh cement on the summit of the ridge. We shall thus have a thin wall steadily growing upward but always crowned by a gigantic coping. From all the cells, both those just commenced and those completed, being thus crowned by a strong coping of wax, the bees can cluster and crawl over the comb without injuring the delicate hexagonal walls. These walls, as Professor Miller has kindly ascertained for me, vary greatly in thickness; being, on an average of twelve measurements made near the border of the comb, 1/352 of an inch in thickness; whereas the basal rhomboidal plates are thicker, nearly in the proportion of three to two, having a mean thickness, from twenty-one measurements, of 1/229 of an inch. By the above singular manner of building, strength is continually given to the comb, with the utmost ultimate economy of wax.

It seems at first to add to the difficulty of understanding how the cells are made, that a multitude of bees all work together; one bee after working a short time at one cell going to another, so that, as Huber has stated, a score of individuals work even at the commencement of the first cell. I was able practically to show this fact, by covering the edges of the hexagonal walls of a single cell, or the extreme margin of the circumferential rim of a growing comb, with an extremely thin layer of melted vermilion wax; and I invariably found that the colour was most delicately diffused by the bees – as delicately as a painter could have done it with his brush – by atoms of the coloured wax having been taken from the spot on which it had been placed, and worked into the growing edges of the cells all round. The work of construction seems to be a sort of balance struck between many bees, all instinctively standing at the same relative distance from each other, all trying to sweep equal spheres, and then building up, or leaving ungnawed, the planes of intersection between these spheres. It was really curious to note in cases of difficulty, as when two pieces of comb met at an angle, how often the bees would pull down and rebuild in different ways the same cell, sometimes recurring to a shape which they had at first rejected.

When bees have a place on which they can stand in their proper positions for working – for instance, on a slip of wood, placed directly under the middle of a comb growing downwards, so that the comb has to be built over one face of the slip – in this case the bees can lay the foundations of one wall of a new hexagon, in its strictly proper place, projecting beyond the other completed cells. It suffices that the bees should be enabled to stand at their proper relative distances from each other and from the walls of the last completed cells, and then, by striking imaginary spheres, they can build up a wall intermediate between two adjoining spheres; but, as far as I have seen, they never gnaw away and finish off the angles of a cell till a large part both of that cell and of the adjoining cells has been built. This capacity in bees of laying down under certain circumstances a rough wall in its proper place between two just-commenced cells, is important, as it bears on a fact, which seems at first subversive of the foregoing theory; namely, that the cells on the extreme margin of wasp-combs are sometimes strictly hexagonal; but I have not space here to enter on this subject. Nor does there seem to me any great difficulty in a single insect (as in the case of a queen-wasp) making hexagonal cells, if she were to work alternately on the inside and outside of two or three cells commenced at the same time, always standing at the proper relative distance from the parts of the cells just begun, sweeping spheres or cylinders, and building up intermediate planes.

As natural selection acts only by the accumulation of slight modifications of structure or instinct, each profitable to the individual under its conditions of life, it may reasonably be asked, how a long and graduated succession of modified architectural instincts, all tending towards the present perfect plan of construction, could have profited the progenitors of the hive bee? I think the answer is not difficult: cells constructed like those of the bee or the wasp gain in strength, and save much in labour and space, and in the materials of which they are constructed. With respect to the formation of wax, it is known that bees are often hard pressed to get sufficient nectar; and I am informed by Mr. Tegetmeier that it has been experimentally proved that from twelve to fifteen pounds of dry sugar are consumed by a hive of bees for the secretion of a pound of wax; so that a prodigious quantity of fluid nectar must be collected and consumed by the bees in a hive for the secretion of the wax necessary for the construction of their combs. Moreover, many bees have to remain idle for many days during the process of secretion. A large store of honey is indispensable to support a large stock of bees during the winter; and the security of the hive is known mainly to depend on a large number of bees being supported. Hence the saving of wax by largely saving honey, and the time consumed in collecting the honey, must be an important element of success to any family of bees. Of course the success of the species may be dependent on the number of its enemies, or parasites, or on quite distinct causes, and so be altogether independent of the quantity of honey which the bees can collect. But let us suppose that this latter circumstance determined, as it probably often has determined, whether a bee allied to our humble-bees could exist in large numbers in any country; and let us further suppose that the community lived through the winter, and consequently required a store of honey: there can in this case be no doubt that it would be an advantage to our imaginary humble-bee if a slight modification of her instincts led her to make her waxen cells near together, so as to intersect a little; for a wall in common even to two adjoining cells would save some little labour and wax. Hence, it would continually be more and more advantageous to our humble-bees, if they were to make their cells more and more regular, nearer together, and aggregated into a mass, like the cells of the Melipona; for in this case a large part of the bounding surface of each cell would serve to bound the adjoining cells, and much labour and wax would be saved. Again, from the same cause, it would be advantageous to the Melipona, if she were to make her cells closer together, and more regular in every way than at present; for then, as we have seen, the spherical surfaces would wholly disappear and be replaced by plane surfaces; and the Melipona would make a comb as perfect as that of the hive bee. Beyond this stage of perfection in architecture, natural selection could not lead; for the comb of the hive bee, as far as we can see, is absolutely perfect in economising labour and wax.

Thus, as I believe, the most wonderful of all known instincts, that of the hive bee, can be explained by natural selection having taken advantage of numerous, successive, slight modifications of simpler instincts; natural selection having, by slow degrees, more and more perfectly led the bees to sweep equal spheres at a given distance from each other in a double layer, and to build up and excavate the wax along the planes of intersection. The bees, of course, no more knowing that they swept their spheres at one particular distance from each other, than they know what are the several angles of the hexagonal prisms and of the basal rhombic plates; the motive power of the process of natural selection having been the construction of cells of due strength and of the proper size and shape for the larvae, this being effected with the greatest possible economy of labour and wax; that individual swarm which thus made the best cells with least labour, and least waste of honey in the secretion of wax, having succeeded best, and having transmitted their newly-acquired economical instincts to new swarms, which in their turn will have had the best chance of succeeding in the struggle for existence.

OBJECTIONS TO THE THEORY OF NATURAL SELECTION AS APPLIED TO INSTINCTS: NEUTER AND STERILE INSECTS

It has been objected to the foregoing view of the origin of instincts that 'the variations of structure and of instinct must have been simultaneous and accurately adjusted to each other, as a modification in the one without an immediate corresponding change in the other would have been fatal.' The force of this objection rests entirely on the assumption that the changes in the instincts and structure are abrupt. To take as an illustration the case of the larger titmouse (Parus major) alluded to in a previous chapter; this bird often holds the seeds of the yew between its feet on a branch, and hammers with its beak till it gets at the kernel. Now what special difficulty would there be in natural selection preserving all the slight individual variations in the shape of the beak, which were better and better adapted to break open the seeds, until a beak was formed, as well constructed for this purpose as that of the nuthatch, at the same time that habit, or compulsion, or spontaneous variations of taste, led the bird to become more and more of a seed-eater? In this case the beak is supposed to be slowly modified by natural selection, subsequently to, but in accordance with, slowly changing habits or taste; but let the feet of the titmouse vary and grow larger from correlation with the beak, or from any other unknown cause, and it is not improbable that such larger feet would lead the bird to climb more and more until it acquired the remarkable climbing instinct and power of the nuthatch. In this case a gradual change of structure is supposed to lead to changed instinctive habits. To take one more case: few instincts are more remarkable than that which leads the swift of the Eastern Islands to make its nest wholly of inspissated saliva. Some birds build their nests of mud, believed to be moistened with saliva; and one of the swifts of North America makes its nest (as I have seen) of sticks agglutinated with saliva, and even with flakes of this substance. Is it then very improbable that the natural selection of individual swifts, which secreted more and more saliva, should at last produce a species with instincts leading it to neglect other materials and to make its nest exclusively of inspissated saliva? And so in other cases. It must, however, be admitted that in many instances we cannot conjecture whether it was instinct or structure which first varied.

No doubt many instincts of very difficult explanation could be opposed to the theory of natural selection – cases, in which we cannot see how an instinct could have originated; cases, in which no intermediate gradations are known to exist; cases of instincts of such trifling importance, that they could hardly have been acted on by natural selection; cases of instincts almost identically the same in animals so remote in the scale of nature that we cannot account for their similarity by inheritance from a common progenitor, and consequently must believe that they were independently acquired through natural selection. I will not here enter on these several cases, but will confine myself to one special difficulty, which at first appeared to me insuperable, and actually fatal to the whole theory.

I allude to the neuters or sterile females in insect communities: for these neuters often differ widely in instinct and in structure from both the males and fertile females, and yet, from being sterile, they cannot propagate their kind.

The subject well deserves to be discussed at great length, but I will here take only a single case, that of working or sterile ants. How the workers have been rendered sterile is a difficulty; but not much greater than that of any other striking modification of structure; for it can be shown that some insects and other articulate animals in a state of nature occasionally become sterile; and if such insects had been social, and it had been profitable to the community that a number should have been annually born capable of work, but incapable of procreation, I can see no especial difficulty in this having been effected through natural selection. But I must pass over this preliminary difficulty. The great difficulty lies in the working ants differing widely from both the males and the fertile females in structure, as in the shape of the thorax, and in being destitute of wings and sometimes of eyes, and in instinct. As far as instinct alone is concerned, the wonderful difference in this respect between the workers and the perfect females would have been better exemplified by the hive bee. If a working ant or other neuter insect had been an ordinary animal, I should have unhesitatingly assumed that all its characters had been slowly acquired through natural selection; namely, by individuals having been born with slight profitable modifications, which were inherited by the offspring, and that these again varied and again were selected, and so onwards. But with the working ant we have an insect differing greatly from its parents, yet absolutely sterile; so that it could never have transmitted successively acquired modifications of structure or instinct to its progeny. It may well be asked how it is possible to reconcile this case with the theory of natural selection?

First, let it be remembered that we have innumerable instances, both in our domestic productions and in those in a state of nature, of all sorts of differences of inherited structure which are correlated with certain ages and with either sex. We have differences correlated not only with one sex, but with that short period when the reproductive system is active, as in the nuptial plumage of many birds, and in the hooked jaws of the male salmon. We have even slight differences in the horns of different breeds of cattle in relation to an artificially imperfect state of the male sex; for oxen of certain breeds have longer horns than the oxen of other breeds, relatively to the length of the horns in both the bulls and cows of these same breeds. Hence, I can see no great difficulty in any character becoming correlated with the sterile condition of certain members of insect communities; the difficulty lies in understanding how such correlated modifications of structure could have been slowly accumulated by natural selection.

This difficulty, though appearing insuperable, is lessened, or, as I believe, disappears, when it is remembered that selection may be applied to the family, as well as to the individual, and may thus gain the desired end. Breeders of cattle wish the flesh and fat to be well marbled together. An animal thus characterized has been slaughtered, but the breeder has gone with confidence to the same stock and has succeeded. Such faith may be placed in the power of selection that a breed of cattle, always yielding oxen with extraordinarily long horns, could, it is probable, be formed by carefully watching which individual bulls and cows, when matched, produced oxen with the longest horns; and yet no one ox would ever have propagated its kind. Here is a better and real illustration: According to M. Verlot, some varieties of the double annual stock, from having been long and carefully selected to the right degree, always produce a large proportion of seedlings bearing double and quite sterile flowers, but they likewise yield some single and fertile plants. These latter, by which alone the variety can be propagated, may be compared with the

Fully mature male Salmon (with hooked jaw), shown with each stage of its development,
from egg and embryo to adult

fertile male and female ants, and the double sterile plants with the neuters of the same community. As with the varieties of the stock, so with social insects, selection has been applied to the family, and not to the individual, for the sake of gaining a serviceable end. Hence, we may conclude that slight modifications of structure or of instinct, correlated with the sterile condition of certain members of the community, have proved advantageous; consequently the fertile males and females have flourished, and transmitted to their fertile offspring a tendency to produce sterile members with the same modifications. This process must have been repeated many times, until that prodigious amount of difference between the fertile and sterile females of the same species has been produced which we see in many social insects.

But we have not as yet touched on the acme of the difficulty; namely, the fact that the neuters of several ants differ, not only from the fertile females and males, but from each other, sometimes to an almost incredible degree, and are thus divided into two or even three castes. The castes, moreover, do not generally graduate into each other, but are perfectly well defined; being as distinct from each other as are any two species of the same genus, or rather as any two genera of the same family. Thus, in Eciton, there are working and soldier neuters, with jaws and instincts extraordinarily different: in Cryptocerus, the workers of one caste alone carry a wonderful sort of shield on their heads, the use of which is quite unknown: in the Mexican Myrmecocystus, the workers of one caste never leave the nest; they are fed by the workers of another caste, and they have an enormously developed abdomen which secretes a sort of honey, supplying the place of that excreted by the aphides, or the domestic cattle as they may be called, which our European ants guard and imprison.

It will indeed be thought that I have an overweening confidence in the principle of natural selection, when I do not admit that such wonderful and well-established facts at once annihilate the theory. In the simpler case of neuter insects all of one caste, which, as I believe, have been rendered different from the fertile males and females through natural selection, we may conclude from the analogy of ordinary variations, that the successive, slight, profitable modifications did not first arise in all the neuters in the same nest, but in some few alone; and that by the survival of the communities with females which produced most neuters having the advantageous modification, all the neuters ultimately came to be thus characterized. According to this view we ought occasionally to find in the same nest neuter-insects, presenting gradations of structure; and this we do find, even not rarely, considering how few neuter-insects out of Europe have been carefully examined. Mr. F. Smith has shown that the neuters of several British ants differ surprisingly from each other in size and sometimes in colour; and that the extreme forms can be linked together by individuals taken out of the same nest: I have myself compared perfect gradations of this kind. It sometimes happens that the larger or the smaller sized workers are the most numerous; or that both large and small are numerous, whilst those of an intermediate size are scanty in numbers. Formica flava has larger and smaller workers, with some few of intermediate size; and, in this species, as Mr. F. Smith has observed, the larger workers have simple eyes (ocelli), which, though small, can be plainly distinguished, whereas the smaller workers have their ocelli rudimentary. Having carefully dissected several specimens of these workers, I can affirm that the eyes are far more rudimentary in the smaller workers than can be accounted for merely by their proportionately lesser size; and I fully believe, though I dare not assert so positively, that the workers of intermediate size have their ocelli in an exactly intermediate condition. So that here we have two bodies of sterile workers in the same nest, differing not only in size, but in their organs of vision, yet connected by some few members in an intermediate condition. I may digress by adding, that if the smaller workers had been the most useful to the community, and those males and females had been continually selected, which produced more and more of the smaller workers, until all the workers were in this condition; we should then have had a species of ant with neuters in nearly the same condition as those of Myrmica. For the workers of Myrmica have not even rudiments of ocelli, though the male and female ants of this genus have well-developed ocelli.

I may give one other case: so confidently did I expect occasionally to find gradations of important structures between the different castes of neuters in the same species, that I gladly availed myself of Mr. F. Smith's offer of numerous specimens from the same nest of the driver ant (Anomma) of West Africa. The reader will perhaps best appreciate the amount of difference in these workers by my giving, not the actual measurements, but a strictly accurate illustration: the difference was the same as if we were to see a set of workmen building a house, of whom many were five feet four inches high, and many sixteen feet high; but we must in addition suppose that the larger workmen had heads four instead of three times as big as those of the smaller men, and jaws nearly five times as big. The jaws, moreover, of the working ants of the several sizes differed wonderfully in shape, and in the form and number of the teeth. But the important fact for us is that, though the workers can be grouped into castes of different sizes, yet they graduate insensibly into each other, as does the widely-different structure of their jaws. I speak confidently on this latter point, as Sir J. Lubbock made drawings for me, with the camera lucida, of the jaws which I dissected from the workers of the several sizes. Mr. Bates, in his interesting 'Naturalist on the Amazons,' has described analogous cases.

With these facts before me, I believe that natural selection, by acting on the fertile ants or parents, could form a species which should regularly produce neuters, all of large size with one form of jaw, or all of small size with widely different jaws; or lastly, and this is the greatest difficulty, one set of workers of one size and structure, and simultaneously another set of workers of a different size and structure; a graduated series having first been formed, as in the case of the driver ant, and then the extreme forms having been produced in greater and greater numbers, through the survival of the parents which generated them, until none with an intermediate structure were produced.

An analogous explanation has been given by Mr. Wallace, of the equally complex case, of certain Malayan butterflies regularly appearing under two or even three distinct female forms; and by Fritz Müller, of certain Brazilian crustaceans likewise appearing under two widely distinct male forms. But this subject need not here be discussed.

I have now explained how, I believe, the wonderful fact of two distinctly defined castes of sterile workers existing in the same nest, both widely different from each other and from their parents, has originated. We can see how useful their production may have been to a social community of ants, on the same principle that the division of labour is useful to civilised man. Ants, however, work by inherited instincts and by inherited organs or tools, whilst man works by acquired knowledge and manufactured instruments. But I must confess, that, with all my faith in natural selection, I should never have anticipated that this principle could have been efficient in so high a degree, had not the case of these neuter insects led me to this conclusion. I have, therefore, discussed this case, at some little but wholly insufficient length, in order to show the power of natural selection, and likewise because this is by far the most serious special difficulty which my theory has encountered. The case, also, is very interesting, as it proves that with animals, as with plants, any amount of modification may be effected by the accumulation of numerous, slight, spontaneous variations, which are in any way profitable, without exercise or habit having been brought into play. For peculiar habits, confined to the workers of sterile females, however long they might be followed, could not possibly affect the males and fertile females, which alone leave descendants. I am surprised that no one has advanced this demonstrative case of neuter insects, against the well-known doctrine of inherited habit, as advanced by Lamarck.

SUMMARY

I have endeavoured in this chapter briefly to show that the mental qualities of our domestic animals vary, and that the variations are inherited. Still more briefly I have attempted to show that instincts vary slightly in a state of nature. No one will dispute that instincts are of the highest importance to each animal. Therefore, there is no real difficulty, under changing conditions of life, in natural selection accumulating to any extent slight modifications of instinct which are in any way useful. In many cases habit or use and disuse have probably come into play. I do not pretend that the facts given in this chapter strengthen in any great degree my theory; but none of the cases of difficulty, to the best of my judgment, annihilate it. On the other hand, the fact that instincts are not always absolutely perfect and are liable to mistakes; – that no instinct can be shown to have been produced for the good of other animals, though animals take advantage of the instincts of others; – that the canon in natural history, of 'Natura non facit saltum,' is applicable to instincts as well as to corporeal structure, and is plainly explicable on the foregoing views, but is otherwise inexplicable, – all tend to corroborate the theory of natural selection.

This theory is also strengthened by some few other facts in regard to instincts; as by that common case of closely allied, but distinct, species, when inhabiting distant parts of the world and living under considerably different conditions of life, yet often retaining nearly the same instincts. For instance, we can understand, on the principle of inheritance, how it is that the thrush of tropical South America lines its nest with mud, in the same peculiar manner as does our British thrush; how it is that the Hornbills of Africa and India have the same extraordinary instinct of plastering up and imprisoning the females in a hole in a tree, with only a small hole left in the plaster through which the males feed them and their young when hatched; how it is that the male wrens (Troglodytes) of North America, build 'cock-nests,' to roost in, like the males of our Kitty-wrens, – a habit wholly unlike that of any other known bird. Finally, it may not be a logical deduction, but to my imagination it is far more satisfactory to look at such instincts as the young cuckoo ejecting its foster-brothers, ants making slaves, the larvae of ichneumonidae feeding within the live bodies of caterpillars, not as specially endowed or created instincts, but as small consequences of one general law leading to the advancement of all organic beings – namely, multiply, vary, let the strongest live and the weakest die.

Chapter IX

🔳 HYBRIDISM

Distinction between the sterility of first crosses and of hybrids–Sterility various in degree, not universal, affected by close interbreeding, removed by domestication–Laws governing the sterility of hybrids– Sterility not a special endowment, but incidental on other differences, not accumulated by natural selection–Causes of the sterility of first crosses and of hybrids–Parallelism between the effects of changed conditions of life and of crossing–Dimorphism and trimorphism–Fertility of varieties when crossed, and of their mongrel offspring not universal–Hybrids and mongrels compared independently of their fertility–Summary.

The view commonly entertained by naturalists is that species, when intercrossed, have been specially endowed with sterility, in order to prevent their confusion. This view certainly seems at first highly probable, for species living together could hardly have been kept distinct had they been capable of freely crossing. The subject is in many ways important for us, more especially as the sterility of species when first crossed, and that of their hybrid offspring, cannot have been acquired, as I shall show, by the preservation of successive profitable degrees of sterility. It is an incidental result of differences in the reproductive systems of the parent-species.

In treating this subject, two classes of facts, to a large extent fundamentally different, have generally been confounded; namely, the sterility of species when first crossed, and the sterility of the hybrids produced from them.

Pure species have of course their organs of reproduction in a perfect condition, yet when intercrossed they produce either few or no offspring. Hybrids, on the other hand, have their reproductive organs functionally impotent, as may be clearly seen in the state of the male element in both plants and animals; though the formative organs themselves are perfect in structure, as far as the microscope reveals. In the first case the two sexual elements which go to form the embryo are perfect; in the second case they are either not at all developed, or are imperfectly developed. This distinction is important, when the cause of the sterility, which is common to the two cases, has to be considered. The distinction probably has been slurred over, owing to the sterility in both cases being looked on as a special endowment, beyond the province of our reasoning powers.

The fertility of varieties, that is of the forms known or believed to be descended from common parents, when crossed, and likewise the fertility of their mongrel offspring, is, with reference to my theory, of equal importance with the sterility of species; for it seems to make a broad and clear distinction between varieties and species.

DEGREES OF STERILITY

First, for the sterility of species when crossed and of their hybrid offspring. It is impossible to study the several memoirs and works of those two conscientious and admirable observers, Kölreuter and Gärtner, who almost devoted their lives to this subject, without being deeply impressed with the high generality of some degree of sterility. Kölreuter makes the rule universal; but then he cuts the knot, for in ten cases in which he found two forms, considered by most authors as distinct species, quite fertile together, he unhesitatingly average number produced by both pure parent-species in a state of nature. But causes of serious error here intervene: a plant, to be hybridised, must be castrated, and, what is often more important, must be secluded in order to prevent pollen being brought to it by insects from other plants. Nearly all the plants experimented on by Gärtner were potted, and were kept in a chamber in his house. That these processes are often injurious to the fertility of a plant cannot be doubted; for Gärtner gives in his table about a score of cases of plants which he castrated, and *artificially fertilised* with their own pollen, and (excluding all cases such as the Leguminosae, in which there is an acknowledged difficulty in the manipulation) half of these twenty plants had their fertility in some degree impaired. Moreover, as Gärtner repeatedly crossed some forms, such as the common red and blue pimpernels (Anagallis arvensis and coerulea), which the best botanists rank as varieties, and found them absolutely sterile, we may doubt whether many species are really so sterile, when intercrossed, as he believed.

It is certain, on the one hand, that the sterility of various species when crossed is so different in degree and graduates away so insensibly, and, on the other hand, that the fertility of pure species is so easily affected by various circumstances, that for all practical purposes it is most difficult to say where perfect fertility ends and sterility begins. I think no better evidence of this can be required than that the two most experienced observers who have ever lived, namely Kölreuter and Gärtner, arrived at diametrically opposite conclusions in regard to some of the very same forms. It is also most instructive to compare – but I have not space here to enter on details – the evidence advanced by our best botanists on the question whether certain doubtful forms should be ranked as species or varieties, with the evidence from fertility adduced by different hybridisers, or by the same observer from experiments made during different years. It can thus be shown that neither sterility nor

Purple Loosestrife, a wetland plant
pollinated by long-tongued insects

ranks them as varieties. Gärtner, also, makes the rule equally universal; and he disputes the entire fertility of Kölreuter's ten cases. But in these and in many other cases, Gärtner is obliged carefully to count the seeds, in order to show that there is any degree of sterility. He always compares the maximum number of seeds produced by two species when first crossed, and the maximum produced by their hybrid offspring, with the fertility affords any certain distinction between species and varieties. The evidence from this source graduates away, and is doubtful in the same degree as is the evidence derived from other constitutional and structural differences.

In regard to the sterility of hybrids in successive generations; though Gärtner was enabled to rear some hybrids, carefully guarding them from a cross with either pure parent, for six or seven, and in one case for ten generations, yet he asserts positively that their fertility never increases, but generally decreases greatly and suddenly. With respect to this decrease, it may first be noticed that when any deviation in structure or constitution is common to both parents, this is often transmitted in an augmented degree to the offspring; and both sexual elements in hybrid plants are already affected in some degree. But I believe that their fertility has been diminished in nearly all these cases by an independent cause, namely, by too close interbreeding. I have made so many experiments and collected so many facts, showing on the one hand that an occasional cross with a distinct individual or variety increases the vigour and fertility of the offspring, and on the other hand that very close interbreeding lessens their vigour and fertility, that I cannot doubt the correctness of this conclusion. Hybrids are seldom raised by experimentalists in great numbers; and as the parent-species, or other allied hybrids, generally grow in the same garden, the visits of insects must be carefully prevented during the flowering season: hence hybrids, if left to themselves, will generally be fertilised during each generation by pollen from the same flower; and this would probably be injurious to their fertility, already lessened by their hybrid origin. I am strengthened in this conviction by a remarkable statement repeatedly made by Gärtner, namely, that if even the less fertile hybrids be artificially fertilised with hybrid pollen of the same kind, their fertility, notwithstanding the frequent ill effects from manipulation, sometimes decidedly increases, and

goes on increasing. Now, in the process of artificial fertilisation, pollen is as often taken by chance (as I know from my own experience) from the anthers of another flower, as from the anthers of the flower itself which is to be fertilised; so that a cross between two flowers, though probably often on the same plant, would be thus effected. Moreover, whenever complicated experiments are in progress, so careful an observer as Gärtner would have castrated his hybrids, and this would have insured in each generation a cross with pollen from a distinct flower, either from the same plant or from another plant of the same hybrid nature. And thus, the strange fact of an increase of fertility in the successive generations of artificially fertilised hybrids, in contrast with those spontaneously self-fertilised, may, as I believe, be accounted for by too close interbreeding having been avoided.

Now let us turn to the results arrived at by a third most experienced hybridiser, namely, the Hon. and Rev. W. Herbert. He is as emphatic in his conclusion that some hybrids are perfectly fertile – as fertile as the pure parent-species – as are Kölreuter and Gärtner that some degree of sterility between distinct species is a universal law of nature. He experimented on some of the very same species as did Gärtner. The difference in their results may, I think, be in part accounted for by Herbert's great horticultural skill, and by his having hot-houses at his command. Of his many important statements I will here give only a single one as an example, namely, that 'every ovule in a pod of Crinum capense fertilised by C. revolutum produced a plant, which I never saw to occur in a case of its natural fecundation.' So that here we have perfect, or even more than commonly perfect fertility, in a first cross between two distinct species.

This case of the Crinum leads me to refer to a singular fact, namely, that individual plants of certain species of Lobelia, Verbascum and Passiflora, can easily be fertilised by the pollen from a distinct species, but not by pollen from the same plant, though this pollen can be proved to be perfectly sound by fertilising other plants or species. In the genus Hippeastrum, in Corydalis as shown by Professor Hildebrand, in various orchids as shown by Mr. Scott and Fritz Müller, all the individuals are in this peculiar condition. So that with some species, certain abnormal individuals, and in other species all the individuals, can actually be hybridised much more readily than they can be fertilised by pollen from the same individual plant! To give one instance, a bulb of Hippeastrum aulicum produced four flowers; three were fertilised by Herbert with their own pollen, and the fourth was subsequently fertilised by the pollen of a compound hybrid descended from three distinct species: the result was that 'the ovaries of the three first flowers soon ceased to grow, and after a few days perished entirely, whereas the pod impregnated by the pollen of the hybrid made vigorous growth and rapid progress to maturity, and bore good seed, which vegetated freely.' Mr. Herbert tried similar experiments during many years, and always with the same result. These cases serve to show on what slight and mysterious causes the lesser or greater fertility of a species sometimes depends.

The practical experiments of horticulturists, though not made with scientific precision, deserve some notice. It is notorious in how complicated a manner the species of Pelargonium, Fuchsia, Calceolaria, Petunia, Rhododendron, etc., have been crossed, yet many of these hybrids seed freely. For instance, Herbert asserts that a hybrid from Calceolaria integrifolia and plantaginea, species most widely dissimilar in general habit, 'reproduces itself as perfectly as if it had been a natural species from the mountains of Chile.' I have taken some pains to ascertain the degree of fertility of some of the complex crosses of Rhododendrons, and I am assured that many of them are perfectly fertile. Mr. C. Noble, for instance, informs me that he raises stocks for grafting from a hybrid between Rhod. ponticum and catawbiense, and that this hybrid 'seeds as

freely as it is possible to imagine.' Had hybrids, when fairly treated, always gone on decreasing in fertility in each successive generation, as Gärtner believed to be the case, the fact would have been notorious to nurserymen. Horticulturists raise large beds of the same hybrid, and such alone are fairly treated, for by insect agency the several individuals are allowed to cross freely with each other, and the injurious influence of close interbreeding is thus prevented. Any one may readily convince himself of the efficiency of insect agency by examining the flowers of the more sterile kinds of hybrid Rhododendrons, which produce no pollen, for he will find on their stigmas plenty of pollen brought from other flowers.

In regard to animals, much fewer experiments have been carefully tried than with plants. If our systematic arrangements can be trusted, that is, if the genera of animals are as distinct from each other as are the genera of plants, then we may infer that animals more widely distinct in the scale of nature can be crossed more easily than in the case of plants; but the hybrids themselves are, I think, more sterile. It should, however, be borne in mind that, owing to few animals breeding freely under confinement, few experiments have been fairly tried: for instance, the canary-bird has been crossed with nine distinct species of finches, but, as not one of these breeds freely in confinement, we have no right to expect that the first crosses between them and the canary, or that their hybrids, should be perfectly fertile. Again, with respect to the fertility in successive generations of the more fertile hybrid animals, I hardly know of an instance in which two families of the same hybrid have been raised at the same time from different parents, so as to avoid the ill effects of close interbreeding. On the contrary, brothers and sisters have usually been crossed in each successive generation, in opposition to the constantly repeated admonition of every breeder. And in this case, it is not at all surprising that the inherent sterility in the hybrids should have gone on increasing.

Although I know of hardly any thoroughly well-authenticated cases of perfectly fertile hybrid animals, I have reason to believe that the hybrids from Cervulus vaginalis and Reevesii, and from Phasianus colchicus with P. torquatus, are perfectly fertile. M. Quatrefages states that the hybrids from two moths (Bombyx cynthia and arrindia) were proved in Paris to be fertile *inter se* for eight generations. It has lately been asserted that two such distinct species as the hare and rabbit, when they can be got to breed together, produce offspring, which are highly fertile when crossed with one of the parent-species. The hybrids from the common and Chinese geese (A. cygnoides), species which are so different that they are generally ranked in distinct genera, have often bred in this country with either pure parent, and in one single instance they have bred *inter se*. This was effected by Mr. Eyton, who raised two hybrids from the same parents, but from different hatches; and from these two birds he raised no less than eight hybrids (grandchildren of the pure geese) from one nest. In India, however, these cross-bred geese must be far more fertile; for I am assured by two eminently capable judges, namely Mr. Blyth and Captain Hutton, that whole flocks of these crossed geese are kept in various parts of the country; and as they are kept for profit, where neither pure parent-species exists, they must certainly be highly or perfectly fertile.

With our domesticated animals, the various races when crossed together are quite fertile; yet in many cases they are descended from two or more wild species. From this fact we must conclude either that the aboriginal parent-species at first produced perfectly fertile hybrids, or that the hybrids subsequently reared under domestication became quite fertile. This latter alternative, which was first propounded by Pallas, seems by far the most probable, and can, indeed, hardly be doubted. It is, for instance, almost certain that our dogs are descended from several wild stocks; yet, with perhaps the exception of certain indigenous domestic dogs of South America, all are quite fertile together; but analogy makes me greatly doubt, whether the several aboriginal species would at first have freely bred together and have produced quite fertile hybrids. So again I have lately acquired decisive evidence that the crossed offspring from the Indian humped and common cattle are *inter se* perfectly fertile; and from the observations by Rütimeyer on their important osteological differences, as well as from those by Mr. Blyth on their differences in habits, voice, constitution, etc., these two forms must be regarded as good and distinct species. The same remarks may be extended to the two chief races of the pig. We must, therefore, either give up the belief of the universal sterility of species when crossed; or we must look at this sterility in animals, not as an indelible characteristic, but as one capable of being removed by domestication.

Finally, considering all the ascertained facts on the intercrossing of plants and animals, it may be concluded that some degree of sterility, both in first crosses and in hybrids, is an extremely general result; but that it cannot, under our present state of knowledge, be considered as absolutely universal.

LAWS GOVERNING THE STERILITY OF FIRST CROSSES AND OF HYBRIDS

We will now consider a little more in detail the laws governing the sterility of first crosses and of hybrids. Our chief object will be to see whether or not these laws indicate that species have been *specially* endowed with this quality, in order to prevent their crossing and blending together in utter confusion. The following conclusions are drawn up chiefly from Gärtner's admirable work on the hybridisation of plants. I have taken much pains to ascertain how far they apply to animals, and, considering how scanty our knowledge is in regard to hybrid animals, I have been surprised to find how generally the same rules apply to both kingdoms.

It has been already remarked, that the degree of fertility, both of first crosses and of hybrids, graduates from zero to perfect fertility. It is surprising in how many curious ways this gradation can be shown; but only the barest outline of the facts can here be given. When pollen from a plant of one family is placed on the stigma of a plant of a distinct family, it exerts no more influence than so much inorganic dust. From this absolute zero of fertility, the pollen of different species applied to the stigma of some one species of the same genus, yields a perfect gradation in the number of seeds produced, up to nearly complete or even quite complete fertility; and, as we have seen, in certain abnormal cases, even to an excess of fertility, beyond that which the plant's own pollen produces. So in hybrids themselves, there are some which never have produced, and probably never would produce, even with the pollen of the pure parents, a single fertile seed: but in some of these cases a first trace of fertility may be detected, by the pollen of one of the pure parent-species causing the flower of the hybrid to wither earlier than it otherwise would have done; and the early withering of the flower is well known to be a sign of incipient fertilisation. From this extreme degree of sterility we have self-fertilised hybrids producing a greater and greater number of seeds up to perfect fertility.

The hybrids raised from two species which are very difficult to cross, and which rarely produce any offspring, are generally very sterile; but the parallelism between the difficulty of making a first cross, and the sterility of the hybrids thus produced – two classes of facts which are generally confounded together – is by no means strict. There are many cases, in which two pure species, as in the genus Verbascum, can be united with unusual facility, and produce numerous hybrid offspring, yet these hybrids are remarkably sterile. On the other hand, there are species which can be crossed very rarely, or with extreme difficulty, but the hybrids, when at last produced, are very fertile. Even within the limits of the same genus, for instance in Dianthus, these two opposite cases occur.

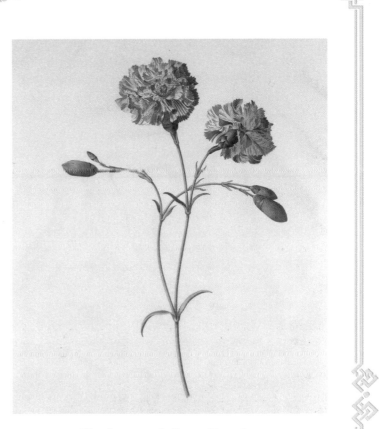

Dianthus carophyllus or Carnation

The fertility, both of first crosses and of hybrids, is more easily affected by unfavourable conditions, than is that of pure species. But the fertility of first crosses is likewise innately variable; for it is not always the same in degree when the same two species are crossed under the same circumstances; it depends in part upon the constitution of the individuals which happen to have been chosen for the experiment. So it is with hybrids, for their degree of fertility is often found to differ greatly in the several individuals raised from seed out of the same capsule and exposed to the same conditions.

By the term systematic affinity is meant, the general resemblance between species in structure and constitution. Now the fertility of first crosses, and of the hybrids produced from them, is largely governed by their systematic affinity. This is clearly shown by hybrids never having been raised between species ranked by systematists in distinct families; and on the other hand, by very closely allied species generally uniting with facility. But the correspondence between systematic affinity and the facility of crossing is by no means strict. A multitude of cases could be given of very closely allied species which will not unite, or only with extreme difficulty; and on the other hand of very distinct species which unite with the utmost facility. In the same family there may be a genus, as Dianthus, in which very many species can most readily be crossed; and another genus, as Silene, in which the most persevering efforts have failed to produce between extremely close species a single hybrid. Even within the limits of the same genus, we meet with this same difference; for instance, the many species of Nicotiana have been more largely crossed than the species of almost any other genus; but Gärtner found that N. acuminata, which is not a particularly distinct species, obstinately failed to fertilise, or to be fertilised, by no less than eight other species of Nicotiana. Many analogous facts could be given.

No one has been able to point out what kind or what amount of difference, in any recognisable character, is sufficient to prevent two species crossing. It can be shown that plants most widely different in habit and general appearance, and having strongly marked differences in every part of the flower, even in the pollen, in the fruit, and in the cotyledons, can be crossed. Annual and perennial plants, deciduous and evergreen trees, plants inhabiting different stations and fitted for extremely different climates, can often be crossed with ease.

By a reciprocal cross between two species, I mean the case, for instance, of a female ass being first crossed by a stallion, and then a mare by a male ass: these two species may then be said to have been reciprocally crossed. There is often the widest possible difference in the facility of making reciprocal crosses. Such cases are highly important, for they prove that the capacity in any two species to cross is often completely independent of their systematic affinity, that is of any difference in their structure or constitution,

excepting in their reproductive systems. The diversity of the result in reciprocal crosses between the same two species was long ago observed by Kölreuter. To give an instance: Mirabilis jalapa can easily be fertilised by the pollen of M. longiflora, and the hybrids thus produced are sufficiently fertile; but Kölreuter tried more than two hundred times, during eight following years, to fertilise reciprocally M. longiflora with the pollen of M. jalapa, and utterly failed. Several other equally striking cases could be given. Thuret has observed the same fact with certain sea-weeds or Fuci. Gärtner, moreover, found that this difference of facility in making reciprocal crosses is extremely common in a lesser degree. He has observed it even between closely related forms (as Matthiola annua and glabra) which many botanists rank only as varieties. It is also a remarkable fact that hybrids raised from reciprocal crosses, though of course compounded of the very same two species, the one species having first been used as the father and then as the mother, though they rarely differ in external characters, yet generally differ in fertility in a small, and occasionally in a high degree.

Several other singular rules could be given from Gärtner: for instance, some species have a remarkable power of crossing with other species; other species of the same genus have a remarkable power of impressing their likeness on their hybrid offspring; but these two powers do not at all necessarily go together. There are certain hybrids which, instead of having, as is usual, an intermediate character between their two parents, always closely resemble one of them; and such hybrids, though externally so like one of their pure parent-species, are with rare exceptions extremely sterile. So again amongst hybrids which are usually intermediate in structure between their parents, exceptional and abnormal individuals sometimes are born, which closely resemble one of their pure parents; and these hybrids are almost always utterly sterile, even when the other hybrids raised from seed from the same capsule have a

Capsicum, from the nightshade family

considerable degree of fertility. These facts show how completely the fertility of a hybrid may be independent of its external resemblance to either pure parent.

Considering the several rules now given, which govern the fertility of first crosses and of hybrids, we see that when forms, which must be considered as good and distinct species, are united, their fertility graduates from zero to perfect fertility, or even to fertility under certain conditions in excess; that their fertility, besides being eminently susceptible to favourable and unfavourable conditions, is innately variable; that it is by no means always the same in degree in the first cross and in the hybrids produced from this cross; that the fertility of hybrids is not related to the degree in which they resemble in external appearance either parent; and lastly, that the facility of making a first cross between any two species is not always governed by their systematic affinity or degree of resemblance to each other. This latter statement is clearly proved by the difference in the result of reciprocal crosses between the same two species, for, according as the one species or the other is used as the father or the mother, there is generally some difference, and occasionally the widest possible difference, in the facility of effecting an union. The hybrids, moreover, produced from reciprocal crosses often differ in fertility.

Now do these complex and singular rules indicate that species have been endowed with sterility simply to prevent their becoming confounded in nature? I think not. For why should the sterility be so extremely different in degree, when various species are crossed, all of which we must suppose it would be equally important to keep from blending together? Why should the degree of sterility be innately variable in the individuals of the same species? Why should some species cross with facility and yet produce very sterile hybrids; and other species cross with extreme difficulty, and yet produce fairly fertile hybrids? Why should there often be so great a

difference in the result of a reciprocal cross between the same two species? Why, it may even be asked, has the production of hybrids been permitted? To grant to species the special power of producing hybrids, and then to stop their further propagation by different degrees of sterility, not strictly related to the facility of the first union between their parents, seems a strange arrangement.

The foregoing rules and facts, on the other hand, appear to me clearly to indicate that the sterility, both of first crosses and of hybrids, is simply incidental or dependent on unknown differences in their reproductive systems; the differences being of so peculiar and limited a nature, that, in reciprocal crosses between the same two species, the male sexual element of the one will often freely act on the female sexual element of the other, but not in a reversed direction. It will be advisable to explain a little more fully, by an example, what I mean by sterility being incidental on other differences, and not a specially endowed quality. As the capacity of one plant to be grafted or budded on another is unimportant for their welfare in a state of nature, I presume that no one will suppose that this capacity is a *specially* endowed quality, but will admit that it is incidental on differences in the laws of growth of the two plants. We can sometimes see the reason why one tree will not take on another from differences in their rate of growth, in the hardness of their wood, in the period of the flow or nature of their sap, etc.; but in a multitude of cases we can assign no reason whatever. Great diversity in the size of two plants, one being woody and the other herbaceous, one being evergreen and the other deciduous, and adaptation to widely different climates, does not always prevent the two grafting together. As in hybridisation, so with grafting, the capacity is limited by systematic affinity, for no one has been able to graft together trees belonging to quite distinct families; and, on the other hand, closely allied species and varieties of the same species, can usually, but not invariably, be grafted with ease. But this capacity, as in hybridisation, is by no means absolutely governed by systematic affinity. Although many distinct genera within the same family have been grafted together, in other cases species of the same genus will not take on each other. The pear can be grafted far more readily on the quince, which is ranked as a distinct genus, than on the apple, which is a member of the same genus. Even different varieties of the pear take with different degrees of facility on the quince; so do different varieties of the apricot and peach on certain varieties of the plum.

As Gärtner found that there was sometimes an innate difference in different *individuals* of the same two species in crossing; so Sagaret believes this to be the case with different individuals of the same two species in being grafted together. As in reciprocal crosses, the facility of effecting an union is often very far from equal, so it sometimes is in grafting. The common gooseberry, for instance, cannot be grafted on the currant, whereas the currant will take, though with difficulty, on the gooseberry.

We have seen that the sterility of hybrids which have their reproductive organs in an imperfect condition, is a different case from the difficulty of uniting two pure species, which have their reproductive organs perfect; yet these two distinct classes of cases run to a large extent parallel. Something analogous occurs in grafting; for Thouin found that three species of Robinia, which seeded freely on their own roots, and which could be grafted with no great difficulty on a fourth species, when thus grafted were rendered barren. On the other hand, certain species of Sorbus, when grafted on other species, yielded twice as much fruit as when on their own roots. We are reminded by this latter fact of the extraordinary cases of Hippeastrum, Passiflora, etc., which seed much more freely when fertilised with the pollen of a distinct species than when fertilised with pollen from the same plant.

We thus see that, although there is a clear and great difference between the mere adhesion of grafted stocks and the union of the male and female elements in the act of reproduction, yet that there is a rude degree of parallelism in the results of grafting and of crossing distinct species. And as we must look at the curious and complex laws governing the facility with which trees can be grafted on each other as incidental on unknown differences in their vegetative systems, so I believe that the still more complex laws governing the facility of first crosses are incidental on unknown differences in their reproductive systems. These differences in both cases follow, to a certain extent, as might have been expected, systematic affinity, by which term every kind of resemblance and dissimilarity between organic beings is attempted to be expressed. The facts by no means seem to indicate that the greater or lesser difficulty of either grafting or crossing various species has been a special endowment; although in the case of crossing, the difficulty is as important for the endurance and stability of specific forms, as in the case of grafting it is unimportant for their welfare.

ORIGIN AND CAUSES OF THE STERILITY OF FIRST CROSSES AND OF HYBRIDS

At one time it appeared to me probable, as it has to others, that the sterility of first crosses and of hybrids might have been slowly acquired through the natural selection of slightly lessened degrees of fertility, which, like any other variation, spontaneously appeared in certain individuals of one variety when crossed with those of another variety. For it would clearly be advantageous to two varieties or incipient species if they could be kept from blending, on the same principle that, when man is selecting at the same time two varieties, it is necessary that he should keep them separate. In the first place, it may be remarked that species inhabiting distinct regions are often sterile when crossed; now it could clearly have been of no advantage to such separated species to have been rendered mutually sterile, and consequently this could not have been effected through natural selection; but it may perhaps be argued, that, if a species was rendered sterile with some one compatriot, sterility with other species would follow as a necessary contingency. In the second place, it is almost as much opposed to the theory of natural selection as to that of special creation, that in reciprocal crosses the male element of one form should have been rendered utterly impotent on a second form, whilst at the same time the male element of this second form is enabled freely to fertilise the first form; for this peculiar state of the reproductive system could hardly have been advantageous to either species.

In considering the probability of natural selection having come into action, in rendering species mutually sterile, the greatest difficulty will be found to lie in the existence of many graduated steps, from slightly lessened fertility to absolute sterility. It may be admitted that it would profit an incipient species, if it were rendered in some slight degree sterile when crossed with its parent form or with some other variety; for thus fewer bastardised and deteriorated offspring would be produced to commingle their blood with the new species in process of formation. But he who will take the trouble to reflect on the steps by which this first degree of sterility could be increased through natural selection to that high degree which is common with so many species, and which is universal with species which have been differentiated to a generic or family rank, will find the subject extraordinarily complex. After mature reflection, it seems to me that this could not have been effected through natural selection. Take the case of any two species which, when crossed, produced few and sterile offspring; now, what is there which could favour the survival of those individuals which happened to be endowed in a slightly higher degree with mutual infertility, and which thus approached by one small step towards absolute sterility? Yet an advance of this kind, if the theory of natural selection be brought to bear, must have incessantly occurred with many species, for a multitude are mutually quite barren. With

sterile neuter insects we have reason to believe that modifications in their structure and fertility have been slowly accumulated by natural selection, from an advantage having been thus indirectly given to the community to which they belonged over other communities of the same species; but an individual animal not belonging to a social community, if rendered slightly sterile when crossed with some other variety, would not thus itself gain any advantage or indirectly give any advantage to the other individuals of the same variety, thus leading to their preservation.

But it would be superfluous to discuss this question in detail: for with plants we have conclusive evidence that the sterility of crossed species must be due to some principle, quite independent of natural selection. Both Gärtner and Kölreuter have proved that in genera including numerous species, a series can be formed from species which when crossed yield fewer and fewer seeds, to species which never produce a single seed, but yet are affected by the pollen of certain other species, for the germen swells. It is here manifestly impossible to select the more sterile individuals, which have already ceased to yield seeds; so that this acme of sterility, when the germen alone is effected, cannot have been gained through selection; and from the laws governing the various grades of sterility being so uniform throughout the animal and vegetable kingdoms, we may infer that the cause, whatever it may be, is the same or nearly the same in all cases.

We will now look a little closer at the probable nature of the differences between species which induce sterility in first crosses and in hybrids. In the case of first crosses, the greater or less difficulty in effecting a union and in obtaining offspring apparently depends on several distinct causes. There must sometimes be a physical impossibility in the male element reaching the ovule, as would be the case with a plant having a pistil too long for the pollen-tubes to reach the ovarium. It has also been observed that when the pollen of one species is placed on the stigma of a distantly allied species, though the pollen-tubes protrude, they do not penetrate the stigmatic surface. Again, the male element may reach the female element, but be incapable of causing an embryo to be developed, as seems to have been the case with some of Thuret's experiments on Fuci. No explanation can be given of these facts, any more than why certain trees cannot be grafted on others. Lastly, an embryo may be developed, and then perish at an early period. This latter alternative has not been sufficiently attended to; but I believe, from observations communicated to me by Mr. Hewitt, who has had great experience in hybridising pheasants and fowls, that the early death of the embryo is a very frequent cause of sterility in first crosses. Mr. Salter has recently given the results of an examination of about 500 eggs produced from various crosses between three species of Gallus and their hybrids; the majority of these eggs had been fertilised; and in the majority of the fertilised eggs, the embryos had either been partially developed and had then perished, or had become nearly mature, but the young chickens had been unable to break through the shell. Of the chickens which were born, more than four-fifths died within the first few days, or at latest weeks, 'without any obvious cause, apparently from mere inability to live;' so that from the 500 eggs only twelve chickens were reared. With plants, hybridized embryos probably often perish in a like manner; at least it is known that hybrids raised from very distinct species are sometimes weak and dwarfed, and perish at an early age; of which fact Max Wichura has recently given some striking cases with hybrid willows. It may be here worth noticing that in some cases of parthenogenesis, the embryos within the eggs of silk moths which had not been fertilised, pass through their early stages of development and then perish like the embryos produced by a cross between distinct species. Until becoming acquainted with these facts, I was unwilling to believe in the frequent early death of hybrid embryos; for hybrids, when once born, are generally healthy and long-lived, as we see in the case of the common mule. Hybrids, however, are differently circumstanced before and after birth: when born and living in a country where their two parents live, they are generally placed under suitable conditions of life. But a hybrid partakes of only half of the nature and constitution of its mother; it may therefore, before birth, as long as it is nourished within its mother's womb, or within the egg or seed produced by the mother, be exposed to conditions in some degree unsuitable, and consequently be liable to perish at an early period; more especially as all very young beings are eminently sensitive to injurious or unnatural conditions of life. But after all, the cause more probably lies in some imperfection in the original act of impregnation, causing the embryo to be imperfectly developed, rather than in the conditions to which it is subsequently exposed.

In regard to the sterility of hybrids, in which the sexual elements are imperfectly developed, the case is somewhat different. I have more than once alluded to a large body of facts showing that, when animals and plants are removed from their natural conditions, they are extremely liable to have their reproductive systems seriously affected. This, in fact, is the great bar to the domestication of animals. Between the sterility thus superinduced and that of hybrids, there are many points of similarity. In both cases the sterility is independent of general health, and is often accompanied by excess of size or great luxuriance. In both cases the sterility occurs in various degrees; in both, the male element is the most liable to be affected; but sometimes the female more than the male. In both, the tendency goes to a certain extent with systematic affinity, for whole groups of animals and plants are rendered impotent by the same unnatural conditions; and whole groups of species tend to produce sterile hybrids. On the other hand, one species in a group will sometimes resist great changes of conditions

Lilium speciosum or Japanese lily, showing reproductive organs (bottom)

with unimpaired fertility; and certain species in a group will produce unusually fertile hybrids. No one can tell till he tries, whether any particular animal will breed under confinement, or any exotic plant seed freely under culture; nor can he tell till he tries, whether any two species of a genus will produce more or less sterile hybrids. Lastly, when organic beings are placed during several generations under conditions not natural to them, they are extremely liable to vary, which seems to be partly due to their reproductive systems having been specially affected, though in a lesser degree than when sterility ensues. So it is with hybrids, for their offspring in successive generations are eminently liable to vary, as every experimentalist has observed.

Thus we see that when organic beings are placed under new and unnatural conditions, and when hybrids are produced by the unnatural crossing of two species, the reproductive system, independently of the general state of health, is affected in a very similar manner. In the one case, the conditions of life have been disturbed, though often in so slight a degree as to be inappreciable by us; in the other case, or that of hybrids, the external conditions have remained the same, but the organisation has been disturbed by two distinct structures and constitutions, including of course the reproductive systems, having been blended into one. For it is scarcely possible that two organisations should be compounded into one, without some disturbance occurring in the development, or periodical action, or mutual relations of the different parts and organs one to another or to the conditions of life. When hybrids are able to breed *inter se*, they transmit to their offspring from generation to generation the same compounded organisation, and hence we need not be surprised that their sterility, though in some degree variable, does not diminish; it is even apt to increase, this being generally the result, as before explained, of too close interbreeding. The above view of the sterility of hybrids being caused by two constitutions being

compounded into one has been strongly maintained by Max Wichura.

It must, however, be owned that we cannot understand, on the above or any other view, several facts with respect to the sterility of hybrids; for instance, the unequal fertility of hybrids produced from reciprocal crosses; or the increased sterility in those hybrids which occasionally and exceptionally resemble closely either pure parent. Nor do I pretend that the foregoing remarks go to the root of the matter: no explanation is offered why an organism, when placed under unnatural conditions, is rendered sterile. All that I have attempted to show is, that in two cases, in some respects allied, sterility is the common result, – in the one case from the conditions of life having been disturbed, in the other case from the organisation having been disturbed by two organisations being compounded into one.

A similar parallelism holds good with an allied yet very different class of facts. It is an old and almost universal belief, founded on a considerable body of evidence, which I have elsewhere given, that slight changes in the conditions of life are beneficial to all living things. We see this acted on by farmers and gardeners in their frequent exchanges of seed, tubers, etc., from one soil or climate to another, and back again. During the convalescence of animals, great benefit is derived from almost any change in their habits of life. Again, both with plants and animals, there is the clearest evidence that a cross between individuals of the same species, which differ to a certain extent, gives vigour and fertility to the offspring; and that close interbreeding continued during several generations between the nearest relations, if these be kept under the same conditions of life, almost always leads to decreased size, weakness, or sterility.

Hence it seems that, on the one hand, slight changes in the conditions of life benefit all organic beings, and on the other hand, that slight crosses, that is, crosses between the males and females of the same species, which have been subjected to

slightly different conditions, or which have slightly varied, give vigour and fertility to the offspring. But, as we have seen, organic beings long habituated to certain uniform conditions under a state of nature, when subjected, as under confinement, to a considerable change in their conditions, very frequently are rendered more or less sterile; and we know that a cross between two forms that have become widely or specifically different, produce hybrids which are almost always in some degree sterile. I am fully persuaded that this double parallelism is by no means an accident or an illusion. He who is able to explain why the elephant, and a multitude of other animals, are incapable of breeding when kept under only partial confinement in their native country, will be able to explain the primary cause of hybrids being so generally sterile. He will at the same time be able to explain how it is that the races of some of our domesticated animals, which have often been subjected to new and not uniform conditions, are quite fertile together, although they are descended from distinct species, which would probably have been sterile if aboriginally crossed. The above two parallel series of facts seem to be connected together by some common but unknown bond, which is essentially related to the principle of life; this principle, according to Mr. Herbert Spencer, being that life depends on, or consists in, the incessant action and reaction of various forces, which, as throughout nature, are always tending towards an equilibrium; and when this tendency is slightly disturbed by any change, the vital forces gain in power.

RECIPROCAL DIMORPHISM AND TRIMORPHISM

This subject may be here briefly discussed, and will be found to throw some light on hybridism. Several plants belonging to distinct orders present two forms, which exist in about equal numbers and which differ in no respect except in their reproductive organs; one form having a long pistil with short stamens, the other a short pistil with long stamens; the two having differently sized pollen-grains. With trimorphic plants there are three forms likewise differing in the lengths of their pistils and stamens, in the size and colour of the pollen-grains, and in some other respects; and as in each of the three forms there are two sets of stamens, the three forms possess altogether six sets of stamens and three kinds of pistils. These organs are so proportioned in length to each other that half the stamens in two of the forms stand on a level with the stigma of the third form. Now I have shown, and the result has been confirmed by other observers, that in order to obtain full fertility with these plants, it is necessary that the stigma of the one form should be fertilised by pollen taken from the stamens of corresponding height in another form. So that with dimorphic species two unions, which may be called legitimate, are fully fertile; and two, which may be called illegitimate, are more or less infertile. With trimorphic species six unions are legitimate, or fully fertile, and twelve are illegitimate, or more or less infertile.

The infertility which may be observed in various dimorphic and trimorphic plants, when they are illegitimately fertilised, that is by pollen taken from stamens not corresponding in height with the pistil, differs much in degree, up to absolute and utter sterility; just in the same manner as occurs in crossing distinct species. As the degree of sterility in the latter case depends in an eminent degree on the conditions of life being more or less favourable, so I have found it with illegitimate unions. It is well known that if pollen of a distinct species be placed on the stigma of a flower, and its own pollen be afterwards, even after a considerable interval of time, placed on the same stigma, its action is so strongly prepotent that it generally annihilates the effect of the foreign pollen; so it is with the pollen of the several forms of the same species, for legitimate pollen is strongly prepotent over illegitimate pollen, when both are placed on the same stigma. I ascertained this by fertilising several flowers, first illegitimately, and twenty-four hours afterwards legitimately, with pollen taken from a peculiarly coloured variety, and all the seedlings were similarly coloured; this shows that the legitimate pollen, though applied twenty-four hours subsequently, had wholly destroyed or prevented the action of the previously applied illegitimate pollen. Again, as in making reciprocal crosses between the same two species, there is occasionally a great difference in the result, so the same thing occurs with trimorphic plants; for instance, the mid-styled form of Lythrum salicaria was illegitimately fertilised with the greatest ease by pollen from the longer stamens of the short-styled form, and yielded many seeds; but the latter form did not yield a single seed when fertilised by the longer stamens of the mid-styled form.

In all these respects, and in others which might be added, the forms of the same undoubted species, when illegitimately united, behave in exactly the same manner as do two distinct species when crossed. This led me carefully to observe during four years many seedlings, raised from several illegitimate unions. The chief result is that these illegitimate plants, as they may be called, are not fully fertile. It is possible to raise from dimorphic species, both long-styled and short-styled illegitimate plants, and from trimorphic plants all three illegitimate forms. These can then be properly united in a legitimate manner. When this is done, there is no apparent reason why they should not yield as many seeds as did their parents when legitimately fertilised. But such is not the case. They are all infertile, in various degrees; some being so utterly and incurably sterile that they did not yield during four seasons a single seed or even seed-capsule. The sterility of these illegitimate plants, when united with each other in a legitimate manner, may be strictly compared with that of hybrids when crossed *inter se*. If, on the other hand, a hybrid is crossed with either pure parent-species, the sterility is usually much lessened: and so it is when an illegitimate

plant is fertilised by a legitimate plant. In the same manner as the sterility of hybrids does not always run parallel with the difficulty of making the first cross between the two parent-species, so that sterility of certain illegitimate plants was unusually great, whilst the sterility of the union from which they were derived was by no means great. With hybrids raised from the same seed-capsule the degree of sterility is innately variable, so it is in a marked manner with illegitimate plants. Lastly, many hybrids are profuse and persistent flowerers, whilst other and more sterile hybrids produce few flowers, and are weak, miserable dwarfs; exactly similar cases occur with the illegitimate offspring of various dimorphic and trimorphic plants.

Altogether there is the closest identity in character and behaviour between illegitimate plants and hybrids. It is hardly an exaggeration to maintain that illegitimate plants are hybrids, produced within the limits of the same species by the improper union of certain forms, whilst ordinary hybrids are produced from an improper union between so-called distinct species. We have also already seen that there is the closest similarity in all respects between first illegitimate unions and first crosses between distinct species. This will perhaps be made more fully apparent by an illustration; we may suppose that a botanist found two well-marked varieties (and such occur) of the long-styled form of the trimorphic Lythrum salicaria, and that he determined to try by crossing whether they were specifically distinct. He would find that they yielded only about one-fifth of the proper number of seed, and that they behaved in all the other above specified respects as if they had been two distinct species. But to make the case sure, he would raise plants from his supposed hybridised seed, and he would find that the seedlings were miserably dwarfed and utterly sterile, and that they behaved in all other respects like ordinary hybrids. He might then maintain that he had actually proved, in accordance with the common view, that his two varieties were as good and as distinct species as any in the world; but he would be completely mistaken.

The facts now given on dimorphic and trimorphic plants are important, because they show us, first, that the physiological test of lessened fertility, both in first crosses and in hybrids, is no safe criterion of specific distinction; secondly, because we may conclude that there is some unknown bond which connects the infertility of illegitimate unions with that of their illegitimate offspring, and we are led to extend the same view to first crosses and hybrids; thirdly, because we find, and this seems to me of especial importance, that two or three forms of the same species may exist and may differ in no respect whatever, either in structure or in constitution, relatively to external conditions, and yet be sterile when united in certain ways. For we must remember that it is the union of the sexual elements of individuals of the same form, for instance, of two long-styled forms, which results in sterility; whilst it is the union of the sexual elements proper to two distinct forms which is fertile. Hence the case appears at first sight exactly the reverse of what occurs, in the ordinary unions of the individuals of the same species and with crosses between distinct species. It is, however, doubtful whether this is really so; but I will not enlarge on this obscure subject.

We may, however, infer as probable from the consideration of dimorphic and trimorphic plants, that the sterility of distinct species when crossed and of their hybrid progeny, depends exclusively on the nature of their sexual elements, and not on any difference in their structure or general constitution. We are also led to this same conclusion by considering reciprocal crosses, in which the male of one species cannot be united, or can be united with great difficulty, with the female of a second species, whilst the converse cross can be effected with perfect facility. That excellent observer, Gärtner, likewise concluded that species when crossed are sterile owing to differences confined to their reproductive systems.

FERTILITY OF VARIETIES WHEN CROSSED, AND OF THEIR MONGREL OFFSPRING, NOT UNIVERSAL

It may be urged as an overwhelming argument that there must be some essential distinction between species and varieties inasmuch as the latter, however much they may differ from each other in external appearance, cross with perfect facility, and yield perfectly fertile offspring. With some exceptions, presently to be given, I fully admit that this is the rule. But the subject is surrounded by difficulties, for, looking to varieties produced under nature, if two forms hitherto reputed to be varieties be found in any degree sterile together, they are at once ranked by most naturalists as species. For instance, the blue and red pimpernel, which are considered by most botanists as varieties, are said by Gärtner to be quite sterile when crossed, and he consequently ranks them as undoubted species. If we thus argue in a circle, the fertility of all varieties produced under nature will assuredly have to be granted.

If we turn to varieties, produced, or supposed to have been produced, under domestication, we are still involved in some doubt. For when it is stated, for instance, that certain South American indigenous domestic dogs do not readily unite with European dogs, the explanation which will occur to everyone, and probably the true one, is that they are descended from aboriginally distinct species. Nevertheless the perfect fertility of so many domestic races, differing widely from each other in appearance, for instance, those of the pigeon, or of the cabbage, is a remarkable fact; more especially when we reflect how many species there are, which, though resembling each other most closely, are utterly sterile when intercrossed. Several considerations, however, render the fertility of domestic varieties less remarkable. In the first place, it may be observed that the amount of external difference between two species is no sure guide to their degree of mutual sterility, so that similar differences in the case of varieties would be no sure guide. It is certain that with species the cause lies exclusively in differences in their sexual constitution. Now the varying conditions to which domesticated animals and cultivated plants have been subjected, have had so little tendency towards modifying the reproductive system in a manner leading to mutual sterility, that we have good grounds for admitting the directly opposite doctrine of Pallas, namely, that such conditions generally eliminate this tendency; so that the domesticated descendants of species, which in their natural state probably would have been in some degree sterile when crossed, become perfectly fertile together. With plants, so far is cultivation from giving a tendency towards sterility between distinct species, that in several well-authenticated cases already alluded to, certain plants have been affected in an opposite manner, for they have become self-impotent, whilst still retaining the capacity of fertilising, and being fertilised by, other species. If the Pallasian doctrine of the elimination of sterility through long-continued domestication be admitted, and it can hardly be rejected, it becomes in the highest degree improbable that similar conditions long-continued should likewise induce this tendency; though in certain cases, with species having a peculiar constitution, sterility might occasionally be thus caused. Thus, as I believe, we can understand why, with domesticated animals, varieties have not been produced which are mutually sterile; and why with plants only a few such cases, immediately to be given, have been observed.

The real difficulty in our present subject is not, as it appears to me, why domestic varieties have not become mutually infertile when crossed, but why this has so generally occurred with natural varieties, as soon as they have been permanently modified in a sufficient degree to take rank as species. We are far from precisely knowing the cause; nor is this surprising, seeing how profoundly ignorant we are in regard to the normal and abnormal action of the reproductive system. But we can see that species, owing to their struggle for existence with numerous competitors, will have been exposed during long periods of time to more uniform conditions, than have domestic varieties; and this may well make a wide difference in the result. For we know how commonly wild animals and plants, when taken from their natural conditions and subjected to captivity, are rendered sterile; and the reproductive functions of organic beings which have always lived under natural conditions would probably in like manner be eminently sensitive to the influence of an unnatural cross. Domesticated productions, on the other hand, which, as shown by the mere fact of their domestication, were not originally highly sensitive to changes in their conditions of life, and which can now generally resist with undiminished fertility repeated changes of conditions, might be expected to produce varieties, which would be little liable to have their reproductive powers injuriously affected by the act of crossing with other varieties which had originated in a like manner.

I have as yet spoken as if the varieties of the same species were invariably fertile when intercrossed. But it is impossible to resist the evidence of the existence of a certain amount of sterility in the few following cases, which I will briefly abstract. The evidence is at least as good as that from which we believe in the sterility of a multitude of species. The evidence is also derived from hostile witnesses, who in all other cases consider fertility and sterility as safe criterions of specific distinction. Gärtner kept, during several years, a dwarf kind of maize with yellow seeds, and a tall variety with red seeds growing near each other in his garden; and although these plants have separated sexes, they never naturally crossed. He then fertilised thirteen flowers of the one kind with pollen of the other; but only a single head produced any seed, and this one head produced only five grains. Manipulation in this case could not have been injurious, as the plants have separated sexes. No one, I believe, has suspected that these varieties of maize are distinct species;

Molène phlomide. Verbascum phlomoides L.

Verbascum phlomoides or orange mullein

and it is important to notice that the hybrid plants thus raised were themselves *perfectly* fertile; so that even Gärtner did not venture to consider the two varieties as specifically distinct.

Girou de Buzareingues crossed three varieties of gourd, which like the maize has separated sexes, and he asserts that their mutual fertilisation is by so much the less easy as their differences are greater. How far these experiments may be trusted, I know not; but the forms experimented on are ranked by Sagaret, who mainly founds his classification by the test of infertility, as varieties, and Naudin has come to the same conclusion.

The following case is far more remarkable, and seems at first incredible; but it is the result of an astonishing number of experiments made during many years on nine species of Verbascum, by so good an observer and so hostile a witness as Gärtner: namely, that the yellow and white varieties when crossed produce less seed than the similarly coloured varieties of the same species. Moreover, he asserts that, when yellow and white varieties of one species are crossed with yellow and white varieties of a *distinct* species, more seed is produced by the crosses between the similarly coloured flowers, than between those which are differently coloured. Mr. Scott also has experimented on the species and varieties of Verbascum; and although unable to confirm Gärtner's results on the crossing of the distinct species, he finds that the dissimilarly coloured varieties of the same species yield fewer seeds, in the proportion of eighty-six to 100, than the similarly coloured varieties. Yet these varieties differ in no respect, except in the colour of their flowers; and one variety can sometimes be raised from the seed of another.

Kölreuter, whose accuracy has been confirmed by every subsequent observer, has proved the remarkable fact that one particular variety of the common tobacco was more fertile than the other varieties, when crossed with a widely distinct species. He experimented on five forms which are commonly reputed to be varieties, and which he tested by the severest trial, namely, by reciprocal crosses, and he found their mongrel offspring perfectly fertile. But one of these five varieties, when used either as the father or mother, and crossed with the Nicotiana glutinosa, always yielded hybrids not so sterile as those which were produced from the four other varieties when crossed with N. glutinosa. Hence the reproductive system of this one variety must have been in some manner and in some degree modified.

From these facts it can no longer be maintained that varieties when crossed are invariably quite fertile. From the great difficulty of ascertaining the infertility of varieties in a state of nature, for a supposed variety, if proved to be infertile in any degree, would almost universally be ranked as a species; from man attending only to external characters in his domestic varieties, and from such varieties not having been exposed for very long periods to uniform conditions of life; from these several considerations we may conclude that fertility does not constitute a fundamental distinction between varieties and species when crossed. The general sterility of crossed species may safely be looked at, not as a special acquirement or endowment, but as incidental on changes of an unknown nature in their sexual elements.

HYBRIDS AND MONGRELS COMPARED, INDEPENDENTLY OF THEIR FERTILITY

Independently of the question of fertility, the offspring of species and of varieties when crossed may be compared in several other respects. Gärtner, whose strong wish it was to draw a distinct line between species and varieties, could find very few, and, as it seems to me, quite unimportant differences between the so-called hybrid offspring of species, and the so-called mongrel offspring of varieties. And, on the other hand, they agree most closely in many important respects.

I shall here discuss this subject with extreme brevity. The most important distinction is, that in

Left: Tobacco plant or Nicotiana *Right:* Lobelia erinus

the first generation mongrels are more variable than hybrids; but Gärtner admits that hybrids from species which have long been cultivated are often variable in the first generation; and I have myself seen striking instances of this fact. Gärtner further admits that hybrids between very closely allied species are more variable than those from very distinct species; and this shows that the difference in the degree of variability graduates away. When mongrels and the more fertile hybrids are propagated for several generations, an extreme amount of variability in the offspring in both cases is notorious; but some few instances of both hybrids and mongrels long retaining a uniform character could be given. The variability, however, in the successive generations of mongrels is, perhaps, greater than in hybrids.

This greater variability in mongrels than in hybrids does not seem at all surprising. For the parents of mongrels are varieties, and mostly domestic varieties (very few experiments having been tried on natural varieties), and this implies that there has been recent variability; which would often continue and would augment that arising from the act of crossing. The slight variability of hybrids in the first generation, in contrast with that in the succeeding generations, is a curious fact and deserves attention. For it bears on the view which I have taken of one of the causes of ordinary variability; namely, that the reproductive system, from being eminently sensitive to changed conditions of life, fails under these circumstances to perform its proper function of producing offspring closely similar in all respects to the parent-form. Now, hybrids in the first generation are descended from species (excluding those long cultivated) which have not had their reproductive systems in any way affected, and they are not variable; but hybrids themselves have their reproductive systems seriously affected, and their descendants are highly variable.

But to return to our comparison of mongrels and hybrids: Gärtner states that mongrels are more liable than hybrids to revert to either parent form; but this, if it be true, is certainly only a difference in degree. Moreover, Gärtner expressly states that the hybrids from long cultivated plants are more subject to reversion than hybrids from species in their natural state; and this probably explains the singular difference in the results arrived at by different observers. Thus Max Wichura doubts whether hybrids ever revert to their parent forms, and he experimented on uncultivated species of willows, whilst Naudin, on the other hand, insists in the strongest terms on the almost universal tendency to reversion in hybrids, and he experimented chiefly on cultivated plants. Gärtner further states that when any two species, although most closely allied to each other, are crossed with a third species, the hybrids are widely different from each other; whereas if two very distinct varieties of one species are crossed with another species, the hybrids do not differ much. But this conclusion, as far as I can make out, is founded on a single experiment; and seems directly opposed to the results of several experiments made by Kölreuter.

Such alone are the unimportant differences which Gärtner is able to point out between hybrid and mongrel plants. On the other hand, the degrees and kinds of resemblance in mongrels and in hybrids to their respective parents, more especially in hybrids produced from nearly related species, follow, according to Gärtner the same laws. When two species are crossed, one has sometimes a prepotent power of impressing its likeness on the hybrid. So I believe it to be with varieties of plants; and with animals, one variety certainly often has this prepotent power over another variety. Hybrid plants produced from a reciprocal cross generally resemble each other closely, and so it is with mongrel plants from a reciprocal cross. Both hybrids and mongrels can be reduced to either pure parent-form, by repeated crosses in successive generations with either parent.

These several remarks are apparently applicable to animals; but the subject is here much complicated, partly owing to the existence of secondary sexual characters; but more especially owing to prepotency in transmitting likeness running more strongly in one sex than in the other, both when one species is crossed with another and when one variety is crossed with another variety. For instance, I think those authors are right who maintain that the ass has a prepotent power over the horse, so that both the mule and the hinny resemble more closely the ass than the horse; but that the prepotency runs more strongly in the male than in the female ass, so that the mule, which is an offspring of the male ass and mare, is more like an ass than is the hinny, which is the offspring of the female ass and stallion.

Much stress has been laid by some authors on the supposed fact, that it is only with mongrels that the offspring are not intermediate in character, but closely resemble one of their parents; but this does sometimes occur with hybrids, yet I grant much less frequently than with mongrels. Looking to the cases which I have collected of cross-bred animals closely resembling one parent, the resemblances seem chiefly confined to characters almost monstrous in their nature, and which have suddenly appeared – such as albinism, melanism, deficiency of tail or horns, or additional fingers and toes; and do not relate to characters which have been slowly acquired through selection. A tendency to sudden reversions to the perfect character of either parent would, also, be much more likely to occur with mongrels, which are descended from varieties often suddenly produced and semi-monstrous in character, than with hybrids, which are descended from species slowly and naturally produced. On the whole, I entirely agree with Dr. Prosper Lucas, who, after arranging an enormous body of facts with respect to animals, comes to the conclusion that the laws of resemblance of the child to its parents are the same, whether the two parents differ little or much from each other, namely, in the union of individuals of the same variety, or of different varieties, or of distinct species.

Independently of the question of fertility and sterility, in all other respects there seems to be a general and close similarity in the offspring of crossed species, and of crossed varieties. If we look at species as having been specially created, and at varieties as having been produced by secondary laws, this similarity would be an astonishing fact. But it harmonises perfectly with the view that there is no essential distinction between species and varieties.

SUMMARY OF CHAPTER

First crosses between forms, sufficiently distinct to be ranked as species, and their hybrids, are very generally, but not universally, sterile. The sterility is of all degrees, and is often so slight that the most careful experimentalists have arrived at diametrically opposite conclusions in ranking forms by this test. The sterility is innately variable in individuals of the same species, and is eminently susceptible to action of favourable and unfavourable conditions. The degree of sterility does not strictly follow systematic affinity, but is governed by several curious and complex laws. It is generally different, and sometimes widely different in reciprocal crosses between the same two species. It is not always equal in degree in a first cross and in the hybrids produced from this cross.

In the same manner as in grafting trees, the capacity in one species or variety to take on another, is incidental on differences, generally of an unknown nature, in their vegetative systems, so in crossing, the greater or less facility of one species to unite with another is incidental on unknown differences in their reproductive systems. There is no more reason to think that species have been specially endowed with various degrees of sterility to prevent their crossing and blending in nature, than to think that trees have been specially endowed with various and somewhat analogous degrees of difficulty in being grafted together in order to prevent their inarching in our forests.

The sterility of first crosses and of their hybrid progeny has not been acquired through natural selection. In the case of first crosses it seems to depend on several circumstances; in some instances in chief part on the early death of the embryo. In the case of hybrids, it apparently depends on their whole organisation having been disturbed by being compounded from two distinct forms; the sterility being closely allied to that which so frequently affects pure species, when exposed to new and unnatural conditions of life. He who will explain these latter cases will be able to explain the sterility of hybrids. This view is strongly supported by a parallelism of another kind: namely, that, firstly, slight changes in the conditions of life add to the vigour and fertility of all organic beings; and secondly, that the crossing of forms, which have been exposed to slightly different conditions of life, or which have varied, favours the size, vigour and fertility of their offspring. The facts given on the sterility of the illegitimate unions of dimorphic and trimorphic plants and of their illegitimate progeny, perhaps render it probable that some unknown bond in all cases connects the degree of fertility of first unions with that of their offspring. The consideration of these facts on dimorphism, as well as of the results of reciprocal crosses, clearly leads to the conclusion that the primary cause of the sterility of crossed species is confined to differences in their sexual elements. But why, in the case of distinct species, the sexual elements should so generally have become more or less modified, leading to their mutual infertility, we do not know; but it seems to stand in some close relation to species having been exposed for long periods of time to nearly uniform conditions of life.

It is not surprising that the difficulty in crossing any two species, and the sterility of their hybrid offspring, should in most cases correspond, even if due to distinct causes: for both depend on the amount of difference between the species which are crossed. Nor is it surprising that the facility of effecting a first cross, and the fertility of the hybrids thus produced, and the capacity of being grafted together – though this latter capacity evidently depends on widely different circumstances – should all run, to a certain extent, parallel with the systematic affinity of the forms subjected to experiment; for systematic affinity includes resemblances of all kinds.

First crosses between forms known to be varieties, or sufficiently alike to be considered as varieties, and their mongrel offspring, are very generally, but not, as is so often stated, invariably fertile. Nor is this almost universal and perfect fertility surprising, when it is remembered how liable we are to argue in a circle with respect to varieties in a state of nature; and when we remember that the greater number of varieties have been produced under domestication by the selection of mere external differences, and that they have not been long exposed to uniform conditions of life. It should also be especially kept in mind, that long-continued domestication tends to eliminate sterility, and is therefore little likely to induce this same quality. Independently of the question of fertility, in all other respects there is the closest general resemblance between hybrids and mongrels, in their variability, in their power of absorbing each other by repeated crosses, and in their inheritance of characters from both parent-forms. Finally, then, although we are as ignorant of the precise cause of the sterility of first crosses and of hybrids as we are why animals and plants removed from their natural conditions become sterile, yet the facts given in this chapter do not seem to me opposed to the belief that species aboriginally existed as varieties.

Chapter X

On the Imperfection of the Geological Record

On the absence of intermediate varieties at the present day–On the nature of extinct intermediate varieties; on their number–On the lapse of time, as inferred from the rate of denudation and of deposition–On the lapse of time as estimated by years–On the poorness of our palaeontological collections–On the intermittence of geological formations–On the denudation of granitic areas–On the absence of intermediate varieties in any one formation–On the sudden appearance of groups of species– On their sudden appearance in the lowest known fossiliferous strata–Antiquity of the habitable earth.

In the sixth chapter I enumerated the chief objections which might be justly urged against the views maintained in this volume. Most of them have now been discussed. One, namely, the distinctness of specific forms, and their not being blended together by innumerable transitional links, is a very obvious difficulty. I assigned reasons why such links do not commonly occur at the present day under the circumstances apparently most favourable for their presence, namely, on an extensive and continuous area with graduated physical conditions. I endeavoured to show, that the life of each species depends in a more important manner on the presence of other already defined organic forms, than on climate, and, therefore, that the really governing conditions of life do not graduate away quite insensibly like heat or moisture. I endeavoured, also, to show that intermediate varieties, from existing in lesser numbers than the forms which they connect, will generally be beaten out and exterminated during the course of further modification and improvement. The main cause, however, of innumerable intermediate links not now occurring everywhere throughout nature, depends on the very process of natural selection, through which new varieties continually take the places of and supplant their parent-forms. But just in proportion as this process of extermination has acted on an enormous scale, so must the number of intermediate varieties, which have formerly existed, be truly enormous. Why then is not every geological formation and every stratum full of such intermediate links? Geology assuredly does not reveal any such finely graduated organic chain; and this, perhaps, is the most obvious and serious objection which can be urged against the theory. The explanation lies, as I believe, in the extreme imperfection of the geological record.

In the first place, it should always be borne in mind what sort of intermediate forms must, on the theory, have formerly existed. I have found it difficult, when looking at any two species, to avoid picturing to myself forms *directly* intermediate between them. But this is a wholly false view; we should always look for forms intermediate between each species and a common but unknown progenitor; and the progenitor will generally have differed in some respects from all its modified descendants. To give a simple illustration: the fantail and pouter pigeons are both descended from the rock-pigeon; if we possessed all the intermediate varieties which have ever existed, we should have an extremely close series between both and the rock-pigeon; but we should have no varieties directly intermediate between the fantail and pouter; none, for instance, combining a tail somewhat expanded with a crop somewhat enlarged, the characteristic features of these two breeds. These two breeds, moreover, have become so much modified, that, if we had no historical or indirect evidence regarding their origin, it would not have been possible to have determined, from a mere comparison of their structure with that of the rock-pigeon, C. livia, whether they had descended from this species or from some other allied form, such as C. oenas.

So, with natural species, if we look to forms very distinct, for instance to the horse and tapir, we have no reason to suppose that links directly intermediate between them ever existed, but between each and an unknown common parent. The common parent will have had in its whole organisation much general resemblance to the tapir and to the horse; but in some points of structure may have differed considerably from both, even perhaps more than they differ from each other. Hence, in all such cases, we should be unable to recognise the parent-form of any two or more species, even if we closely compared the structure of the parent with that of its modified descendants, unless at the same time we had a nearly perfect chain of the intermediate links.

It is just possible, by the theory, that one of two living forms might have descended from the other; for instance, a horse from a tapir; and in this case *direct* intermediate links will have existed between them. But such a case would imply that one form had remained for a very long period unaltered, whilst its descendants had undergone a vast amount of change; and the principle of competition between organism and organism, between child and parent, will render this a very rare event; for in all cases the new and improved forms of life tend to supplant the old and unimproved forms.

By the theory of natural selection all living species have been connected with the parent-species of each genus, by differences not greater than we see between the natural and domestic varieties of the same species at the present day; and these parent-species, now generally extinct,

have in their turn been similarly connected with more ancient forms; and so on backwards, always converging to the common ancestor of each great class. So that the number of intermediate and transitional links, between all living and extinct species, must have been inconceivably great. But assuredly, if this theory be true, such have lived upon the earth.

ON THE LAPSE OF TIME, AS INFERRED FROM THE RATE OF DEPOSITION AND EXTENT OF DENUDATION

Independently of our not finding fossil remains of such infinitely numerous connecting links, it may be objected that time cannot have sufficed for so great an amount of organic change, all changes having been effected slowly. It is hardly possible for me to recall to the reader who is not a practical geologist, the facts leading the mind feebly to comprehend the lapse of time. He who can read Sir Charles Lyell's grand work on the Principles of Geology, which the future historian will recognise as having produced a revolution in natural science, and yet does not admit how vast have been the past periods of time, may at once close this volume. Not that it suffices to study the Principles of Geology, or to read special treatises by different observers on separate formations, and to mark how each author attempts to give an inadequate idea of the duration of each formation, or even of each stratum. We can best gain some idea of past time by knowing the agencies at work; and learning how deeply the surface of the land has been denuded, and how much sediment has been deposited. As Lyell has well remarked, the extent and thickness of our sedimentary formations are the result and the measure of the denudation which the earth's crust has elsewhere undergone. Therefore a man should examine for himself the great piles of superimposed strata, and watch the rivulets bringing down mud, and the waves wearing away the sea-cliffs, in order to comprehend something about the duration of past time, the monuments of which we see all around us.

TAPIRUS DOWI.

The Tapir, found in Guatemala, Nicaragua and Costa Rica

It is good to wander along the coast, when formed of moderately hard rocks, and mark the process of degradation. The tides in most cases reach the cliffs only for a short time twice a day, and the waves eat into them only when they are charged with sand or pebbles; for there is good evidence that pure water effects nothing in wearing away rock. At last the base of the cliff is undermined, huge fragments fall down, and these remaining fixed, have to be worn away atom by atom, until after being reduced in size they can be rolled about by the waves, and then they are more quickly ground into pebbles, sand, or mud. But how often do we see along the bases of retreating cliffs rounded boulders, all thickly clothed by marine productions, showing how little they are abraded and how seldom they are rolled about! Moreover, if we follow for a few miles any line of rocky cliff, which is undergoing degradation, we find that it is only here and there, along a short length or round a promontory, that the cliffs are at the present time suffering. The appearance of the surface and the vegetation show that elsewhere years have elapsed since the waters washed their base.

We have, however, recently learned from the observations of Ramsay, in the van of many excellent observers – of Jukes, Geikie, Croll and others, that subaerial degradation is a much more important agency than coast-action, or the power of the waves. The whole surface of the land is exposed to the chemical action of the air and of the rain-water, with its dissolved carbonic acid, and in colder countries to frost; the disintegrated matter is carried down even gentle slopes during heavy rain, and to a greater extent than might be supposed, especially in arid districts, by the wind; it is then transported by the streams and rivers, which, when rapid deepen their channels, and triturate the fragments. On a rainy day, even in a gently undulating country, we see the effects of subaerial degradation in the muddy rills which flow down every slope. Messrs. Ramsay and Whitaker have shown, and the observation is a most striking one,

that the great lines of escarpment in the Wealden district and those ranging across England, which formerly were looked at as ancient sea-coasts, cannot have been thus formed, for each line is composed of one and the same formation, whilst our sea-cliffs are everywhere formed by the intersection of various formations. This being the case, we are compelled to admit that the escarpments owe their origin in chief part to the rocks of which they are composed having resisted subaerial denudation better than the surrounding surface; this surface consequently has been gradually lowered, with the lines of harder rock left projecting. Nothing impresses the mind with the vast duration of time, according to our ideas of time, more forcibly than the conviction thus gained that subaerial agencies, which apparently have so little power, and which seem to work so slowly, have produced great results.

When thus impressed with the slow rate at which the land is worn away through subaerial and littoral action, it is good, in order to appreciate the past duration of time, to consider, on the one hand, the masses of rock which have been removed over many extensive areas, and on the other hand the thickness of our sedimentary formations. I remember having been much struck when viewing volcanic islands, which have been worn by the waves and pared all round into perpendicular cliffs of one or two thousand feet in height; for the gentle slope of the lava-streams, due to their formerly liquid state, showed at a glance how far the hard, rocky beds had once extended into the open ocean. The same story is told still more plainly by faults, – those great cracks along which the strata have been upheaved on one side, or thrown down on the other, to the height or depth of thousands of feet; for since the crust cracked, and it makes no great difference whether the upheaval was sudden, or, as most geologists now believe, was slow and effected by many starts, the surface of the land has been so completely planed down that no trace of these vast dislocations is externally visible. The Craven fault, for instance, extends for upward of thirty miles, and along this line the vertical displacement of the strata varies from 600 to 3,000 feet. Professor Ramsay has published an account of a downthrow in Anglesea of 2,300 feet; and he informs me that he fully believes that there is one in Merionethshire of 12,000 feet; yet in these cases there is nothing on the surface of the land to show such prodigious movements; the pile of rocks on either side of the crack having been smoothly swept away.

On the other hand, in all parts of the world the piles of sedimentary strata are of wonderful thickness. In the Cordillera, I estimated one mass of conglomerate at ten thousand feet; and although conglomerates have probably been accumulated at a quicker rate than finer sediments, yet from being formed of worn and rounded pebbles, each of which bears the stamp of time, they are good to show how slowly the mass must have been heaped together. Professor Ramsay has given me the maximum thickness, from actual measurement in most cases, of the successive formations in *different* parts of Great Britain; and this is the result:–

	Feet
Palaeozoic strata (not including igneous beds)	57,154
Secondary strata	13,190
Tertiary strata	2,240

– making altogether 72,584 feet; that is, very nearly thirteen and three-quarters British miles. Some of these formations, which are represented in England by thin beds, are thousands of feet in thickness on the Continent. Moreover, between each successive formation we have, in the opinion of most geologists, blank periods of enormous length. So that the lofty pile of sedimentary rocks in Britain gives but an inadequate idea of the time which has elapsed during their accumulation. The consideration of these various facts impresses the mind almost in the same manner as does the vain endeavour to grapple with the idea of eternity.

Nevertheless this impression is partly false. Mr. Croll, in an interesting paper, remarks that we do not err 'in forming too great a conception of the length of geological periods,' but in estimating them by years. When geologists look at large and complicated phenomena, and then at the figures representing several million years, the two produce a totally different effect on the mind, and the figures are at once pronounced too small. In regard to subaerial denudation, Mr. Croll shows, by calculating the known amount of sediment annually brought down by certain rivers, relatively to their areas of drainage, that 1,000 feet of solid rock, as it became gradually disintegrated, would thus be removed from the mean level of the whole area in the course of six million years. This seems an astonishing result, and some considerations lead to the suspicion that it may be too large, but if halved or quartered it is still very surprising. Few of us, however, know what a million really means: Mr. Croll gives the following illustration: Take a narrow strip of paper, eighty-three feet four inches in length, and stretch it along the wall of a large hall; then mark off at one end the tenth of an inch. This tenth of an inch will represent one hundred years, and the entire strip a million years. But let it be borne in mind, in relation to the subject of this work, what a hundred years implies, represented as it is by a measure utterly insignificant in a hall of the above dimensions. Several eminent breeders, during a single lifetime, have so largely modified some of the higher animals, which propagate their kind much more slowly than most of the lower animals, that they have formed what well deserves to be called a new sub-breed. Few men have attended with due care to any one strain for more than half a century, so that a hundred years represents the work of two breeders in succession. It is not to be supposed that species in a state of nature ever change so quickly as domestic animals under the guidance of methodical selection. The comparison would be in every way fairer with the effects which follow from unconscious selection, that is, the preservation of the most useful or beautiful animals, with no intention of modifying the breed; but by this process of unconscious selection, various breeds have been sensibly changed in the course of two or three centuries.

Species, however, probably change much more slowly, and within the same country only a few change at the same time. This slowness follows from all the inhabitants of the same country being already so well adapted to each other, that new places in the polity of nature do not occur until after long intervals, due to the occurrence of physical changes of some kind, or through the immigration of new forms. Moreover, variations or individual differences of the right nature, by which some of the inhabitants might be better fitted to their new places under the altered circumstance, would not always occur at once. Unfortunately we have no means of determining, according to the standard of years, how long a period it takes to modify a species; but to the subject of time we must return.

ON THE POORNESS OF PALAEONTOLOGICAL COLLECTIONS

Now let us turn to our richest museums, and what a paltry display we behold! That our collections are imperfect is admitted by every one. The remark of that admirable palaeontologist, Edward Forbes, should never be forgotten, namely, that very many fossil species are known and named from single and often broken specimens, or from a few specimens collected on some one spot. Only a small portion of the surface of the earth has been geologically explored, and no part with sufficient care, as the important discoveries made every year in Europe prove. No organism wholly soft can be preserved. Shells and bones decay and disappear when left on the bottom of the sea, where sediment is not accumulating. We probably take a quite erroneous view, when we assume that sediment is being deposited over nearly the whole bed of the sea, at a rate sufficiently quick to embed and preserve fossil remains. Throughout an

enormously large proportion of the ocean, the bright blue tint of the water bespeaks its purity. The many cases on record of a formation conformably covered, after an immense interval of time, by another and later formation, without the underlying bed having suffered in the interval any wear and tear, seem explicable only on the view of the bottom of the sea not rarely lying for ages in an unaltered condition. The remains which do become embedded, if in sand or gravel, will, when the beds are upraised, generally be dissolved by the percolation of rain water charged with carbonic acid. Some of the many kinds of animals which live on the beach between high and low water mark seem to be rarely preserved. For instance, the several species of the Chthamalinae (a sub-family of sessile cirripedes) coat the rocks all over the world in infinite numbers: they are all strictly littoral, with the exception of a single Mediterranean species, which inhabits deep water, and this has been found fossil in Sicily, whereas not one other species has hitherto been found in any tertiary formation: yet it is known that the genus Chthamalus existed during the Chalk period. Lastly, many great deposits, requiring a vast length of time for their accumulation, are entirely destitute of organic remains, without our being able to assign any reason: one of the most striking instances is that of the Flysch formation, which consists of shale and sandstone, several thousand, occasionally even six thousand feet in thickness, and extending for at least 300 miles from Vienna to Switzerland; and although this great mass has been most carefully searched, no fossils, except a few vegetable remains, have been found.

With respect to the terrestrial productions which lived during the Secondary and Palaeozoic periods, it is superfluous to state that our evidence is fragmentary in an extreme degree. For instance, until recently not a land shell was known belonging to either of these vast periods, with the exception of one species discovered by Sir C. Lyell and Dr. Dawson in the carboniferous strata of

North America; but now land-shells have been found in the lias. In regard to mammiferous remains, a glance at the historical table published in Lyell's Manual, will bring home the truth, how accidental and rare is their preservation, far better than pages of detail. Nor is their rarity surprising, when we remember how large a proportion of the bones of tertiary mammals have been discovered either in caves or in lacustrine deposits; and that not a cave or true lacustrine bed is known belonging to the age of our secondary or palaeozoic formations.

But the imperfection in the geological record largely results from another and more important cause than any of the foregoing; namely, from the several formations being separated from each other by wide intervals of time. This doctrine has been emphatically admitted by many geologists and palaeontologists, who, like E. Forbes, entirely disbelieve in the change of species. When we see the formations tabulated in written works, or when we follow them in nature, it is difficult to avoid believing that they are closely consecutive. But we know, for instance, from Sir R. Murchison's great work on Russia, what wide gaps there are in that country between the superimposed formations; so it is in North America, and in many other parts of the world. The most skilful geologist, if his attention had been confined exclusively to these large territories, would never have suspected that during the periods which were blank and barren in his own country, great piles of sediment, charged with new and peculiar forms of life, had elsewhere been accumulated. And if, in every separate territory, hardly any idea can be formed of the length of time which has elapsed between the consecutive formations, we may infer that this could nowhere be ascertained. The frequent and great changes in the mineralogical composition of consecutive formations, generally implying great changes in the geography of the surrounding lands, whence the sediment was derived, accord with the belief of vast intervals of time having elapsed between each formation.

We can, I think, see why the geological formations of each region are almost invariably intermittent; that is, have not followed each other in close sequence. Scarcely any fact struck me more when examining many hundred miles of the South American coasts, which have been upraised several hundred feet within the recent period, than the absence of any recent deposits sufficiently extensive to last for even a short geological period. Along the whole west coast, which is inhabited by a peculiar marine fauna, tertiary beds are so poorly developed that no record of several successive and peculiar marine faunas will probably be preserved to a distant age. A little reflection will explain why, along the rising coast of the western side of South America, no extensive formations with recent or tertiary remains can anywhere be found, though the supply of sediment must for ages have been great, from the enormous degradation of the coast rocks and from the muddy streams entering the sea. The explanation, no doubt, is that the littoral and sub-littoral deposits are continually worn away, as soon as they are brought up by the slow and gradual rising of the land within the grinding action of the coast-waves.

We may, I think, conclude that sediment must be accumulated in extremely thick, solid, or extensive masses, in order to withstand the incessant action of the waves, when first upraised and during subsequent oscillations of level, as well as the subsequent subaerial degradation. Such thick and extensive accumulations of sediment may be formed in two ways; either in profound depths of the sea, in which case the bottom will not be inhabited by so many and such varied forms of life as the more shallow seas; and the mass when upraised will give an imperfect record of the organisms which existed in the neighbourhood during the period of its accumulation. Or sediment may be deposited to any thickness and extent over a shallow bottom, if it continue slowly to subside. In this latter case, as long as the rate of subsidence and supply of sediment nearly balance each other, the sea will remain shallow and favourable for many and varied forms, and thus a rich fossiliferous formation, thick enough, when upraised, to resist a large amount of denudation, may be formed.

I am convinced that nearly all our ancient formations, which are throughout the greater part of their thickness *rich in fossils*, have thus been formed during subsidence. Since publishing my views on this subject in 1845, I have watched the progress of geology, and have been surprised to note how author after author, in treating of this or that great formation, has come to the conclusion that it was accumulated during subsidence. I may add, that the only ancient tertiary formation on the west coast of South America, which has been bulky enough to resist such degradation as it has as yet suffered, but which will hardly last to a distant geological age, was deposited during a downward oscillation of level, and thus gained considerable thickness.

All geological facts tell us plainly that each area has undergone numerous slow oscillations of level, and apparently these oscillations have affected wide spaces. Consequently, formations rich in fossils and sufficiently thick and extensive to resist subsequent degradation, will have been formed over wide spaces during periods of subsidence, but only where the supply of sediment was sufficient to keep the sea shallow and to embed and preserve the remains before they had time to decay. On the other hand, as long as the bed of the sea remained stationary, *thick* deposits cannot have been accumulated in the shallow parts, which are the most favourable to life. Still less can this have happened during the alternate periods of elevation; or, to speak more accurately, the beds which were then accumulated will generally have been destroyed by being upraised and brought within the limits of the coast-action.

These remarks apply chiefly to littoral and sublittoral deposits. In the case of an extensive and shallow sea, such as that within a large part of the Malay Archipelago, where the depth varies from thirty or forty to sixty fathoms, a widely extended formation might be formed during a period of elevation, and yet not suffer excessively from denudation during its slow upheaval; but the thickness of the formation could not be great, for owing to the elevatory movement it would be less than the depth in which it was formed; nor would the deposit be much consolidated, nor be capped by overlying formations, so that it would run a good chance of being worn away by atmospheric degradation and by the action of the sea during subsequent oscillations of level. It has, however, been suggested by Mr. Hopkins, that if one part of the area, after rising and before being denuded, subsided, the deposit formed during the rising movement, though not thick, might afterwards become protected by fresh accumulations, and thus be preserved for a long period.

Mr. Hopkins also expresses his belief that sedimentary beds of considerable horizontal extent have rarely been completely destroyed. But all geologists, excepting the few who believe that our present metamorphic schists and plutonic rocks once formed the primordial nucleus of the globe, will admit that these latter rocks have been stripped of their covering to an enormous extent. For it is scarcely possible that such rocks could have been solidified and crystallised whilst uncovered; but if the metamorphic action occurred at profound depths of the ocean, the former protecting mantle of rock may not have been very thick. Admitting then that gneiss, mica-schist, granite, diorite, etc., were once necessarily covered up, how can we account for the naked and extensive areas of such rocks in many parts of the world, except on the belief that they have subsequently been completely denuded of all overlying strata? That such extensive areas do exist cannot be doubted: the granitic region of Parime is described by Humboldt as being at least nineteen times as large as Switzerland. South of the Amazon, Boue colours an area composed of rocks of this nature as equal to that of Spain, France, Italy, part of Germany, and the British

Islands, all conjoined. This region has not been carefully explored, but from the concurrent testimony of travellers, the granitic area is very large: thus Von Eschwege gives a detailed section of these rocks, stretching from Rio de Janeiro for 260 geographical miles inland in a straight line; and I travelled for 150 miles in another direction, and saw nothing but granitic rocks. Numerous specimens, collected along the whole coast, from near Rio de Janeiro to the mouth of the Plata, a distance of 1,100 geographical miles, were examined by me, and they all belonged to this class. Inland, along the whole northern bank of the Plata, I saw, besides modern tertiary beds, only one small patch of slightly metamorphosed rock, which alone could have formed a part of the original capping of the granitic series. Turning to a well-known region, namely, to the United States and Canada, as shown in Professor H.D. Rogers' beautiful map, I have estimated the areas by cutting out and weighing the paper, and I find that the metamorphic (excluding the 'semi-metamorphic') and granite rocks exceed, in the proportion of 19 to 12.5, the whole of the newer Palaeozoic formations. In many regions the metamorphic and granite rocks would be found much more widely extended than they appear to be, if all the sedimentary beds were removed which rest unconformably on them, and which could not have formed part of the original mantle under which they were crystallised. Hence, it is probable that in some parts of the world whole formations have been completely denuded, with not a wreck left behind.

One remark is here worth a passing notice. During periods of elevation the area of the land and of the adjoining shoal parts of the sea will be increased and new stations will often be formed: – all circumstances favourable, as previously explained, for the formation of new varieties and species; but during such periods there will generally be a blank in the geological record. On the other hand, during subsidence, the inhabited area and number of inhabitants will decrease (excepting on the shores of a continent when first broken up into an archipelago), and consequently during subsidence, though there will be much extinction, few new varieties or species will be formed; and it is during these very periods of subsidence that the deposits which are richest in fossils have been accumulated.

ON THE ABSENCE OF NUMEROUS INTERMEDIATE VARIETIES IN ANY SINGLE FORMATION

From these several considerations it cannot be doubted that the geological record, viewed as a whole, is extremely imperfect; but if we confine our attention to any one formation, it becomes much more difficult to understand why we do not therein find closely graduated varieties between the allied species which lived at its commencement and at its close. Several cases are on record of the same species presenting varieties in the upper and lower parts of the same formation; thus, Trautschold gives a number of instances with Ammonites; and Hilgendorf has described a most curious case of ten graduated forms of Planorbis multiformis in the successive beds of a fresh-water formation in Switzerland. Although each formation has indisputably required a vast number of years for its deposition, several reasons can be given why each should not commonly include a graduated series of links between the species which lived at its commencement and close; but I cannot assign due proportional weight to the following considerations.

Although each formation may mark a very long lapse of years, each probably is short compared with the period requisite to change one species into another. I am aware that two palaeontologists, whose opinions are worthy of much deference, namely Bronn and Woodward, have concluded that the average duration of each formation is twice or thrice as long as the average duration of specific forms. But insuperable difficulties, as it seems to me, prevent us from coming to any just conclusion on this head.

Top: assorted fossils *Bottom:* sedimentary strata

formation, the probability is that it only then first immigrated into that area. It is well known, for instance, that several species appear somewhat earlier in the palaeozoic beds of North America than in those of Europe; time having apparently been required for their migration from the American to the European seas. In examining the latest deposits, in various quarters of the world, it has everywhere been noted, that some few still existing species are common in the deposit, but have become extinct in the immediately surrounding sea; or, conversely, that some are now abundant in the neighbouring sea, but are rare or absent in this particular deposit. It is an excellent lesson to reflect on the ascertained amount of migration of the inhabitants of Europe during the glacial epoch, which forms only a part of one whole geological period; and likewise to reflect on the changes of level, on the extreme change of climate, and on the great lapse of time, all included within this same glacial period. Yet it may be doubted whether, in any quarter of the world, sedimentary deposits, *including fossil remains*, have gone on accumulating within the same area during the whole of this period. It is not, for instance, probable that sediment was deposited during the whole of the glacial period near the mouth of the Mississippi, within that limit of depth at which marine animals can best flourish: for we know that great geographical changes occurred in other parts of America during this space of time. When such beds as were deposited in shallow water near the mouth of the Mississippi during some part of the glacial period shall have been upraised, organic remains will probably first appear and disappear at different levels, owing to the migrations of species and to geographical changes. And in the distant future, a geologist, examining these beds, would be tempted to conclude that the average duration of life of the embedded fossils had been less than that of the glacial period, instead of having been really far greater, that is, extending from before the glacial epoch to the present day.

When we see a species first appearing in the middle of any formation, it would be rash in the extreme to infer that it had not elsewhere previously existed. So again, when we find a species disappearing before the last layers have been deposited, it would be equally rash to suppose that it then became extinct. We forget how small the area of Europe is compared with the rest of the world; nor have the several stages of the same formation throughout Europe been correlated with perfect accuracy.

We may safely infer that with marine animals of all kinds there has been a large amount of migration due to climatal and other changes; and when we see a species first appearing in any

In order to get a perfect gradation between two forms in the upper and lower parts of the same formation, the deposit must have gone on continuously accumulating during a long period, sufficient for the slow process of modification; hence, the deposit must be a very thick one; and the species undergoing change must have lived in the same district throughout the whole time. But we have seen that a thick formation, fossiliferous throughout its entire thickness, can accumulate only during a period of subsidence; and to keep the depth approximately the same, which is necessary that the same marine species may live on the same space, the supply of sediment must nearly counterbalance the amount of subsidence. But this same movement of subsidence will tend to submerge the area whence the sediment is derived, and thus diminish the supply, whilst the downward movement continues. In fact, this nearly exact balancing between the supply of sediment and the amount of subsidence is probably a rare contingency; for it has been observed by more than one palaeontologist that very thick deposits are usually barren of organic remains, except near their upper or lower limits.

It would seem that each separate formation, like the whole pile of formations in any country, has generally been intermittent in its accumulation. When we see, as is so often the case, a formation composed of beds of widely different mineralogical composition, we may reasonably suspect that the process of deposition has been more or less interrupted. Nor will the closest inspection of a formation give us any idea of the length of time which its deposition may have consumed. Many instances could be given of beds, only a few feet in thickness, representing formations which are elsewhere thousands of feet in thickness, and which must have required an enormous period for their accumulation; yet no one ignorant of this fact would have even suspected the vast lapse of time represented by the thinner formation. Many cases could be given of the lower beds of a formation having been

upraised, denuded, submerged, and then re-covered by the upper beds of the same formation, – facts, showing what wide, yet easily overlooked, intervals have occurred in its accumulation. In other cases we have the plainest evidence in great fossilised trees, still standing upright as they grew, of many long intervals of time and changes of level during the process of deposition, which would not have been suspected, had not the trees been preserved: thus Sir C. Lyell and Dr. Dawson found carboniferous beds 1,400 feet thick in Nova Scotia, with ancient root-bearing strata, one above the other, at no less than sixty-eight different levels. Hence, when the same species occurs at the bottom, middle, and top of a formation, the probability is that it has not lived on the same spot during the whole period of deposition, but has disappeared and reappeared, perhaps many times, during the same geological period. Consequently if it were to undergo a considerable amount of modification during the deposition of any one geological formation, a section would not include all the fine intermediate gradations which must on our theory have existed, but abrupt, though perhaps slight, changes of form.

It is all-important to remember that naturalists have no golden rule by which to distinguish species and varieties; they grant some little variability to each species, but when they meet with a somewhat greater amount of difference between any two forms, they rank both as species, unless they are enabled to connect them together by the closest intermediate gradations; and this, from the reasons just assigned, we can seldom hope to effect in any one geological section. Supposing B and C to be two species, and a third, A, to be found in an older and underlying bed; even if A were strictly intermediate between B and C, it would simply be ranked as a third and distinct species, unless at the same time it could be closely connected by intermediate varieties with either one or both forms. Nor should it be forgotten, as before explained, that A might be the actual progenitor of

B and C, and yet would not necessarily be strictly intermediate between them in all respects. So that we might obtain the parent-species and its several modified descendants from the lower and upper beds of the same formation, and unless we obtained numerous transitional gradations, we should not recognise their blood-relationship, and should consequently rank them as distinct species.

It is notorious on what excessively slight differences many palaeontologists have founded their species; and they do this the more readily if the specimens come from different sub-stages of the same formation. Some experienced conchologists are now sinking many of the very fine species of D'Orbigny and others into the rank of varieties; and on this view we do find the kind of evidence of change which on the theory we ought to find. Look again at the later tertiary deposits, which include many shells believed by the majority of naturalists to be identical with existing species; but some excellent naturalists, as Agassiz and Pictet, maintain that all these tertiary species are specifically distinct, though the distinction is admitted to be very slight; so that here, unless we believe that these eminent naturalists have been misled by their imaginations, and that these late tertiary species really present no difference whatever from their living representatives, or unless we admit, in opposition to the judgment of most naturalists, that these tertiary species are all truly distinct from the recent, we have evidence of the frequent occurrence of slight modifications of the kind required. If we look to rather wider intervals of time, namely, to distinct but consecutive stages of the same great formation, we find that the embedded fossils, though universally ranked as specifically different, yet are far more closely related to each other than are the species found in more widely separated formations; so that here again we have undoubted evidence of change in the direction required by the theory; but to this latter subject I shall return in the following chapter.

With animals and plants that propagate rapidly and do not wander much, there is reason to suspect, as we have formerly seen, that their varieties are generally at first local; and that such local varieties do not spread widely and supplant their parent-form until they have been modified and perfected in some considerable degree. According to this view, the chance of discovering in a formation in any one country all the early stages of transition between any two forms, is small, for the successive changes are supposed to have been local or confined to some one spot. Most marine animals have a wide range; and we have seen that with plants it is those which have the widest range, that oftenest present varieties, so that, with shells and other marine animals, it is probable that those which had the widest range, far exceeding the limits of the known geological formations in Europe, have oftenest given rise, first to local varieties and ultimately to new species; and this again would greatly lessen the chance of our being able to trace the stages of transition in any one geological formation.

It is a more important consideration, leading to the same result, as lately insisted on by Dr. Falconer, namely, that the period during which each species underwent modification, though long as measured by years, was probably short in comparison with that during which it remained without undergoing any change.

It should not be forgotten, that at the present day, with perfect specimens for examination, two forms can seldom be connected by intermediate varieties, and thus proved to be the same species, until many specimens are collected from many places; and with fossil species this can rarely be done. We shall, perhaps, best perceive the improbability of our being enabled to connect species by numerous, fine, intermediate, fossil links, by asking ourselves whether, for instance, geologists at some future period will be able to prove that our different breeds of cattle, sheep, horses, and dogs are descended from a single stock or from several aboriginal stocks; or, again, whether certain sea shells inhabiting the shores of North America, which are ranked by some conchologists as distinct species from their European representatives, and by other conchologists as only varieties, are really varieties, or are, as it is called, specifically distinct. This could be effected by the future geologist only by his discovering in a fossil state numerous intermediate gradations; and such success is improbable in the highest degree.

It has been asserted over and over again, by writers who believe in the immutability of species, that geology yields no linking forms. This assertion, as we shall see in the next chapter, is certainly erroneous. As Sir J. Lubbock has remarked, 'Every species is a link between other allied forms.' If we take a genus having a score of species, recent and extinct, and destroy four-fifths of them, no one doubts that the remainder will stand much more distinct from each other. If the extreme forms in the genus happen to have been thus destroyed, the genus itself will stand more distinct from other allied genera. What geological research has not revealed, is the former existence of infinitely numerous gradations, as fine as existing varieties, connecting together nearly all existing and extinct species. But this ought not to be expected; yet this has been repeatedly advanced as a most serious objection against my views.

It may be worth while to sum up the foregoing remarks on the causes of the imperfection of the geological record under an imaginary illustration. The Malay Archipelago is about the size of Europe from the North Cape to the Mediterranean, and from Britain to Russia; and therefore equals all the geological formations which have been examined with any accuracy, excepting those of the United States of America. I fully agree with Mr. Godwin-Austen, that the present condition of the Malay Archipelago, with its numerous large islands separated by wide and shallow seas, probably represents the former state of Europe, whilst most of our formations were accumulating. The Malay Archipelago is one of

the richest regions in organic beings; yet if all the species were to be collected which have ever lived there, how imperfectly would they represent the natural history of the world!

But we have every reason to believe that the terrestrial productions of the archipelago would be preserved in an extremely imperfect manner in the formations which we suppose to be there accumulating. Not many of the strictly littoral animals, or of those which lived on naked submarine rocks, would be embedded; and those embedded in gravel or sand would not endure to a distant epoch. Wherever sediment did not accumulate on the bed of the sea, or where it did not accumulate at a sufficient rate to protect organic bodies from decay, no remains could be preserved.

Formations rich in fossils of many kinds, and of thickness sufficient to last to an age as distant in futurity as the secondary formations lie in the past, would generally be formed in the archipelago only during periods of subsidence. These periods of subsidence would be separated from each other by immense intervals of time, during which the area would be either stationary or rising; whilst rising, the fossiliferous formations on the steeper shores would be destroyed, almost as soon as accumulated, by the incessant coast-action, as we now see on the shores of South America. Even throughout the extensive and shallow seas within the archipelago, sedimentary beds could hardly be accumulated of great thickness during the periods of elevation, or become capped and protected by subsequent deposits, so as to have a good chance of enduring to a very distant future. During the periods of subsidence, there would probably be much extinction of life; during the periods of elevation, there would be much variation, but the geological record would then be less perfect.

It may be doubted whether the duration of any one great period of subsidence over the whole or part of the archipelago, together with a contemporaneous accumulation of sediment, would *exceed* the average duration of the same specific forms; and these contingencies are indispensable for the preservation of all the transitional gradations between any two or more species. If such gradations were not all fully preserved, transitional varieties would merely appear as so many new, though closely allied species. It is also probable that each great period of subsidence would be interrupted by oscillations of level, and that slight climatical changes would intervene during such lengthy periods; and in these cases the inhabitants of the archipelago would migrate, and no closely consecutive record of their modifications could be preserved in any one formation.

Very many of the marine inhabitants of the archipelago now range thousands of miles beyond its confines; and analogy plainly leads to the belief that it would be chiefly these far-ranging species, though only some of them, which would oftenest produce new varieties; and the varieties would at first be local or confined to one place, but if possessed of any decided advantage, or when further modified and improved, they would slowly spread and supplant their parent-forms. When such varieties returned to their ancient homes, as they would differ from their former state in a nearly uniform, though perhaps extremely slight degree, and as they would be found embedded in slightly different sub-stages of the same formation, they would, according to the principles followed by many palaeontologists, be ranked as new and distinct species.

If then there be some degree of truth in these remarks, we have no right to expect to find, in our geological formations, an infinite number of those fine transitional forms, which, on our theory, have connected all the past and present species of the same group into one long and branching chain of life. We ought only to look for a few links, and such assuredly we do find – some more distantly, some more closely, related to each other; and these links, let them be ever so close, if found in different stages of the same formation, would, by many palaeontologists, be

ranked as distinct species. But I do not pretend that I should ever have suspected how poor was the record in the best preserved geological sections, had not the absence of innumerable transitional links between the species which lived at the commencement and close of each formation, pressed so hardly on my theory.

ON THE SUDDEN APPEARANCE OF WHOLE GROUPS OF ALLIED SPECIES

The abrupt manner in which whole groups of species suddenly appear in certain formations, has been urged by several palaeontologists – for instance, by Agassiz, Pictet, and Sedgwick – as a fatal objection to the belief in the transmutation of species. If numerous species, belonging to the same genera or families, have really started into life at once, the fact would be fatal to the theory of evolution through natural selection. For the development by this means of a group of forms, all of which are descended from some one progenitor, must have been an extremely slow process; and the progenitors must have lived long before their modified descendants. But we continually overrate the perfection of the geological record, and falsely infer, because certain genera or families have not been found beneath a certain stage, that they did not exist before that stage. In all cases positive palaeontological evidence may be implicitly trusted; negative evidence is worthless, as experience has so often shown. We continually forget how large the world is, compared with the area over which our geological formations have been carefully examined; we forget that groups of species may elsewhere have long existed, and have slowly multiplied, before they invaded the ancient archipelagoes of Europe and the United States. We do not make due allowance for the enormous intervals of time which have elapsed between our consecutive formations, – longer perhaps in many cases than the time required for the accumulation of each formation. These intervals will have given time for the multiplication of species from some one parent-form: and in the succeeding formation, such groups or species will appear as if suddenly created.

I may here recall a remark formerly made, namely, that it might require a long succession of ages to adapt an organism to some new and peculiar line of life, for instance, to fly through the air; and consequently that the transitional forms would often long remain confined to some one region; but that, when this adaptation had once been effected, and a few species had thus acquired a great advantage over other organisms, a comparatively short time would be necessary to produce many divergent forms, which would spread rapidly and widely throughout the world. Professor Pictet, in his excellent Review of this work, in commenting on early transitional forms, and taking birds as an illustration, cannot see how the successive modifications of the anterior limbs of a supposed prototype could possibly have been of any advantage. But look at the penguins of the Southern Ocean; have not these birds their front limbs in this precise intermediate state of 'neither true arms nor true wings?' Yet these birds hold their place victoriously in the battle for life; for they exist in infinite numbers and of many kinds. I do not suppose that we here see the real transitional grades through which the wings of birds have passed; but what special difficulty is there in believing that it might profit the modified descendants of the penguin, first to become enabled to flap along the surface of the sea like the logger-headed duck, and ultimately to rise from its surface and glide through the air?

I will now give a few examples to illustrate the foregoing remarks, and to show how liable we are to error in supposing that whole groups of species have suddenly been produced. Even in so short an interval as that between the first and second editions of Pictet's great work on Palaeontology, published in 1844-46 and in 1853-57, the conclusions on the first appearance and disappearance of several groups of animals have been considerably modified; and a third edition

would require still further changes. I may recall the well-known fact that in geological treatises, published not many years ago, mammals were always spoken of as having abruptly come in at the commencement of the tertiary series. And now one of the richest known accumulations of fossil mammals belongs to the middle of the secondary series; and true mammals have been discovered in the new red sandstone at nearly the commencement of this great series. Cuvier used to urge that no monkey occurred in any tertiary stratum; but now extinct species have been discovered in India, South America and in Europe, as far back as the Miocene stage. Had it not been for the rare accident of the preservation of footsteps in the new red sandstone of the United States, who would have ventured to suppose that no less than at least thirty different bird-like animals, some of gigantic size, existed during that period? Not a fragment of bone has been discovered in these beds. Not long ago, palaeontologists maintained that the whole class of birds came suddenly into existence during the Eocene period; but now we know, on the authority of Professor Owen, that a bird certainly lived during the deposition of the upper greensand; and still more recently, that strange bird, the Archeopteryx, with a long lizard-like tail, bearing a pair of feathers on each joint, and with its wings furnished with two free claws, has been discovered in the oolitic slates of Solenhofen. Hardly any recent discovery shows more forcibly than this how little we as yet know of the former inhabitants of the world.

I may give another instance, which, from having passed under my own eyes has much struck me. In a memoir on Fossil Sessile Cirripedes, I stated that, from the large number of existing and extinct tertiary species; from the extraordinary abundance of the individuals of many species all over the world, from the Arctic regions to the equator, inhabiting various zones of depths, from the upper tidal limits to fifty fathoms; from the perfect manner in which specimens are preserved in the oldest tertiary beds; from the case with which even a fragment of a valve can be recognised; from all these circumstances, I inferred that, had sessile cirripedes existed during the secondary periods, they would certainly have been preserved and discovered; and as not one species had then been discovered in beds of this age, I concluded that this great group had been suddenly developed at the commencement of the tertiary series. This was a sore trouble to me, adding, as I then thought, one more instance of the abrupt appearance of a great group of species. But my work had hardly been published, when a skilful palaeontologist, M. Bosquet, sent me a drawing of a perfect specimen of an unmistakable sessile cirripede, which he had himself extracted from the chalk of Belgium. And, as if to make the case as striking as possible, this cirripede was a Chthamalus, a very common, large, and ubiquitous genus, of which not one species has as yet been found even in any tertiary stratum. Still more recently, a Pyrgoma, a member of a distinct sub-family of sessile cirripedes, has been discovered by Mr. Woodward in the upper chalk; so that we now have abundant evidence of the existence of this group of animals during the secondary period.

The case most frequently insisted on by palaeontologists of the apparently sudden appearance of a whole group of species, is that of the teleostean fishes, low down, according to Agassiz, in the Chalk period. This group includes the large majority of existing species. But certain Jurassic and Triassic forms are now commonly admitted to be teleostean; and even some palaeozoic forms have thus been classed by one high authority. If the teleosteans had really appeared suddenly in the northern hemisphere at the commencement of the chalk formation, the fact would have been highly remarkable; but it would not have formed an insuperable difficulty, unless it could likewise have been shown that at the same period the species were suddenly and simultaneously developed in other quarters of the world. It is almost superfluous to remark that

hardly any fossil-fish are known from south of the equator; and by running through Pictet's Palaeontology it will be seen that very few species are known from several formations in Europe. Some few families of fish now have a confined range; the teleostean fishes might formerly have had a similarly confined range, and after having been largely developed in some one sea, have spread widely. Nor have we any right to suppose that the seas of the world have always been so freely open from south to north as they are at present. Even at this day, if the Malay Archipelago were converted into land, the tropical parts of the Indian Ocean would form a large and perfectly enclosed basin, in which any great group of marine animals might be multiplied; and here they would remain confined, until some of the species became adapted to a cooler climate, and were enabled to double the southern capes of Africa or Australia, and thus reach other and distant seas.

From these considerations, from our ignorance of the geology of other countries beyond the confines of Europe and the United States, and from the revolution in our palaeontological knowledge effected by the discoveries of the last dozen years, it seems to me to be about as rash to dogmatize on the succession of organic forms throughout the world, as it would be for a naturalist to land for five minutes on a barren point in Australia, and then to discuss the number and range of its productions.

ON THE SUDDEN APPEARANCE OF GROUPS OF ALLIED SPECIES IN THE LOWEST KNOWN FOSSILIFEROUS STRATA

There is another and allied difficulty, which is much more serious. I allude to the manner in which species belonging to several of the main divisions of the animal kingdom suddenly appear in the lowest known fossiliferous rocks. Most of the arguments which have convinced me that all the existing species of the same group are descended from a single progenitor, apply with equal force to the earliest known species. For instance, it cannot be doubted that all the Cambrian and Silurian trilobites are descended from some one crustacean, which must have lived long before the Cambrian age, and which probably differed greatly from any known animal. Some of the most ancient animals, as the Nautilus, Lingula, etc., do not differ much from living species; and it cannot on our theory be supposed, that these old species were the progenitors of all the species belonging to the same groups which have subsequently appeared, for they are not in any degree intermediate in character.

Consequently, if the theory be true, it is indisputable that

The Pearly Nautilus, with the left half of its shell removed

before the lowest Cambrian stratum was deposited long periods elapsed, as long as, or probably far longer than, the whole interval from the Cambrian age to the present day; and that during these vast periods the world swarmed with living creatures. Here we encounter a formidable objection; for it seems doubtful whether the earth, in a fit state for the habitation of living creatures, has lasted long enough. Sir W. Thompson concludes that the consolidation of the crust can hardly have occurred less than twenty or more than four hundred million years ago, but probably not less than ninety-eight or more than two hundred million years. These very wide limits show how doubtful the data are; and other elements may have hereafter to be introduced into the problem. Mr. Croll estimates that about sixty

million years have elapsed since the Cambrian period, but this, judging from the small amount of organic change since the commencement of the Glacial epoch, appears a very short time for the many and great mutations of life, which have certainly occurred since the Cambrian formation; and the previous one hundred and forty million years can hardly be considered as sufficient for the development of the varied forms of life which already existed during the Cambrian period. It is, however, probable, as Sir William Thompson insists, that the world at a very early period was subjected to more rapid and violent changes in its physical conditions than those now occurring; and such changes would have tended to induce changes at a corresponding rate in the organisms which then existed.

To the question why we do not find rich fossiliferous deposits belonging to these assumed earliest periods prior to the Cambrian system, I can give no satisfactory answer. Several eminent geologists, with Sir R. Murchison at their head, were until recently convinced that we beheld in the organic remains of the lowest Silurian stratum the first dawn of life. Other highly competent judges, as Lyell and E. Forbes, have disputed this conclusion. We should not forget that only a small portion of the world is known with accuracy. Not very long ago M. Barrande added another and lower stage, abounding with new and peculiar species, beneath the then known Silurian system; and now, still lower down in the Lower Cambrian formation, Mr Hicks has found South Wales beds rich in trilobites, and containing various molluscs and annelids. The presence of phosphatic nodules and bituminous matter, even in some of the lowest azoic rocks, probably indicates life at these periods; and the existence of the Eozoon in the Laurentian formation of Canada is generally admitted. There are three great series of strata beneath the Silurian system in Canada, in the lowest of which the Eozoon is found. Sir W. Logan states that their 'united thickness may possibly far surpass that of all the succeeding rocks, from the base of the palaeozoic series to the present time. We are thus carried back to a period so remote, that the appearance of the so-called primordial fauna (of Barrande) may by some be considered as a comparatively modern event.' The Eozoon belongs to the most lowly organised of all classes of animals, but is highly organised for its class; it existed in countless numbers, and, as Dr. Dawson has remarked, certainly preyed on other minute organic beings, which must have lived in great numbers. Thus the words, which I wrote in 1859, about the existence of living beings long

before the Cambrian period, and which are almost the same with those since used by Sir W. Logan, have proved true. Nevertheless, the difficulty of assigning any good reason for the absence of vast piles of strata rich in fossils beneath the Cambrian system is very great. It does not seem probable that the most ancient beds have been quite worn away by denudation, or that their fossils have been wholly obliterated by metamorphic action, for if this had been the case we should have found only small remnants of

Portion of a branching colony of Annelids

the formations next succeeding them in age, and these would always have existed in a partially metamorphosed condition. But the descriptions which we possess of the Silurian deposits over immense territories in Russia and in North America, do not support the view that the older a formation is the more invariably it has suffered extreme denudation and metamorphism.

The case at present must remain inexplicable; and may be truly urged as a valid argument against the views here entertained. To show that it may hereafter receive some explanation, I will give the following hypothesis. From the nature of the organic remains which do not appear to have inhabited profound depths, in the several formations of Europe and of the United States; and from the amount of sediment, miles in thickness, of which the formations are composed, we may infer that from first to last large islands or tracts of land, whence the sediment was derived, occurred in the neighbourhood of the now existing continents of Europe and North America. This same view has since been maintained by Agassiz and others. But we do not know what was the state of things in the intervals between the several successive formations; whether Europe and the United States during these intervals existed as dry land, or as a submarine surface near land, on which sediment was not deposited, or as the bed of an open and unfathomable sea.

Looking to the existing oceans, which are thrice as extensive as the land, we see them studded with many islands; but hardly one truly oceanic island (with the exception of New Zealand, if this can be called a truly oceanic island) is as yet known to afford even a remnant of any palaeozoic or secondary formation. Hence, we may perhaps infer, that during the palaeozoic and secondary periods, neither continents nor continental islands existed where our oceans now extend; for had they existed, palaeozoic and secondary formations would in all probability have been accumulated from sediment derived from their wear and tear; and would have been at least

partially upheaved by the oscillations of level, which must have intervened during these enormously long periods. If, then, we may infer anything from these facts, we may infer that, where our oceans now extend, oceans have extended from the remotest period of which we have any record; and on the other hand, that where continents now exist, large tracts of land have existed, subjected, no doubt, to great oscillations of level, since the Cambrian period. The coloured map appended to my volume on Coral Reefs, led me to conclude that the great oceans are still mainly areas of subsidence, the great archipelagoes still areas of oscillations of level, and the continents areas of elevation. But we have no reason to assume that things have thus remained from the beginning of the world. Our continents seem to have been formed by a preponderance, during many oscillations of level, of the force of elevation. But may not the areas of preponderant movement have changed in the lapse of ages? At a period long antecedent to the Cambrian epoch, continents may have existed where oceans are now spread out, and clear and open oceans may have existed where our continents now stand. Nor should we be justified in assuming that if, for instance, the bed of the Pacific Ocean were now converted into a continent we should there find sedimentary formations, in recognisable condition, older than the Cambrian strata, supposing such to have been formerly deposited; for it might well happen that strata which had subsided some miles nearer to the centre of the earth, and which had been pressed on by an enormous weight of superincumbent water, might have undergone far more metamorphic action than strata which have always remained nearer to the surface. The immense areas in some parts of the world, for instance in South America, of naked metamorphic rocks, which must have been heated under great pressure, have always seemed to me to require some special explanation; and we may perhaps believe that we see in these large areas the many formations long anterior to the Cambrian epoch in a completely metamorphosed and denuded condition.

The several difficulties here discussed, namely, that, though we find in our geological formations many links between the species which now exist and which formerly existed, we do not find infinitely numerous fine transitional forms closely joining them all together. The sudden manner in which several groups of species first appear in our European formations, the almost entire absence, as at present known, of formations rich in fossils beneath the Cambrian strata, are all undoubtedly of the most serious nature. We see this in the fact that the most eminent palaeontologists, namely, Cuvier, Agassiz, Barrande, Pictet, Falconer, E. Forbes, etc., and all our greatest geologists, as Lyell, Murchison, Sedgwick, etc., have unanimously, often vehemently, maintained the immutability of species. But Sir Charles Lyell now gives the

support of his high authority to the opposite side, and most geologists and palaeontologists are much shaken in their former belief. Those who believe that the geological record is in any degree perfect, will undoubtedly at once reject the theory. For my part, following out Lyell's metaphor, I look at the geological record as a history of the world imperfectly kept and written in a changing dialect. Of this history we possess the last volume alone, relating only to two or three countries. Of this volume, only here and there a short chapter has been preserved, and of each page, only here and there a few lines. Each word of the slowly-changing language, more or less different in the successive chapters, may represent the forms of life, which are entombed in our consecutive formations, and which falsely appear to have been abruptly introduced. On this view the difficulties above discussed are greatly diminished or even disappear.

Chapter XI

ON THE GEOLOGICAL SUCCESSION OF ORGANIC BEINGS

On the slow and successive appearance of new species—On their different rates of change—Species once lost do not reappear—Groups of species follow the same general rules in their appearance and disappearance as do single species—On extinction—On simultaneous changes in the forms of life throughout the world—On the affinities of extinct species to each other and to living species—On the state of development of ancient forms—On the succession of the same types within the same areas—Summary of preceding and present chapters.

Let us now see whether the several facts and laws relating to the geological succession of organic beings accord best with the common view of the immutability of species, or with that of their slow and gradual modification, through variation and natural selection.

New species have appeared very slowly, one after another, both on the land and in the waters. Lyell has shown that it is hardly possible to resist the evidence on this head in the case of the several tertiary stages; and every year tends to fill up the blanks between the stages, and to make the proportion between the lost and existing forms more gradual. In some of the most recent beds, though undoubtedly of high antiquity if measured by years, only one or two species are extinct, and only one or two are new, having appeared there for the first time, either locally, or, as far as we know, on the face of the earth. The secondary formations are more broken; but, as Bronn has remarked, neither the appearance nor disappearance of the many species embedded in each formation has been simultaneous.

Species belonging to different genera and classes have not changed at the same rate, or in the same degree. In the older tertiary beds a few living shells may still be found in the midst of a multitude of extinct forms. Falconer has given a striking instance of a similar fact, for an existing crocodile is associated with many lost mammals and reptiles in the sub-Himalayan deposits. The Silurian Lingula differs but little from the living species of this genus; whereas most of the other Silurian molluscs and all the crustaceans have changed greatly. The productions of the land seem to have changed at a quicker rate than those of the sea, of which a striking instance has been observed in Switzerland. There is some reason to believe that organisms high in the scale, change more quickly than those that are low: though there are exceptions to this rule. The amount of organic change, as Pictet has remarked, is not the same in each successive so-called formation. Yet if we compare any but the most closely related formations, all the species will be found to have undergone some change. When a species has once disappeared from the face of the earth, we have no reason to believe that the same identical form ever reappears. The strongest apparent exception to this latter rule is that of the so-called 'colonies' of M. Barrande, which intrude for a period in the midst of an older formation, and then allow the pre-existing fauna to reappear; but Lyell's explanation, namely, that it is a case of temporary migration from a distinct geographical province, seems satisfactory.

These several facts accord well with our theory, which includes no fixed law of development, causing all the inhabitants of an area to change abruptly, or simultaneously, or to an equal degree. The process of modification must be slow, and will generally affect only a few species at the same time; for the variability of each species is independent of that of all others. Whether such variations or individual differences as may arise will be accumulated through natural selection in a greater or less degree, thus causing a greater or less amount of permanent modification, will depend on many complex contingencies – on the variations being of a beneficial nature, on the freedom of intercrossing, on the slowly changing physical conditions of the country, on the immigration of new colonists, and on the nature of the other inhabitants with which the varying species come into competition. Hence it is by no means surprising that one species should retain the same identical form much longer than others; or, if changing, should change in a less degree. We find similar relations between the existing inhabitants of distinct countries; for instance, the land shells and coleopterous insects of Madeira have come to differ considerably from their nearest allies on the continent of Europe, whereas the marine shells and birds have remained unaltered. We can perhaps understand the apparently quicker rate of change in terrestrial and in more highly organised productions compared with marine and lower productions, by the more complex relations of the higher beings to their organic and inorganic conditions of life, as explained in a former chapter. When many of the inhabitants of any area have become modified and improved, we can understand, on the principle of competition, and from the all-important relations of organism to organism in the struggle for life, that any form which did not become in some degree modified and improved, would be liable to extermination. Hence, we see why all the species in the same region do at last, if we look to long enough intervals of time, become modified; for otherwise they would become extinct.

In members of the same class the average amount of change, during long and equal periods of time, may, perhaps, be nearly the same; but as the accumulation of enduring formations, rich in fossils, depends on great masses of sediment being deposited on subsiding areas, our formations have been almost necessarily accumulated at wide and irregularly intermittent intervals of time, consequently the amount of organic change exhibited by the fossils embedded in consecutive formations is not equal. Each formation, on this view, does not mark a new and complete act of creation, but only an occasional scene, taken almost at hazard, in an ever slowly changing drama.

We can clearly understand why a species when once lost should never reappear, even if the very same conditions of life, organic and inorganic, should recur. For though the offspring of one species might be adapted (and no doubt this has occurred in innumerable instances) to fill the place of another species in the economy of nature, and thus supplant it; yet the two forms – the old and the new – would not be identically the same; for both would almost certainly inherit different characters from their distinct progenitors; and organisms already differing would vary in a different manner. For instance, it is possible, if all our fantail pigeons were destroyed, that fanciers might make a new breed hardly distinguishable from the present breed; but if the parent rock-pigeon were likewise destroyed, and under nature we have every reason to believe that parent-forms are generally supplanted and exterminated by their improved offspring, it is incredible that a fantail, identical with the existing breed, could be raised from any other species of pigeon, or even from any other well established race of the domestic pigeon, for the successive variations would almost certainly be in some degree different, and the newly-formed variety would probably inherit from its progenitor some characteristic differences.

Groups of species, that is, genera and families, follow the same general rules in their appearance and disappearance as do single species, changing more or less quickly, and in a greater or lesser degree. A group, when it has once disappeared, never reappears; that is, its existence, as long as it lasts, is continuous. I am aware that there are some apparent exceptions to this rule, but the exceptions are surprisingly few, so few that E. Forbes, Pictet, and Woodward (though all strongly opposed to such views as I maintain) admit its truth; and the rule strictly accords with the theory. For all the species of the same group, however long it may have lasted, are the modified descendants one from the other, and all from a common progenitor. In the genus Lingula, for instance, the species which have successively appeared at all ages must have been connected by an unbroken series of generations, from the lowest Silurian stratum to the present day.

We have seen in the last chapter that whole groups of species sometimes falsely appear to have been abruptly developed; and I have attempted to give an explanation of this fact, which if true would be fatal to my views. But such cases are certainly exceptional; the general rule being a gradual increase in number, until the group reaches its maximum, and then, sooner or later, a gradual decrease. If the number of the species included within a genus, or the number of the genera within a family, be represented by a vertical line of varying thickness, ascending through the successive geological formations, in which the species are found, the line will sometimes falsely appear to begin at its lower end, not in a sharp point, but abruptly; it then gradually thickens upwards, often keeping of equal thickness for a space, and ultimately thins out in the upper beds, marking the decrease and final extinction of the species. This gradual increase in number of the species of a group is strictly conformable with the theory; for the species of the same genus, and the genera of the same family, can increase only slowly and progressively; the process of modification and the production of a number of allied forms

necessarily being a slow and gradual process, one species first giving rise to two or three varieties, these being slowly converted into species, which in their turn produce by equally slow steps other varieties and species, and so on, like the branching of a great tree from a single stem, till the group becomes large.

ON EXTINCTION

We have as yet only spoken incidentally of the disappearance of species and of groups of species. On the theory of natural selection, the extinction of old forms and the production of new and improved forms are intimately connected together. The old notion of all the inhabitants of the earth having been swept away by catastrophes at successive periods is very generally given up, even by those geologists, as Elie de Beaumont, Murchison, Barrande, etc., whose general views would naturally lead them to this conclusion. On the contrary, we have every reason to believe, from the study of the tertiary formations, that species and groups of species gradually disappear, one after another, first from one spot, then from another, and finally from the world. In some few cases, however, as by the breaking of an isthmus and the consequent irruption of a multitude of new inhabitants into an adjoining sea, or by the final subsidence of an island, the process of extinction may have been rapid. Both single species and whole groups of species last for very unequal periods; some groups, as we have seen, have endured from the earliest known dawn of life to the present day; some have disappeared before the close of the palaeozoic period. No fixed law seems to determine the length of time during which any single species or any single genus endures. There is reason to believe that the extinction of a whole group of species is generally a slower process than their production: if their appearance and disappearance be represented, as before, by a vertical line of varying thickness the line is found to taper more gradually at its upper end, which marks the progress of extermination,

Shell of an Ammonite, fossilized remains of a marine animal

by the Spaniards into South America, has run wild over the whole country and has increased in numbers at an unparalleled rate, I asked myself what could so recently have exterminated the former horse under conditions of life apparently so favourable. But my astonishment was groundless. Professor Owen soon perceived that the tooth, though so like that of the existing horse, belonged to an extinct species. Had this horse been still living, but in some degree rare, no naturalist would have felt the least surprise at its rarity; for rarity is the attribute of a vast number of species of all classes, in all countries. If we ask ourselves why this or that species is rare, we answer that something is

than at its lower end, which marks the first appearance and the early increase in number of the species. In some cases, however, the extermination of whole groups, as of ammonites, towards the close of the secondary period, has been wonderfully sudden.

The extinction of species has been involved in the most gratuitous mystery. Some authors have even supposed that, as the individual has a definite length of life, so have species a definite duration. No one can have marvelled more than I have done at the extinction of species. When I found in La Plata the tooth of a horse embedded with the remains of Mastodon, Megatherium, Toxodon and other extinct monsters, which all co-existed with still living shells at a very late geological period, I was filled with astonishment; for, seeing that the horse, since its introduction

unfavourable in its conditions of life; but what that something is, we can hardly ever tell. On the supposition of the fossil horse still existing as a rare species, we might have felt certain, from the analogy of all other mammals, even of the slow-breeding elephant, and from the history of the naturalisation of the domestic horse in South America, that under more favourable conditions it would in a very few years have stocked the whole continent. But we could not have told what the unfavourable conditions were which checked its increase, whether some one or several contingencies, and at what period of the horse's life, and in what degree they severally acted. If the conditions had gone on, however slowly, becoming less and less favourable, we assuredly should not have perceived the fact, yet the fossil horse would certainly have become rarer and

Skeleton of a Megatherium (meaning 'great beast'). It was one of a genus of giant ground sloths rarer, and finally extinct; – its place being seized on by some more successful competitor.

It is most difficult always to remember that the increase of every living creature is constantly being checked by unperceived hostile agencies; and that these same unperceived agencies are amply sufficient to cause rarity, and finally extinction. So little is this subject understood, that I have heard surprise repeatedly expressed at such great monsters as the Mastodon and the more ancient Dinosaurians having become extinct; as if mere bodily strength gave victory in the battle of life. Mere size, on the contrary, would in some cases determine, as has been remarked by Owen, quicker extermination, from the greater amount of requisite food. Before man inhabited India or Africa, some cause must have checked the continued increase of the existing elephant. A highly capable judge, Dr. Falconer, believes that it is chiefly insects which, from incessantly harassing and weakening the elephant in India, check its increase; and this was Bruce's conclusion with respect to the African elephant in Abyssinia. It is certain that insects and blood-sucking bats determine the existence of the larger naturalised quadrupeds in several parts of South America.

We see in many cases in the more recent tertiary formations that rarity precedes extinction; and we know that this has been the progress of events with those animals which have been exterminated, either locally or wholly, through man's agency. I may repeat what I published in 1845, namely, that to admit that species generally become rare before they become extinct – to feel no surprise at the rarity of a species, and yet to marvel greatly when the species ceases to exist, is much the same as to admit that sickness in the individual is the forerunner of death – to feel no surprise at sickness, but, when the sick man dies, to wonder and to suspect that he died by some deed of violence.

The theory of natural selection is grounded on the belief that each new variety and ultimately each new species, is produced and maintained by

having some advantage over those with which it comes into competition; and the consequent extinction of less-favoured forms almost inevitably follows. It is the same with our domestic productions: when a new and slightly improved variety has been raised, it at first supplants the less improved varieties in the same neighbourhood; when much improved it is transported far and near, like our short-horn cattle, and takes the place of other breeds in other countries. Thus the appearance of new forms and the disappearance of old forms, both those naturally and artificially produced, are bound together. In flourishing groups, the number of new specific forms which have been produced within a given time has at some periods probably been greater than the number of the old specific forms which have been exterminated; but we know that species have not gone on indefinitely increasing, at least during the later geological epochs, so that, looking to later times, we may believe that the production of new forms has caused the extinction of about the same number of old forms.

The competition will generally be most severe, as formerly explained and illustrated by examples, between the forms which are most like each other in all respects. Hence the improved and modified descendants of a species will generally cause the extermination of the parent-species; and if many new forms have been developed from any one species, the nearest allies of that species, i.e. the species of the same genus, will be the most liable to extermination. Thus, as I believe, a number of new species descended from one species, that is a new genus, comes to supplant an old genus, belonging to the same family. But it must often have happened that a new species belonging to some one group has seized on the place occupied by a species belonging to a distinct group, and thus have caused its extermination. If many allied forms be developed from the successful intruder, many will have to yield their places; and it will generally be of the great and almost extinct group of Ganoid fishes still inhabit our fresh waters. Therefore, the utter extinction of a group is generally, as we have seen, a slower process than its production.

With respect to the apparently sudden extermination of whole families or orders, as of Trilobites at the close of the palaeozoic period, and of Ammonites at the close of the secondary period, we must remember what has been already said on the probable wide intervals of time between our consecutive formations; and in these intervals there may have been much slow extermination. Moreover, when, by sudden immigration or by unusually rapid development, many species of a new group have taken possession of an area, many of the older species will have been exterminated in a correspondingly rapid manner; and the forms which thus yield their places will commonly be allied, for they will partake of the same inferiority in common.

Thus, as it seems to me, the manner in which single species and whole groups of species become extinct accords well with the theory of natural selection. We need not marvel at extinction; if we must marvel, let it be at our presumption in imagining for a moment that we understand the many complex contingencies on which the existence of each species depends. If we forget for an instant that each species tends to increase inordinately, and that some check is always in action, yet seldom perceived by us, the whole economy of nature will be utterly obscured. Whenever we can precisely say why this species is more abundant in individuals than that; why this species and not another can be naturalised in a given country; then, and not until then, we may justly feel surprise why we cannot account for the extinction of any particular species or group of species.

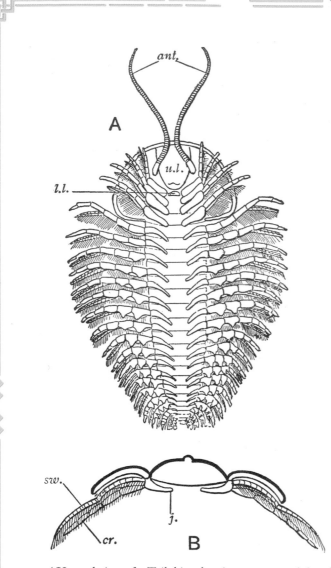

A Ventral view of a Trilobite showing numerous jointed limbs: *ant.* antennae *u.l.* upper lip *l.l.* lower lip
B Cross-section showing limb regions:
j. jaw *sw.* and *cr.* swimming and creeping leg branches

the allied forms, which will suffer from some inherited inferiority in common. But whether it be species belonging to the same or to a distinct class, which have yielded their places to other modified and improved species, a few of the sufferers may often be preserved for a long time, from being fitted to some peculiar line of life, or from inhabiting some distant and isolated station, where they will have escaped severe competition. For instance, some species of Trigonia, a great genus of shells in the secondary formations, survive in the Australian seas; and a few members

ON THE FORMS OF LIFE CHANGING ALMOST SIMULTANEOUSLY THROUGHOUT THE WORLD

Scarcely any palaeontological discovery is more striking than the fact that the forms of life change almost simultaneously throughout the world. Thus our European Chalk formation can be recognised in many distant regions, under the most different climates, where not a fragment of the mineral chalk itself can be found; namely, in North America, in equatorial South America, in Tierra del Fuego, at the Cape of Good Hope, and in the peninsula of India. For at these distant points, the organic remains in certain beds present an unmistakable resemblance to those of the Chalk. It is not that the same species are met with; for in some cases not one species is identically the same, but they belong to the same families, genera, and sections of genera, and sometimes are similarly characterised in such trifling points as mere superficial sculpture. Moreover, other forms, which are not found in the Chalk of Europe, but which occur in the formations either above or below, occur in the same order at these distant points of the world. In the several successive palaeozoic formations of Russia, Western Europe and North America, a similar parallelism in the forms of life has been observed by several authors; so it is, according to Lyell, with the European and North American tertiary deposits. Even if the few fossil species which are common to the Old and New Worlds were kept wholly out of view, the general parallelism in the successive forms of life, in the palaeozoic and tertiary stages, would still be manifest, and the several formations could be easily correlated.

These observations, however, relate to the marine inhabitants of the world: we have not sufficient data to judge whether the productions of the land and of fresh water at distant points change in the same parallel manner. We may doubt whether they have thus changed: if the Megatherium, Mylodon, Macrauchenia, and Toxodon had been brought to Europe from La Plata, without any information in regard to their geological position, no one would have suspected that they had co-existed with sea shells all still living; but as these anomalous monsters co-existed with the Mastodon and Horse, it might at least have been inferred that they had lived during one of the later tertiary stages.

When the marine forms of life are spoken of as having changed simultaneously throughout the world, it must not be supposed that this expression relates to the same year, or even to the same century, or even that it has a very strict geological sense; for if all the marine animals now living in Europe, and all those that lived in Europe during the pleistocene period (a very remote period as measured by years, including the whole glacial epoch) were compared with those now existing in South America or in Australia, the most skilful naturalist would hardly be able to say whether the present or the pleistocene inhabitants of Europe resembled most closely those of the southern hemisphere. So, again, several highly competent observers maintain that the existing productions of the United States are more closely related to those which lived in Europe during certain late tertiary stages, than to the present inhabitants of Europe; and if this be so, it is evident that fossiliferous beds now deposited on the shores of North America would hereafter be liable to be classed with somewhat older European beds. Nevertheless, looking to a remotely future epoch, there can be little doubt that all the more modern *marine* formations, namely, the upper pliocene, the pleistocene and strictly modern beds of Europe, North and South America, and Australia, from containing fossil remains in some degree allied, and from not including those forms which are found only in the older underlying deposits, would be correctly ranked as simultaneous in a geological sense.

The fact of the forms of life changing simultaneously in the above large sense, at distant parts of the world, has greatly struck those admirable observers, MM. de Verneuil and d'Archiac. After referring to the parallelism of the palaeozoic forms of life in various parts of Europe, they add, 'If struck by this strange sequence, we turn our attention to North America, and there discover a series of analogous phenomena, it will appear certain that all these modifications of species, their extinction, and the introduction of new ones, cannot be owing to mere changes in marine currents or other causes more or less local and temporary, but depend on general laws which govern the whole animal kingdom.' M. Barrande has made forcible remarks to precisely the same effect. It is, indeed, quite futile to look to changes of currents, climate, or other physical conditions, as the cause of these great mutations in the forms of life throughout the world, under the most different climates. We must, as Barrande has remarked, look to some special law. We shall see this more clearly when we treat of the present distribution of organic beings, and find how slight is the relation between the physical conditions of various countries and the nature of their inhabitants.

This great fact of the parallel succession of the forms of life throughout the world, is explicable on the theory of natural selection. New species are formed by having some advantage over older forms; and the forms, which are already dominant, or have some advantage over the other forms in their own country, give birth to the greatest number of new varieties or incipient species. We have distinct evidence on this head, in the plants which are dominant, that is, which are commonest and most widely diffused, producing the greatest number of new varieties. It is also natural that the dominant, varying and far-spreading species, which have already invaded, to a certain extent, the territories of other species, should be those which would have the best chance of spreading still further, and of giving rise in new countries to other new varieties and species. The process of diffusion would often be very slow, depending on climatal and geographical changes, on strange accidents, and on the gradual

acclimatization of new species to the various climates through which they might have to pass, but in the course of time the dominant forms would generally succeed in spreading and would ultimately prevail. The diffusion would, it is probable, be slower with the terrestrial inhabitants of distinct continents than with the marine inhabitants of the continuous sea. We might therefore expect to find, as we do find, a less strict degree of parallelism in the succession of the productions of the land than with those of the sea.

Thus, as it seems to me, the parallel, and, taken in a large sense, simultaneous, succession of the same forms of life throughout the world, accords well with the principle of new species having been formed by dominant species spreading widely and varying; the new species thus produced being themselves dominant, owing to their having had some advantage over their already dominant parents, as well as over other species; and again spreading, varying, and producing new forms. The old forms which are beaten and which yield their places to the new and victorious forms, will generally be allied in groups, from inheriting some inferiority in common; and, therefore, as new and improved groups spread throughout the world, old groups disappear from the world; and the succession of forms everywhere tends to correspond both in their first appearance and final disappearance.

There is one other remark connected with this subject worth making. I have given my reasons for believing that most of our great formations, rich in fossils, were deposited during periods of subsidence; and that blank intervals of vast duration, as far as fossils are concerned, occurred during the periods when the bed of the sea was either stationary or rising, and likewise when sediment was not thrown down quickly enough to embed and preserve organic remains. During these long and blank intervals I suppose that the inhabitants of each region underwent a considerable amount of modification and extinction, and that there was much migration from other parts of the world. As we have reason to believe that large areas are affected by the same movement, it is probable that strictly contemporaneous formations have often been accumulated over very wide spaces in the same quarter of the world; but we are very far from having any right to conclude that this has invariably been the case, and that large areas have invariably been affected by the same movements. When two formations have been deposited in two regions during nearly, but not exactly, the same period, we should find in both, from the causes explained in the foregoing paragraphs, the same general succession in the forms of life; but the species would not exactly correspond; for there will have been a little more time in the one region than in the other for modification, extinction, and immigration.

I suspect that cases of this nature occur in Europe. Mr. Prestwich, in his admirable Memoirs on the eocene deposits of England and France, is able to draw a close general parallelism between the successive stages in the two countries; but when he compares certain stages in England with those in France, although he finds in both a curious accordance in the numbers of the species belonging to the same genera, yet the species themselves differ in a manner very difficult to account for considering the proximity of the two areas, unless, indeed, it be assumed that an isthmus separated two seas inhabited by distinct, but contemporaneous faunas. Lyell has made similar observations on some of the later tertiary formations. Barrande, also, shows that there is a striking general parallelism in the successive Silurian deposits of Bohemia and Scandinavia; nevertheless he finds a surprising amount of difference in the species. If the several formations in these regions have not been deposited during the same exact periods, – a formation in one region often corresponding with a blank interval in the other, – and if in both regions the species have gone on slowly changing during the accumulation of the several formations and during

the long intervals of time between them; in this case the several formations in the two regions could be arranged in the same order, in accordance with the general succession of the forms of life, and the order would falsely appear to be strictly parallel; nevertheless the species would not all be the same in the apparently corresponding stages in the two regions.

ON THE AFFINITIES OF EXTINCT SPECIES TO EACH OTHER, AND TO LIVING FORMS

Let us now look to the mutual affinities of extinct and living species. All fall into a few grand classes; and this fact is at once explained on the principle of descent. The more ancient any form is, the more, as a general rule, it differs from living forms. But, as Buckland long ago remarked, extinct species can all be classed either in still existing groups, or between them. That the extinct forms of life help to fill up the intervals between existing genera, families, and orders, is certainly true; but as this statement has often been ignored or even denied, it may be well to make some remarks on this subject, and to give some instances. If we confine our attention either to the living or to the extinct species of the same class, the series is far less perfect than if we combine both into one general system. In the writings of Professor Owen we continually meet with the expression of generalised forms, as applied to extinct animals; and in the writings of Agassiz, of prophetic or synthetic types; and these terms imply that such forms are, in fact, intermediate or connecting links. Another distinguished palaeontologist, M. Gaudry, has shown in the most striking manner that many of the fossil mammals discovered by him in Attica serve to break down the intervals between existing genera. Cuvier ranked the Ruminants and Pachyderms as two of the most distinct orders of mammals; but so many fossil links have been disentombed that Owen has had to alter the whole classification, and has placed certain pachyderms in the same sub-order with ruminants; for example, he dissolves by

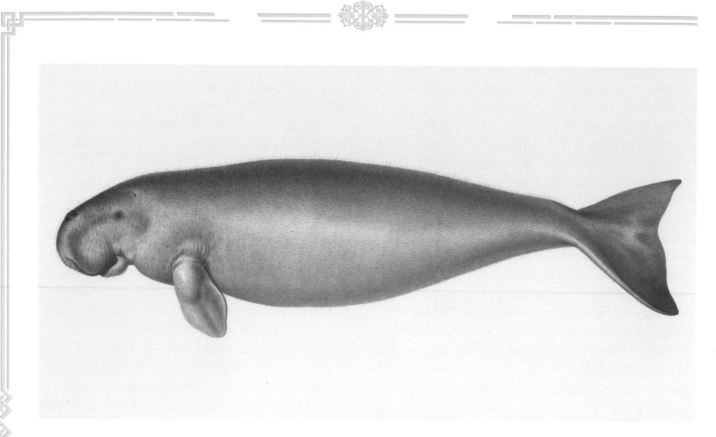

The Halicore Dugong (Sea Cow)

gradations the apparently wide interval between the pig and the camel. The Ungulata or hoofed quadrupeds are now divided into the even-toed or odd-toed divisions; but the Macrauchenia of South America connects to a certain extent these two grand divisions. No one will deny that the Hipparion is intermediate between the existing horse and certain other ungulate forms. What a wonderful connecting link in the chain of mammals is the Typotherium from South America, as the name given to it by Professor Gervais expresses, and which cannot be placed in any existing order. The Sirenia form a very distinct group of the mammals, and one of the most remarkable peculiarities in existing dugong and lamentin is the entire absence of hind limbs, without even a rudiment being left; but the extinct Halitherium had, according to Professor Flower, an ossified thigh-bone 'articulated to a well-defined acetabulum in the pelvis,' and it thus makes some approach to ordinary hoofed quadrupeds, to which the Sirenia are in other respects allied. The cetaceans or whales are widely different from all other mammals, but the tertiary Zeuglodon and Squalodon, which have been placed by some naturalists in an order by themselves, are considered by Professor Huxley to be undoubtedly cetaceans, 'and to constitute connecting links with the aquatic carnivora.'

Even the wide interval between birds and reptiles has been shown by the naturalist just quoted to be partially bridged over in the most unexpected manner, on the one hand, by the ostrich and extinct Archeopteryx, and on the other hand by the Compsognathus, one of the Dinosaurians – that group which includes the most gigantic of all terrestrial reptiles. Turning to the Invertebrata, Barrande asserts, a higher authority could not be named, that he is every day taught that, although palaeozoic animals can certainly be classed under existing groups, yet that at this ancient period the groups were not so distinctly separated from each other as they now are.

Some writers have objected to any extinct species, or group of species, being considered as intermediate between any two living species, or groups of species. If by this term it is meant that an extinct form is directly intermediate in all its characters between two living forms or groups, the objection is probably valid. But in a natural classification many fossil species certainly stand between living species, and some extinct genera between living genera, even between genera belonging to distinct families. The most common case, especially with respect to very distinct groups, such as fish and reptiles, seems to be that, supposing them to be distinguished at the present day by a score of characters, the ancient members are separated by a somewhat lesser number of characters, so that the two groups formerly made a somewhat nearer approach to each other than they now do.

It is a common belief that the more ancient a form is, by so much the more it tends to connect by some of its characters groups now widely separated from each other. This remark no doubt must be restricted to those groups which have undergone much change in the course of geological ages; and it would be difficult to prove the truth of the proposition, for every now and then even a living animal, as the Lepidosiren, is discovered having affinities directed towards very distinct groups. Yet if we compare the older Reptiles and Batrachians, the older Fish, the older Cephalopods, and the eocene Mammals, with the recent members of the same classes, we must admit that there is truth in the remark.

Let us see how far these several facts and inferences accord with the theory of descent with modification. As the subject is somewhat complex, I must request the reader to turn to the diagram in the fourth chapter [see page 86]. We may suppose that the numbered letters in italics represent genera, and the dotted lines diverging from them the species in each genus. The diagram is much too simple, too few genera and too few species being given, but this is unimportant for us. The

horizontal lines may represent successive geological formations, and all the forms beneath the uppermost line may be considered as extinct. The three existing genera, a^{14}, q^{14}, p^{14}, will form a small family; b^{14} and f^{14}, a closely allied family or sub-family; and o^{14}, i^{14}, m^{14}, a third family. These three families, together with the many extinct genera on the several lines of descent diverging from the parent form (A) will form an order; for all will have inherited something in common from their ancient progenitor. On the principle of the continued tendency to divergence of character, which was formerly illustrated by this diagram, the more recent any form is the more it will generally differ from its ancient progenitor. Hence, we can understand the rule that the most ancient fossils differ most from existing forms. We must not, however, assume that divergence of character is a necessary contingency; it depends solely on the descendants from a species being thus enabled to seize on many and different places in the economy of nature. Therefore it is quite possible, as we have seen in the case of some Silurian forms, that a species might go on being slightly modified in relation to its slightly altered conditions of life, and yet retain throughout a vast period the same general characteristics. This is represented in the diagram by the letter F^{14}.

All the many forms, extinct and recent, descended from (A), make, as before remarked, one order; and this order, from the continued effects of extinction and divergence of character, has become divided into several sub-families and families, some of which are supposed to have perished at different periods, and some to have endured to the present day.

By looking at the diagram we can see that if many of the extinct forms supposed to be embedded in the successive formations, were discovered at several points low down in the series, the three existing families on the uppermost line would be rendered less distinct from each other. If, for instance, the genera a^1, a^5, a^{10}, f^8, m^3, m^6, m^9, were disinterred, these three families would be so closely linked together that they probably would have to be united into one great family, in nearly the same manner as has occurred with ruminants and certain pachyderms. Yet he who objected to consider as intermediate the extinct genera, which thus link together the living genera of three families, would be partly justified, for they are intermediate, not directly, but only by a long and circuitous course through many widely different forms. If many extinct forms were to be discovered above one of the middle horizontal lines or geological formations – for instance, above No. VI – but none from beneath this line, then only two of the families (those on the left hand a^{14}, etc., and b^{14}, etc.) would have to be united into one; and there would remain two families which would be less distinct from each other than they were before the discovery of the fossils. So again, if the three families formed of eight genera (a^{14} to m^{14}), on the uppermost line, be supposed to differ from each other by half-a-dozen important characters, then the families which existed at a period marked VI would certainly have differed from each other by a less number of characters; for they would at this early stage of descent have diverged in a less degree from their common progenitor. Thus it comes that ancient and extinct genera are often in a greater or less degree intermediate in character between their modified descendants, or between their collateral relations.

Under nature the process will be far more complicated than is represented in the diagram; for the groups will have been more numerous; they will have endured for extremely unequal lengths of time, and will have been modified in various degrees. As we possess only the last volume of the geological record, and that in a very broken condition, we have no right to expect, except in rare cases, to fill up the wide intervals in the natural system, and thus to unite distinct families or orders. All that we have a right to expect is, that those groups which have, within known geological periods, undergone much

modification, should in the older formations make some slight approach to each other; so that the older members should differ less from each other in some of their characters than do the existing members of the same groups; and this by the concurrent evidence of our best palaeontologists is frequently the case.

Thus, on the theory of descent with modification, the main facts with respect to the mutual affinities of the extinct forms of life to each other and to living forms, are explained in a satisfactory manner. And they are wholly inexplicable on any other view.

On this same theory, it is evident that the fauna during any one great period in the earth's history will be intermediate in general character between that which preceded and that which succeeded it. Thus the species which lived at the sixth great stage of descent in the diagram are the modified offspring of those which lived at the fifth stage, and are the parents of those which became still more modified at the seventh stage; hence they could hardly fail to be nearly intermediate in character between the forms of life above and below. We must, however, allow for the entire extinction of some preceding forms, and in any one region for the immigration of new forms from other regions, and for a large amount of modification during the long and blank intervals between the successive formations. Subject to these allowances, the fauna of each geological period undoubtedly is intermediate in character, between the preceding and succeeding faunas. I need give only one instance, namely, the manner in which the fossils of the Devonian system, when this system was first discovered, were at once recognised by palaeontologists as intermediate in character between those of the overlying carboniferous and underlying Silurian systems. But each fauna is not necessarily exactly intermediate, as unequal intervals of time have elapsed between consecutive formations.

It is no real objection to the truth of the statement that the fauna of each period as a whole is nearly intermediate in character between the preceding and succeeding faunas, that certain genera offer exceptions to the rule. For instance, the species of mastodons and elephants, when arranged by Dr. Falconer in two series, – in the first place according to their mutual affinities, and in the second place according to their periods of existence, – do not accord in arrangement. The species extreme in character are not the oldest or the most recent; nor are those which are intermediate in character, intermediate in age. But supposing for an instant, in this and other such cases, that the record of the first appearance and disappearance of the species was complete, which is far from the case, we have no reason to believe that forms successively produced necessarily endure for corresponding lengths of time. A very ancient form may occasionally have lasted much longer than a form elsewhere subsequently produced, especially in the case of terrestrial productions inhabiting separated districts. To compare small things with great; if the principal living and extinct races of the domestic pigeon were arranged in serial affinity, this arrangement would not closely accord with the order in time of their production, and even less with the order of their disappearance; for the parent rock-pigeon still lives; and many varieties between the rock-pigeon and the carrier have become extinct; and carriers which are extreme in the important character of length of beak originated earlier than short-beaked tumblers, which are at the opposite end of the series in this respect.

Closely connected with the statement, that the organic remains from an intermediate formation are in some degree intermediate in character, is the fact, insisted on by all palaeontologists, that fossils from two consecutive formations are far more closely related to each other, than are the fossils from two remote formations. Pictet gives as a well-known instance, the general resemblance of the organic remains from the several stages of the Chalk formation, though the species are distinct in each stage. This fact alone, from its generality, seems to have shaken Professor Pictet in his belief in the immutability of species. He who is acquainted with the distribution of existing species over the globe, will not attempt to account for the close resemblance of distinct species in closely consecutive formations, by the physical conditions of the ancient areas having remained nearly the same. Let it be remembered that the forms of life, at least those inhabiting the sea, have changed almost simultaneously throughout the world, and therefore under the most different climates and conditions. Consider the prodigious vicissitudes of climate during the pleistocene period, which includes the whole glacial epoch, and note how little the specific forms of the inhabitants of the sea have been affected.

On the theory of descent, the full meaning of the fossil remains from closely consecutive formations, being closely related, though ranked as distinct species, is obvious. As the accumulation of each formation has often been interrupted, and as long blank intervals have intervened between successive formations, we ought not to expect to find, as I attempted to show in the last chapter, in any one or in any two formations, all the intermediate varieties between the species which appeared at the commencement and close of these periods: but we ought to find after intervals, very long as measured by years, but only moderately long as measured geologically, closely allied forms, or, as they have been called by some authors, representative species; and these assuredly we do find. We find, in short, such evidence of the slow and scarcely sensible mutations of specific forms, as we have the right to expect.

ON THE STATE OF DEVELOPMENT OF ANCIENT COMPARED WITH LIVING FORMS

We have seen in the fourth chapter that the degree of differentiation and specialisation of the parts in organic beings, when arrived at maturity, is the best standard, as yet suggested, of their degree of perfection or highness. We have also seen that, as the specialisation of parts is an advantage to each being, so natural selection will tend to render the organisation of each being more specialised and perfect, and in this sense higher; not but that it may leave many creatures with simple and unimproved structures fitted for simple conditions of life, and in some cases will even degrade or simplify the organisation, yet leaving such degraded beings better fitted for their new walks of life. In another and more general manner, new species become superior to their predecessors; for they have to beat in the struggle for life all the older forms, with which they come into close competition. We may therefore conclude that if under a nearly similar climate the eocene inhabitants of the world could be put into competition with the existing inhabitants, the former would be beaten and exterminated by the latter, as would the secondary by the eocene, and the palaeozoic by the secondary forms. So that by this fundamental test of victory in the battle for life, as well as by the standard of the specialisation of organs, modern forms ought, on the theory of natural selection, to stand higher than ancient forms. Is this the case? A large majority of palaeontologists would answer in the affirmative; and it seems that this answer must be admitted as true, though difficult of proof.

It is no valid objection to this conclusion, that certain Brachiopods have been but slightly modified from an extremely remote geological epoch; and that certain land and fresh-water shells have remained nearly the same, from the time when, as far as is known, they first appeared. It is not an insuperable difficulty that Foraminifera have not, as insisted on by Dr. Carpenter, progressed in organisation since even the Laurentian epoch; for some organisms would have to remain fitted for simple conditions of life, and what could be better fitted for this end than these lowly organised Protozoa? Such objections as the above would be fatal to my view, if it included

advance in organisation as a necessary contingent. They would likewise be fatal, if the above Foraminifera, for instance, could be proved to have first come into existence during the Laurentian epoch, or the above Brachiopods during the Cambrian formation; for in this case, there would not have been time sufficient for the development of these organisms up to the standard which they had then reached. When advanced up to any given point, there is no necessity, on the theory of natural selection, for their further continued process; though they will, during each successive age, have to be slightly modified, so as to hold their places in relation to slight changes in their conditions. The foregoing objections hinge on the question whether we really know how old the world is, and at what period the various forms of life first appeared; and this may well be disputed.

The problem whether organisation on the whole has advanced is in many ways excessively intricate. The geological record, at all times imperfect, does not extend far enough back to show with unmistakable clearness that within the known history of the world organisation has largely advanced. Even at the present day, looking to members of the same class, naturalists are not unanimous which forms ought to be ranked as highest: thus, some look at the selaceans or sharks, from their approach in some important points of structure to reptiles, as the highest fish; others look at the teleosteans as the highest. The ganoids stand intermediate between the selaceans and teleosteans; the latter at the present day are largely preponderant in number; but formerly selaceans and ganoids alone existed; and in this case, according to the standard of highness chosen, so will it be said that fishes have advanced or retrograded in organisation. To attempt to compare members of distinct types in the scale of highness seems hopeless; who will decide whether a cuttle-fish be higher than a bee – that insect which the great Von Baer believed to be 'in fact more highly organised than a fish, although upon

A Blue Shark or Squalus glaucus.

another type?' In the complex struggle for life it is quite credible that crustaceans, not very high in their own class, might beat cephalopods, the highest molluscs; and such crustaceans, though not highly developed, would stand very high in the scale of invertebrate animals, if judged by the most decisive of all trials – the law of battle. Beside these inherent difficulties in deciding which forms are the most advanced in organisation, we ought not solely to compare the highest members of a class at any two periods – though undoubtedly this is one and perhaps the most important element in striking a balance – but we ought to compare all the members, high and low, at two periods. At an ancient epoch the highest and lowest molluscoidal animals, namely, cephalopods and brachiopods, swarmed in numbers; at the present time both groups are greatly reduced, whilst others, intermediate in organisation, have largely increased; consequently some naturalists maintain that molluscs were formerly more highly developed than at present; but a stronger case can be made out on the opposite side, by considering the vast reduction of brachiopods, and the fact that our existing

cephalopods, though few in number, are more highly organised than their ancient representatives. We ought also to compare the relative proportional numbers, at any two periods, of the high and low classes throughout the world: if, for instance, at the present day fifty thousand kinds of vertebrate animals exist, and if we knew that at some former period only ten thousand kinds existed, we ought to look at this increase in number in the highest class, which implies a great displacement of lower forms, as a decided advance in the organisation of the world. We thus see how hopelessly difficult it is to compare with perfect fairness, under such extremely complex relations, the standard of organisation of the imperfectly-known faunas of successive periods.

We shall appreciate this difficulty more clearly by looking to certain existing faunas and floras. From the extraordinary manner in which European productions have recently spread over New Zealand, and have seized on places which must have been previously occupied by the indigenes, we must believe, that if all the animals and plants of Great Britain were set free in New Zealand, a multitude of British forms would in the course of time become thoroughly naturalized there, and would exterminate many of the natives. On the other hand, from the fact that hardly a single inhabitant of the southern hemisphere has become wild in any part of Europe, we may well doubt whether, if all the productions of New Zealand were set free in Great Britain, any considerable number would be enabled to seize on places now occupied by our native plants and animals. Under this point of view, the productions of Great Britain stand much higher in the scale than those of New Zealand. Yet the most skilful naturalist, from an examination of the species of the two countries, could not have foreseen this result.

Agassiz and several other highly competent judges insist that ancient animals resemble to a certain extent the embryos of recent animals belonging to the same classes; and that the geological succession of extinct forms is nearly parallel with the embryological development of existing forms. This view accords admirably well with our theory. In a future chapter I shall attempt to show that the adult differs from its embryo, owing to variations having supervened at a not early age, and having been inherited at a corresponding age. This process, whilst it leaves the embryo almost unaltered, continually adds, in the course of successive generations, more and more difference to the adult. Thus the embryo comes to be left as a sort of picture, preserved by nature, of the former and less modified condition of the species. This view may be true, and yet may never be capable of proof. Seeing, for instance, that the oldest known mammals, reptiles, and fishes strictly belong to their proper classes, though some of these old forms are in a slight degree less distinct from each other than are the typical members of the same groups at the present day, it would be vain to look for animals having the common embryological character of the Vertebrata, until beds rich in fossils are discovered far beneath the lowest Cambrian strata – a discovery of which the chance is small.

ON THE SUCCESSION OF THE SAME TYPES WITHIN THE SAME AREAS, DURING THE LATER TERTIARY PERIODS

Mr. Clift many years ago showed that the fossil mammals from the Australian caves were closely allied to the living marsupials of that continent. In South America, a similar relationship is manifest, even to an uneducated eye, in the gigantic pieces of armour, like those of the armadillo, found in several parts of La Plata; and Professor Owen has shown in the most striking manner that most of the fossil mammals, buried there in such numbers, are related to South American types. This relationship is even more clearly seen in the wonderful collection of fossil bones made by MM. Lund and Clausen in the caves of Brazil. I was so much impressed with these facts that I strongly insisted, in 1839 and 1845, on this 'law of the succession of types,' – on 'this wonderful relationship in the same continent between the dead and the living.' Professor Owen has subsequently extended the same generalisation to the mammals of the Old World. We see the same law in this author's restorations of the extinct and gigantic birds of New Zealand. We see it also in the birds of the caves of Brazil. Mr. Woodward has shown that the same law holds good with sea shells, but, from the wide distribution of most molluscs, it is not well displayed by them. Other cases could be added, as the relation between the extinct and living land shells of Madeira; and between the extinct and living brackish water shells of the Aralo-Caspian Sea.

Now, what does this remarkable law of the succession of the same types within the same areas mean? He would be a bold man who, after comparing the present climate of Australia and of parts of South America, under the same latitude, would attempt to account, on the one hand through dissimilar physical conditions, for the dissimilarity of the inhabitants of these two continents; and, on the other hand through similarity of conditions, for the uniformity of the same types in each continent during the later tertiary periods. Nor can it be pretended that it is an immutable law that marsupials should have been chiefly or solely produced in Australia; or that Edentata and other American types should have been solely produced in South America. For we know that Europe in ancient times was peopled by numerous marsupials; and I have shown in the publications above alluded to, that in America the law of distribution of terrestrial mammals was formerly different from what it now is. North America formerly partook strongly of the present character of the southern half of the continent; and the southern half was formerly more closely allied, than it is at present, to the northern half. In a similar manner we know, from Falconer and Cautley's discoveries, that Northern India was formerly more closely related in its mammals to Africa than it is at the present time. Analogous facts could be given in relation to the distribution of marine animals.

On the theory of descent with modification, the great law of the long enduring, but not immutable, succession of the same types within the same areas, is at once explained; for the inhabitants of each quarter of the world will obviously tend to leave in that quarter, during the next succeeding period of time, closely allied though in some degree modified descendants. If the inhabitants of one continent formerly differed greatly from those of another continent, so will their modified descendants still differ in nearly the same manner and degree. But after very long intervals of time, and after great geographical changes, permitting much intermigration, the feebler will yield to the more dominant forms, and there will be nothing immutable in the distribution of organic beings.

It may be asked in ridicule whether I suppose that the megatherium and other allied huge monsters, which formerly lived in South America, have left behind them the sloth, armadillo, and anteater, as their degenerate descendants. This cannot for an instant be admitted. These huge animals have become wholly extinct, and have left no progeny. But in the caves of Brazil there are many extinct species which are closely allied in size and in all other characters to the species still living in South America; and some of these fossils may have been the actual progenitors of the living species. It must not be forgotten that, on our theory, all the species of the same genus are the descendants of some one species; so that, if six genera, each having eight species, be found in one geological formation, and in a succeeding formation there be six other allied or representative genera, each with the same number of species, then we may conclude that generally only one species of each of the older genera has left modified descendants, which constitute the new genera containing the several species; the other seven species of each old genus having died out and left no progeny. Or, and this will be a far commoner case, two or three species in two or three alone of the six older genera will be the

The Armadillo (left) and the Anteater (right)

parents of the new genera: the other species and the other old genera having become utterly extinct. In failing orders, with the genera and species decreasing in numbers as is the case with the Edentata of South America, still fewer genera and species will leave modified blood-descendants.

SUMMARY OF THE PRECEDING AND PRESENT CHAPTERS

I have attempted to show that the geological record is extremely imperfect; that only a small portion of the globe has been geologically explored with care; that only certain classes of organic beings have been largely preserved in a fossil state; that the number both of specimens and of species, preserved in our museums, is absolutely as nothing compared with the number of generations which must have passed away even during a single formation; that, owing to subsidence being almost necessary for the accumulation of deposits rich in fossil species of many kinds, and thick enough to outlast future degradation, great intervals of time must have elapsed between most of our successive formations; that there has probably been more extinction during the periods of subsidence, and more variation during the periods of elevation, and during the latter the record will have been least perfectly kept; that each single formation has not been continuously deposited; that the duration of each formation is probably short compared with the average duration of specific forms; that migration has played an important part in the first appearance of new forms in any one area and formation; that widely ranging species are those which have varied most frequently, and have oftenest given rise to new species; that varieties have at first been local; and lastly, although each species must have passed through numerous transitional stages, it is probable that the periods, during which each underwent modification, though many and long as measured by years, have been short in comparison with the periods during which each remained in an unchanged condition. These causes, taken conjointly, will to a large extent explain why –

though we do find many links – we do not find interminable varieties, connecting together all extinct and existing forms by the finest graduated steps. It should also be constantly borne in mind that any linking variety between two forms, which might be found, would be ranked, unless the whole chain could be perfectly restored, as a new and distinct species; for it is not pretended that we have any sure criterion by which species and varieties can be discriminated.

He who rejects this view of the imperfection of the geological record, will rightly reject the whole theory. For he may ask in vain where are the numberless transitional links which must formerly have connected the closely allied or representative species, found in the successive stages of the same great formation? He may disbelieve in the immense intervals of time which must have elapsed between our consecutive formations; he may overlook how important a part migration has played, when the formations of any one great region, as those of Europe, are considered; he may urge the apparent, but often falsely apparent, sudden coming in of whole groups of species. He may ask where are the remains of those infinitely numerous organisms which must have existed long before the Cambrian system was deposited? We now know that at least one animal did then exist; but I can answer this last question only by supposing that where our oceans now extend they have extended for an enormous period, and where our oscillating continents now stand they have stood since the commencement of the Cambrian system; but that, long before that epoch, the world presented a widely different aspect; and that the older continents, formed of formations older than any known to us, exist now only as remnants in a metamorphosed condition, or lie still buried under the ocean.

Passing from these difficulties, the other great leading facts in palaeontology agree admirably with the theory of descent with modification through variation and natural selection. We can thus understand how it is that new species come in slowly and successively; how species of different classes do not necessarily change together, or at the same rate, or in the same degree; yet in the long run that all undergo modification to some extent. The extinction of old forms is the almost inevitable consequence of the production of new forms. We can understand why, when a species has once disappeared, it never reappears. Groups of species increase in numbers slowly, and endure for unequal periods of time; for the process of modification is necessarily slow, and depends on many complex contingencies. The dominant species belonging to large and dominant groups tend to leave many modified descendants, which form new sub-groups and groups. As these are formed, the species of the less vigorous groups, from their inferiority inherited from a common progenitor, tend to become extinct together, and to leave no modified offspring on the face of the earth. But the utter extinction of a whole group of species has sometimes been a slow process, from the survival of a few descendants, lingering in protected and isolated situations. When a group has once wholly disappeared, it does not reappear; for the link of generation has been broken.

We can understand how it is that dominant forms which spread widely and yield the greatest number of varieties tend to people the world with allied, but modified, descendants; and these will generally succeed in displacing the groups which are their inferiors in the struggle for existence. Hence, after long intervals of time, the productions of the world appear to have changed simultaneously.

We can understand how it is that all the forms of life, ancient and recent, make together a few grand classes. We can understand, from the continued tendency to divergence of character, why the more ancient a form is, the more it generally differs from those now living. Why ancient and extinct forms often tend to fill up gaps between existing forms, sometimes blending two groups, previously classed as distinct into one; but more commonly bringing them only a little closer together. The more ancient a form is, the more often it stands in some degree intermediate between groups now distinct; for the more ancient a form is, the more nearly it will be related to, and consequently resemble, the common progenitor of groups, since become widely divergent. Extinct forms are seldom directly intermediate between existing forms; but are intermediate only by a long and circuitous course through other extinct and different forms. We can clearly see why the organic remains of closely consecutive formations are closely allied; for they are closely linked together by generation. We can clearly see why the remains of an intermediate formation are intermediate in character.

The inhabitants of the world at each successive period in its history have beaten their predecessors in the race for life, and are, in so far, higher in the scale, and their structure has generally become more specialised; and this may account for the common belief held by so many palaeontologists, that organisation on the whole has progressed. Extinct and ancient animals resemble to a certain extent the embryos of the more recent animals belonging to the same classes, and this wonderful fact receives a simple explanation according to our views. The succession of the same types of structure within the same areas during the later geological periods ceases to be mysterious, and is intelligible on the principle of inheritance.

If, then, the geological record be as imperfect as many believe, and it may at least be asserted that the record cannot be proved to be much more perfect, the main objections to the theory of natural selection are greatly diminished or disappear. On the other hand, all the chief laws of palaeontology plainly proclaim, as it seems to me, that species have been produced by ordinary generation: old forms having been supplanted by new and improved forms of life, the products of Variation and the Survival of the Fittest.

Chapter XII

GEOGRAPHICAL DISTRIBUTION

Present distribution cannot be accounted for by differences in physical conditions–Importance of barriers–Affinity of the productions of the same continent–Centres of creation–Means of dispersal by changes of climate and of the level of the land, and by occasional means–Dispersal during the Glacial period–Alternate Glacial periods in the North and South.

In considering the distribution of organic beings over the face of the globe, the first great fact which strikes us is, that neither the similarity nor the dissimilarity of the inhabitants of various regions can be wholly accounted for by climatal and other physical conditions. Of late, almost every author who has studied the subject has come to this conclusion. The case of America alone would almost suffice to prove its truth; for if we exclude the arctic and northern temperate parts, all authors agree that one of the most fundamental divisions in geographical distribution is that between the New and Old Worlds; yet if we travel over the vast American continent, from the central parts of the United States to its extreme southern point, we meet with the most diversified conditions; humid districts, arid deserts, lofty mountains, grassy plains, forests, marshes, lakes and great rivers, under almost every temperature. There is hardly a climate or condition in the Old World which cannot be paralleled in the New – at least so closely as the same species generally require. No doubt small areas can be pointed out in the Old World hotter than any in the New World; but these are not inhabited by a fauna different from that of the surrounding districts; for it is rare to find a group of organisms confined to a small area, of which the conditions are peculiar in only a slight degree.

Notwithstanding this general parallelism in the conditions of the Old and New Worlds, how widely different are their living productions!

In the southern hemisphere, if we compare large tracts of land in Australia, South Africa, and western South America, between latitudes 25° and 35°, we shall find parts extremely similar in all their conditions, yet it would not be possible to point out three faunas and floras more utterly dissimilar. Or, again, we may compare the productions of South America south of latitude 35° with those north of 25°, which consequently are separated by a space of ten degrees of latitude, and are exposed to considerably different conditions; yet they are incomparably more closely related to each other than they are to the productions of Australia or Africa under nearly the same climate. Analogous facts could be given with respect to the inhabitants of the sea.

A second great fact which strikes us in our general review is, that barriers of any kind, or obstacles to free migration, are related in a close and important manner to the differences between the productions of various regions. We see this in the great differences in nearly all the terrestrial productions of the New and Old Worlds, excepting in the northern parts, where the land almost joins, and where, under a slightly different climate, there might have been free migration for the northern temperate forms, as there now is for the strictly arctic productions. We see the same fact in the great difference between the inhabitants of Australia, Africa, and South America under the same latitude; for these countries are almost as much isolated from each other as is possible. On each continent, also, we see the same fact; for on the opposite sides of lofty and continuous mountain-ranges, of great deserts and even of large rivers, we find different productions; though as mountain-chains, deserts, etc., are not as impassable, or likely to have endured so long, as the oceans separating continents, the differences are very inferior in degree to those characteristic of distinct continents.

Turning to the sea, we find the same law. The marine inhabitants of the eastern and western shores of South America are very distinct, with extremely few shells, crustacea, or echinodermata in common; but Dr. Günther has recently shown that about thirty per cent of the fishes are the same on the opposite sides of the isthmus of Panama; and this fact has led naturalists to believe that the isthmus was formerly open. Westward of the shores of America, a wide space of open ocean extends, with not an island as a halting-place for emigrants; here we have a barrier of another kind, and as soon as this is passed we meet in the eastern islands of the Pacific with another and totally distinct fauna. So that three marine faunas range northward and southward in parallel lines not far from each other, under corresponding climate; but from being separated from each other by impassable barriers, either of land or open sea, they are almost wholly distinct. On the other hand, proceeding still further westward from the eastern islands of the tropical parts of the Pacific, we encounter no impassable barriers, and we have innumerable islands as halting-places, or continuous coasts, until, after travelling over a hemisphere, we come to the shores of Africa; and over this vast space we meet with no well-defined and distinct marine faunas. Although so few marine animals are common to the above-named

three approximate faunas of Eastern and Western America and the eastern Pacific islands, yet many fishes range from the Pacific into the Indian Ocean, and many shells are common to the eastern islands of the Pacific and the eastern shores of Africa on almost exactly opposite meridians of longitude.

A third great fact, partly included in the foregoing statement, is the affinity of the productions of the same continent or of the same sea, though the species themselves are distinct at different points and stations. It is a law of the widest generality, and every continent offers innumerable instances. Nevertheless, the naturalist, in travelling, for instance, from north to south, never fails to be struck by the manner in which successive groups of beings, specifically distinct, though nearly related, replace each other. He hears from closely allied, yet distinct kinds of birds, notes nearly similar, and sees their nests similarly constructed, but not quite alike, with eggs coloured in nearly the same manner. The plains near the Straits of Magellan are inhabited by one species of Rhea (American ostrich), and northward the plains of La Plata by another species of the same genus; and not by a true ostrich or emu, like those inhabiting Africa and Australia under the same latitude. On these same plains of La Plata we see the agouti and bizcacha, animals having nearly the same habits as our hares and rabbits, and belonging to the same order of rodents, but they plainly display an American type of structure. We ascend the lofty peaks of the Cordillera, and we find an alpine species of bizcacha; we look to the waters, and we do not find the beaver or muskrat, but the coypu and capybara, rodents of the South American type. Innumerable other instances could be given. If we look to the islands off the American shore, however much they may differ in geological structure, the inhabitants are essentially American, though they may be all peculiar species. We may look back to past ages, as shown in the last chapter, and we find American types then

Rhea Darwinii.

The South American Rhea, a flightless bird with three-toed feet that enable it to run quite swiftly prevailing on the American continent and in the American seas. We see in these facts some deep organic bond, throughout space and time, over the same areas of land and water, independently of physical conditions. The naturalist must be dull who is not led to inquire what this bond is.

The bond is simply inheritance, that cause which alone, as far as we positively know, produces organisms quite like each other, or, as we see in the case of varieties, nearly alike. The dissimilarity of the inhabitants of different regions may be attributed to modification through variation and natural selection, and probably in a subordinate degree to the definite influence of different physical conditions. The degrees of dissimilarity will depend on the migration of the more dominant forms of life from one region into another having been more or less effectually prevented, at periods more or less remote; – on the nature and number of the former immigrants; – and on the action of the inhabitants on each other in leading to the preservation of different modifications; the relation of organism to organism in the struggle for life being, as I have already often remarked, the most important of all relations. Thus the high importance of barriers comes into play by checking migration; as does time for the slow process of modification through natural selection. Widely-ranging species, abounding in individuals, which have already triumphed over many competitors in their own widely extended homes, will have the best chance of seizing on new places, when they spread out into new countries. In their new homes they will be exposed to new conditions, and will frequently undergo further modification and improvement; and thus they will become still further victorious, and will produce groups of modified descendants. On this principle of inheritance with modification we can understand how it is that sections of genera, whole genera, and even families, are confined to the same areas, as is so commonly and notoriously the case.

There is no evidence, as was remarked in the last chapter, of the existence of any law of necessary development. As the variability of each species is an independent property, and will be taken advantage of by natural selection, only so far as it profits each individual in its complex struggle for life, so the amount of modification in different species will be no uniform quantity. If a number of species, after having long competed with each other in their old home, were to migrate in a body into a new and afterwards isolated country, they would be little liable to modification; for neither migration nor isolation in themselves effect anything. These principles come into play only by bringing organisms into new relations with each other and in a lesser degree with the surrounding physical conditions. As we have seen in the last chapter that some forms have retained nearly the same character from an enormously remote geological period, so certain species have migrated over vast spaces, and have not become greatly or at all modified.

According to these views, it is obvious that the several species of the same genus, though inhabiting the most distant quarters of the world, must originally have proceeded from the same source, as they are descended from the same progenitor. In the case of those species which have undergone, during whole geological periods, little modification, there is not much difficulty in believing that they have migrated from the same region; for during the vast geographical and climatical changes which have supervened since ancient times, almost any amount of migration is possible. But in many other cases, in which we have reason to believe that the species of a genus have been produced within comparatively recent times, there is great difficulty on this head. It is also obvious that the individuals of the same species, though now inhabiting distant and isolated regions, must have proceeded from one spot, where their parents were first produced: for, as has been explained, it is incredible that individuals identically the same should have been produced from parents specifically distinct.

SINGLE CENTRES OF SUPPOSED CREATION

We are thus brought to the question which has been largely discussed by naturalists, namely, whether species have been created at one or more points of the earth's surface. Undoubtedly there are many cases of extreme difficulty in understanding how the same species could possibly have migrated from some one point to the several distant and isolated points, where now found. Nevertheless the simplicity of the view that each species was first produced within a single region captivates the mind. He who rejects it, rejects the *vera causa* of ordinary generation with subsequent migration, and calls in the agency of a miracle. It is universally admitted, that in most cases the area inhabited by a species is continuous; and that when a plant or animal inhabits two points so distant from each other, or with an interval of such a nature, that the space could not have been easily passed over by migration, the fact is given as something remarkable and exceptional. The incapacity of migrating across a wide sea is more clear in the case of terrestrial mammals than perhaps with any other organic beings; and, accordingly, we find no inexplicable instances of the same mammals inhabiting distant points of the world. No geologist feels any difficulty in Great Britain possessing the same quadrupeds with the rest of Europe, for they were no doubt once united. But if the same species can be produced at two separate points, why do we not find a single mammal common to Europe and Australia or South America? The conditions of life are nearly the same, so that a multitude of European animals and plants have become naturalised in America and Australia; and some of the aboriginal plants are identically the same at these distant points of the northern and southern hemispheres? The answer, as I believe, is, that mammals have not been able to migrate, whereas some plants, from their varied means of dispersal, have migrated across the wide and broken interspaces. The great and striking influence of barriers of all kinds, is intelligible only on the view that the great

majority of species have been produced on one side, and have not been able to migrate to the opposite side. Some few families, many subfamilies, very many genera, a still greater number of sections of genera, are confined to a single region; and it has been observed by several naturalists that the most natural genera, or those genera in which the species are most closely related to each other, are generally confined to the same country, or if they have a wide range that their range is continuous. What a strange anomaly it would be if a directly opposite rule were to prevail when we go down one step lower in the series, namely to the individuals of the same species, and these had not been, at least at first, confined to some one region!

Hence, it seems to me, as it has to many other naturalists, that the view of each species having been produced in one area alone, and having subsequently migrated from that area as far as its powers of migration and subsistence under past and present conditions permitted, is the most probable. Undoubtedly many cases occur in which we cannot explain how the same species could have passed from one point to the other. But the geographical and climatical changes which have certainly occurred within recent geological times, must have rendered discontinuous the formerly continuous range of many species. So that we are reduced to consider whether the exceptions to continuity of range are so numerous, and of so grave a nature, that we ought to give up the belief, rendered probable by general considerations, that each species has been produced within one area, and has migrated thence as far as it could. It would be hopelessly tedious to discuss all the exceptional cases of the same species, now living at distant and separated points; nor do I for a moment pretend that any explanation could be offered of many instances. But, after some preliminary remarks, I will discuss a few of the most striking classes of facts, namely, the existence of the same species on the summits of distant mountain ranges, and at distant points in the arctic and antarctic regions; and secondly (in the following chapter), the wide distribution of fresh water productions; and thirdly, the occurrence of the same terrestrial species on islands and on the nearest mainland, though separated by hundreds of miles of open sea. If the existence of the same species at distant and isolated points of the earth's surface can in many instances be explained on the view of each species having migrated from a single birthplace; then, considering our ignorance with respect to former climatical and geographical changes, and to the various occasional means of transport, the belief that a single birthplace is the law seems to me incomparably the safest.

In discussing this subject we shall be enabled at the same time to consider a point equally important for us, namely, whether the several species of a genus which must on our theory all be descended from a common progenitor, can have migrated, undergoing modification during their migration from some one area. If, when most of the species inhabiting one region are different from those of another region, though closely allied to them, it can be shown that migration from the one region to the other has probably occurred at some former period, our general view will be much strengthened; for the explanation is obvious on the principle of descent with modification. A volcanic island, for instance, upheaved and formed at the distance of a few hundreds of miles from a continent, would probably receive from it in the course of time a few colonists, and their descendants, though modified, would still be related by inheritance to the inhabitants of that continent. Cases of this nature are common, and are, as we shall hereafter see, inexplicable on the theory of independent creation. This view of the relation of the species of one region to those of another, does not differ much from that advanced by Mr. Wallace, who concludes that 'every species has come into existence coincident both in space and time with a pre-existing closely allied species.' And it is now well known that he attributes this coincidence to descent with modification.

The question of single or multiple centres of creation differs from another though allied question, namely, whether all the individuals of the same species are descended from a single pair, or single hermaphrodite, or whether, as some authors suppose, from many individuals simultaneously created. With organic beings which never intercross, if such exist, each species, must be descended from a succession of modified varieties, that have supplanted each other, but have never blended with other individuals or varieties of the same species, so that, at each successive stage of modification, all the individuals of the same form will be descended from a single parent. But in the great majority of cases, namely, with all organisms which habitually unite for each birth, or which occasionally intercross, the individuals of the same species inhabiting the same area will be kept nearly uniform by intercrossing; so that many individuals will go on simultaneously changing, and the whole amount of modification at each stage will not be due to descent from a single parent. To illustrate what I mean: our English race horses differ from the horses of every other breed; but they do not owe their difference and superiority to descent from any single pair, but to continued care in the selecting and training of many individuals during each generation.

Before discussing the three classes of facts, which I have selected as presenting the greatest amount of difficulty on the theory of 'single centres of creation,' I must say a few words on the means of dispersal.

MEANS OF DISPERSAL
Sir C. Lyell and other authors have ably treated this subject. I can give here only the briefest abstract of the more important facts. Change of climate must have had a powerful influence on migration. A region now impassable to certain organisms from the nature of its climate, might have been a high road for migration, when the climate was different. I shall, however, presently have to discuss this branch of the subject in some

detail. Changes of level in the land must also have been highly influential: a narrow isthmus now separates two marine faunas; submerge it, or let it formerly have been submerged, and the two faunas will now blend together, or may formerly have blended. Where the sea now extends, land may at a former period have connected islands or possibly even continents together, and thus have allowed terrestrial productions to pass from one to the other. No geologist disputes that great mutations of level have occurred within the period of existing organisms. Edward Forbes insisted that all the islands in the Atlantic must have been recently connected with Europe or Africa, and Europe likewise with America. Other authors have thus hypothetically bridged over every ocean, and united almost every island with some mainland. If, indeed, the arguments used by Forbes are to be trusted, it must be admitted that scarcely a single island exists which has not recently been united to some continent. This view cuts the Gordian knot of the dispersal of the same species to the most distant points, and removes many a difficulty; but to the best of my judgment we are not authorized in admitting such enormous geographical changes within the period of existing species. It seems to me that we have abundant evidence of great oscillations in the level of the land or sea; but not of such vast changes in the position and extension of our continents, as to have united them within the recent period to each other and to the several intervening oceanic islands. I freely admit the former existence of many islands, now buried beneath the sea, which may have served as halting places for plants and for many animals during their migration. In the coral-producing oceans such sunken islands are now marked by rings of coral or atolls standing over them. Whenever it is fully admitted, as it will some day be, that each species has proceeded from a single birthplace, and when in the course of time we know something definite about the means of distribution, we shall be enabled to speculate with security on the former extension of the land. But I

do not believe that it will ever be proved that within the recent period most of our continents which now stand quite separate, have been continuously, or almost continuously united with each other, and with the many existing oceanic islands. Several facts in distribution – such as the great difference in the marine faunas on the opposite sides of almost every continent, – the close relation of the tertiary inhabitants of several lands and even seas to their present inhabitants, – the degree of affinity between the mammals inhabiting islands with those of the nearest continent, being in part determined (as we shall hereafter see) by the depth of the intervening ocean, – these and other such facts are opposed to the admission of such prodigious geographical revolutions within the recent period, as are necessary on the view advanced by Forbes and admitted by his followers. The nature and relative proportions of the inhabitants of oceanic islands are likewise opposed to the belief of their former continuity of continents. Nor does the almost universally volcanic composition of such islands favour the admission that they are the wrecks of sunken continents; if they had originally existed as continental mountain ranges, some at least of the islands would have been formed, like other mountain summits, of granite, metamorphic schists, old fossiliferous and other rocks, instead of consisting of mere piles of volcanic matter.

I must now say a few words on what are called accidental means, but which more properly should be called occasional means of distribution. I shall here confine myself to plants. In botanical works, this or that plant is often stated to be ill adapted for wide dissemination; but the greater or less facilities for transport across the sea may be said to be almost wholly unknown. Until I tried, with Mr. Berkeley's aid, a few experiments, it was not even known how far seeds could resist the injurious action of sea-water. To my surprise I found that out of eighty-seven kinds, sixty-four germinated after an immersion of twenty-eight days, and a few survived an immersion of 137 days. It deserves notice that certain orders were far more injured than others: nine Leguminosae were tried, and, with one exception, they resisted the salt water badly; seven species of the allied orders, Hydrophyllaceae and Polemoniaceae, were all killed by a month's immersion. For convenience sake I chiefly tried small seeds without the capsules or fruit; and as all of these sank in a few days, they could not have been floated across wide spaces of the sea, whether or not they were injured by salt water. Afterwards I tried some larger fruits, capsules, etc., and some of these floated for a long time. It is well known what a difference there is in the buoyancy of green and seasoned timber; and it occurred to me that floods would often wash into the sea dried plants or branches with seed-capsules or fruit attached to them. Hence I was led to dry the stems and branches of ninety-four plants with ripe fruit, and to place them on sea-water. The majority sank quickly, but some which, whilst green, floated for a very short time, when dried floated much longer; for instance, ripe hazelnuts sank immediately, but when dried they floated for ninety days, and afterwards when planted germinated; an asparagus plant with ripe berries floated for twenty-three days, when dried it floated for eighty-five days, and the seeds afterwards germinated: the ripe seeds of Helosciadium sank in two days, when dried they floated for above ninety days, and afterwards germinated. Altogether, out of the ninety-four dried plants, eighteen floated for above twenty-eight days; and some of the eighteen floated for a very much longer period. So that as 64/87 kinds of seeds germinated after an immersion of twenty-eight days; and as 18/94 distinct species with ripe fruit (but not all the same species as in the foregoing experiment) floated, after being dried, for above twenty-eight days, we may conclude, as far as anything can be inferred from these scanty facts, that the seeds of 14/100 kinds of plants of any country might be floated by sea-currents during twenty-eight days, and would retain their power of germination. In Johnston's Physical Atlas, the average rate of the several Atlantic currents is thirty-three miles per diem (some currents running at the rate of sixty miles per diem); on this average, the seeds of 14/100 plants belonging to one country might be floated across 924 miles of sea to another country; and when stranded, if blown by an inland gale to a favourable spot, would germinate.

Subsequently to my experiments, M. Martens tried similar ones, but in a much better manner, for he placed the seeds in a box in the actual sea, so that they were alternately wet and exposed to the air like really floating plants. He tried ninety-eight seeds, mostly different from mine, but he chose many large fruits, and likewise seeds, from plants which live near the sea; and this would have favoured both the average length of their flotation and their resistance to the injurious action of the salt water. On the other hand, he did not previously dry the plants or branches with the fruit; and this, as we have seen, would have caused some of them to have floated much longer. The result was that 18/98 of his seeds of different kinds floated for forty-two days, and were then capable of germination. But I do not doubt that plants exposed to the waves would float for a less time than those protected from violent movement as in our experiments. Therefore, it would perhaps be safer to assume that the seeds of about 10/100 plants of a flora, after having been dried, could be floated across a space of sea 900 miles in width, and would then germinate. The fact of the larger fruits often floating longer than the small, is interesting; as plants with large seeds or fruit which, as Alph. de Candolle has shown, generally have restricted ranges, could hardly be transported by any other means.

Seeds may be occasionally transported in another manner. Drift timber is thrown up on most islands, even on those in the midst of the widest oceans; and the natives of the coral islands in the Pacific procure stones for their tools, solely from the roots of drifted trees, these stones being a valuable royal tax. I find that when irregularly shaped stones are embedded in the roots of trees, small parcels of earth are very frequently enclosed in their interstices and behind them, so perfectly that not a particle could be washed away during the longest transport: out of one small portion of earth thus *completely* enclosed by the roots of an oak about fifty years old, three dicotyledonous plants germinated: I am certain of the accuracy of this observation. Again, I can show that the carcasses of birds, when floating on the sea, sometimes escape being immediately devoured; and many kinds of seeds in the crops of floating birds long retain their vitality: peas and vetches, for instance, are killed by even a few days' immersion in sea-water; but some taken out of the crop of a pigeon, which had floated on artificial sea-water for thirty days, to my surprise nearly all germinated.

Living birds can hardly fail to be highly effective agents in the transportation of seeds. I could give many facts showing how frequently birds of many kinds are blown by gales to vast distances across the ocean. We may safely assume that under such circumstances their rate of flight would often be thirty-five miles an hour; and some authors have given a far higher estimate. I have never seen an instance of nutritious seeds passing through the intestines of a bird; but hard seeds of fruit pass uninjured through even the digestive organs of a turkey. In the course of two months, I picked up in my garden twelve kinds of seeds, out of the excrement of small birds, and these seemed perfect, and some of them, which were tried, germinated. But the following fact is more important: the crops of birds do not secrete gastric juice, and do not, as I know by trial, injure in the least the germination of seeds; now, after a bird has found and devoured a large supply of food, it is positively asserted that all the grains do not pass into the gizzard for twelve or even eighteen hours. A bird in this interval might easily be blown to the distance of five hundred miles, and hawks are known to look out for tired birds, and the contents of their torn crops might thus readily get

scattered. Some hawks and owls bolt their prey whole, and after an interval of from twelve to twenty hours, disgorge pellets, which, as I know from experiments made in the Zoological Gardens, include seeds capable of germination. Some seeds of the oat, wheat, millet, canary, hemp, clover, and beet germinated after having been from twelve to twenty-one hours in the stomachs of different birds of prey; and two seeds of beet grew after having been thus retained for two days and fourteen hours. Fresh-water fish, I find, eat seeds of many land and water plants; fish are frequently devoured by birds, and thus the seeds might be transported from place to place. I forced many kinds of seeds into the stomachs of dead fish, and then gave their bodies to fishing-eagles, storks, and pelicans; these birds, after an interval of many hours, either rejected the seeds in pellets or passed them in their excrement; and several of these seeds retained the power of germination. Certain seeds, however, were always killed by this process.

Locusts are sometimes blown to great distances from the land. I myself caught one 370 miles from the coast of Africa, and have heard of others caught at greater distances. The Rev. R.T. Lowe informed Sir C. Lyell that in November, 1844, swarms of locusts visited the island of Madeira. They were in countless numbers, as thick as the flakes of snow in the heaviest snowstorm, and extended upward as far as could be seen with a telescope. During two or three days they slowly careered round and round in an immense ellipse, at least five or six miles in diameter, and at night alighted on the taller trees, which were completely coated with them. They then disappeared over the sea, as suddenly as they had appeared, and have not since visited the island. Now, in parts of Natal it is believed by some farmers, though on insufficient evidence, that injurious seeds are introduced into their grass-land in the dung left by the great flights of locusts which often visit that country. In consequence of this belief Mr. Weale sent me in a

The Locust, the 'migratory grasshopper'

letter a small packet of the dried pellets, out of which I extracted under the microscope several seeds, and raised from them seven grass plants, belonging to two species, of two genera. Hence a swarm of locusts, such as that which visited Madeira, might readily be the means of introducing several kinds of plants into an island lying far from the mainland.

Although the beaks and feet of birds are generally clean, earth sometimes adheres to them: in one case I removed sixty-one grains, and in another case twenty-two grains of dry argillaceous earth from the foot of a partridge, and in the earth there was a pebble as large as the seed of a vetch. Here is a better case: the leg of a woodcock was sent to me by a friend, with a little cake of dry earth attached to the shank, weighing only nine grains; and this contained a seed of the toad-rush (Juncus bufonius) which germinated and flowered. Mr. Swaysland, of Brighton, who during the last forty years has paid close attention to our migratory birds, informs me that he has often shot wagtails (Motacillae), wheatears, and whinchats (Saxicolae), on their first arrival on our shores, before they had alighted; and he has several times noticed little cakes of earth attached to their feet. Many facts could be given showing how generally soil is charged with seeds. For instance, Professor Newton sent me the leg of a red-legged partridge (Caccabis rufa) which had been wounded and could not fly, with a ball of hard earth adhering to it, and weighing six and a half ounces. The earth had been kept for three years, but when broken, watered and placed under a bell glass, no less than eighty-two plants sprung from it: these consisted of twelve monocotyledons, including the common oat, and at least one kind of grass, and of seventy dicotyledons, which consisted, judging from the young leaves, of at least three distinct species. With such facts before us, can we doubt that the many birds which are annually blown by gales across great spaces of ocean, and which annually migrate – for instance, the millions of quails across the Mediterranean – must occasionally transport a few seeds embedded in dirt adhering to their feet or beaks? But I shall have to recur to this subject.

As icebergs are known to be sometimes loaded with earth and stones, and have even carried brushwood, bones, and the nest of a land bird, it can hardly be doubted that they must occasionally, as suggested by Lyell, have transported seeds from one part to another of the arctic and antarctic regions; and during the Glacial period from one part of the now temperate regions to another. In the Azores, from the large number of plants common to Europe, in comparison with the species on the other islands of the Atlantic, which stand nearer to the mainland, and (as remarked by Mr. H.C. Watson) from their somewhat northern character, in comparison with the latitude, I suspected that these islands had been partly stocked by ice-borne seeds during the Glacial epoch. At my request Sir C. Lyell wrote to M. Hartung to inquire whether he had observed erratic boulders on these islands, and he answered that he had found large fragments of granite and other rocks, which do not occur in the archipelago. Hence we may safely infer that icebergs formerly landed their rocky burdens on the shores of these mid-ocean islands, and it is at least possible that they may have brought thither the seeds of northern plants.

Considering that these several means of transport, and that other means, which without doubt remain to be discovered, have been in action year after year for tens of thousands of years, it would, I think, be a marvellous fact if many plants had not thus become widely transported. These means of transport are sometimes called accidental, but this is not strictly correct: the currents of the sea are not accidental, nor is the direction of prevalent gales of wind. It should be observed that scarcely any means of transport would carry seeds for very great distances; for seeds do not retain their vitality when exposed for a great length of time to the action of sea water; nor could they be long carried in the crops or intestines of birds. These means, however, would suffice for occasional transport across tracts of sea some hundred miles in breadth, or from island to island, or from a continent to a neighbouring island, but not from one distant continent to another. The floras of distant continents would not by such means become mingled; but would remain as distinct as they now are. The currents, from their course, would never bring seeds from North America to Britain, though they might and do bring seeds from the West Indies to our western shores, where, if not killed by their very long immersion in salt water, they could not endure our climate. Almost every year, one or two land birds are blown across the whole Atlantic Ocean, from North America to the western shores of Ireland and England; but seeds could be transported by these rare wanderers only by one means, namely, by dirt adhering to their feet or beaks, which is in itself a rare accident. Even in this case, how small would be the chance of a seed falling on favourable soil, and coming to maturity! But it would be a great error to argue that because a well-stocked island, like Great Britain, has not, as far as is known (and it would be very difficult to prove this), received within the last few centuries, through occasional means of transport, immigrants from Europe or any other continent, that a poorly-stocked island, though standing more remote from the mainland, would not receive colonists by similar means. Out of a hundred kinds of seeds or animals transported to an island, even if far less well-stocked than Britain, perhaps not more than one would be so well fitted to its new home, as to become naturalised. But this is no valid argument against what would be effected by occasional means of transport, during the long lapse of geological time, whilst the island was being upheaved, and before it had become fully stocked with inhabitants. On almost bare land, with few or no destructive insects or birds living there, nearly every seed which chanced to arrive, if fitted for the climate, would germinate and survive.

DISPERSAL DURING THE GLACIAL PERIOD

The identity of many plants and animals, on mountain-summits, separated from each other by hundreds of miles of lowlands, where Alpine species could not possibly exist, is one of the most striking cases known of the same species living at distant points, without the apparent possibility of their having migrated from one point to the other. It is indeed a remarkable fact to see so many plants of the same species living on the snowy regions of the Alps or Pyrenees, and in the extreme northern parts of Europe; but it is far more remarkable, that the plants on the White Mountains, in the United States of America, are all the same with those of Labrador, and nearly all the same, as we hear from Asa Gray, with those on the loftiest mountains of Europe. Even as long ago as 1747, such facts led Gmelin to conclude that the same species must have been independently created at many distinct points; and we might have remained in this same belief, had not Agassiz and others called vivid attention to the Glacial period, which, as we shall immediately see, affords a simple explanation of these facts. We have evidence of almost every conceivable kind, organic and inorganic, that, within a very recent geological period, central Europe and North America suffered under an arctic climate. The ruins of a house burnt by fire do not tell their tale more plainly than do the mountains of Scotland and Wales, with their scored flanks, polished surfaces, and perched boulders, of the icy streams with

which their valleys were lately filled. So greatly has the climate of Europe changed, that in Northern Italy, gigantic moraines, left by old glaciers, are now clothed by the vine and maize. Throughout a large part of the United States, erratic boulders and scored rocks plainly reveal a former cold period.

The former influence of the glacial climate on the distribution of the inhabitants of Europe, as explained by Edward Forbes, is substantially as follows. But we shall follow the changes more readily, by supposing a new glacial period slowly to come on, and then pass away, as formerly occurred. As the cold came on, and as each more southern zone became fitted for the inhabitants of the north, these would take the places of the former inhabitants of the temperate regions. The latter, at the same time would travel further and further southward, unless they were stopped by barriers, in which case they would perish. The mountains would become covered with snow and ice, and their former Alpine inhabitants would descend to the plains. By the time that the cold had reached its maximum, we should have an arctic fauna and flora, covering the central parts of Europe, as far south as the Alps and Pyrenees, and even stretching into Spain. The now temperate regions of the United States would likewise be covered by arctic plants and animals and these would be nearly the same with those of Europe; for the present circumpolar inhabitants, which we suppose to have everywhere travelled southward, are remarkably uniform round the world.

As the warmth returned, the arctic forms would retreat northward, closely followed up in their retreat by the productions of the more temperate regions. And as the snow melted from the bases of the mountains, the arctic forms would seize on the cleared and thawed ground, always ascending, as the warmth increased and the snow still further disappeared, higher and higher, whilst their brethren were pursuing their northern journey. Hence, when the warmth had fully returned, the same species, which had lately lived together on the European and North American lowlands, would again be found in the arctic regions of the Old and New Worlds, and on many isolated mountain summits far distant from each other.

Thus we can understand the identity of many plants at points so immensely remote as the mountains of the United States and those of Europe. We can thus also understand the fact that the Alpine plants of each mountain range are more especially related to the arctic forms living due north or nearly due north of them: for the first migration when the cold came on, and the re-migration on the returning warmth, would generally have been due south and north. The Alpine plants, for example, of Scotland, as remarked by Mr. H.C. Watson, and those of the

Alpine flora

Pyrenees, as remarked by Ramond, are more especially allied to the plants of northern Scandinavia; those of the United States to Labrador; those of the mountains of Siberia to the arctic regions of that country. These views, grounded as they are on the perfectly well-ascertained occurrence of a former Glacial period, seem to me to explain in so satisfactory a manner the present distribution of the Alpine and arctic productions of Europe and America, that when in other regions we find the same species on distant mountain summits, we may almost conclude, without other evidence, that a colder climate formerly permitted their migration across the intervening lowlands, now become too warm for their existence.

As the arctic forms moved first southward and afterwards backward to the north, in unison with the changing climate, they will not have been exposed during their long migrations to any great diversity of temperature; and as they all migrated in a body together, their mutual relations will not have been much disturbed. Hence, in accordance with the principles inculcated in this volume, these forms will not have been liable to much modification. But with the Alpine productions, left isolated from the moment of the returning warmth, first at the bases and ultimately on the summits of the mountains, the case will have been somewhat different; for it is not likely that all the same arctic species will have been left on mountain ranges far distant from each other, and have survived there ever since; they will also, in all probability, have become mingled with ancient Alpine species, which must have existed on the mountains before the commencement of the Glacial epoch, and which during the coldest period will have been temporarily driven down to the plains; they will, also, have been subsequently exposed to somewhat different climatical influences. Their mutual relations will thus have been in some degree disturbed; consequently they will have been liable to modification; and they have been modified; for if we compare the present

Alpine plants and animals of the several great European mountain ranges, one with another, though many of the species remain identically the same, some exist as varieties, some as doubtful forms or sub-species and some as distinct yet closely allied species representing each other on the several ranges.

In the foregoing illustration, I have assumed that at the commencement of our imaginary Glacial period, the arctic productions were as uniform round the polar regions as they are at the present day. But it is also necessary to assume that many sub-arctic and some few temperate forms were the same round the world, for some of the species which now exist on the lower mountain slopes and on the plains of North America and Europe are the same; and it may be asked how I account for this degree of uniformity of the sub-arctic and temperate forms round the world, at the commencement of the real Glacial period. At the present day, the sub-arctic and northern temperate productions of the Old and New Worlds are separated from each other by the whole Atlantic Ocean and by the northern part of the Pacific. During the Glacial period, when the inhabitants of the Old and New Worlds lived further southwards than they do at present, they must have been still more completely separated from each other by wider spaces of ocean; so that it may well be asked how the same species could then or previously have entered the two continents. The explanation, I believe, lies in the nature of the climate before the commencement of the Glacial period. At this, the newer Pliocene period, the majority of the inhabitants of the world were specifically the same as now, and we have good reason to believe that the climate was warmer than at the present day. Hence, we may suppose that the organisms which now live under latitude 60°, lived during the Pliocene period further north, under the Polar Circle, in latitude 66-67°; and that the present arctic productions then lived on the broken land still nearer to the pole. Now, if we look at a terrestrial globe, we see under the Polar Circle

that there is almost continuous land from western Europe through Siberia, to eastern America. And this continuity of the circumpolar land, with the consequent freedom under a more favourable climate for intermigration, will account for the supposed uniformity of the sub-arctic and temperate productions of the Old and New Worlds, at a period anterior to the Glacial epoch.

Believing, from reasons before alluded to, that our continents have long remained in nearly the same relative position, though subjected to great oscillations of level, I am strongly inclined to extend the above view, and to infer that during some earlier and still warmer period, such as the older Pliocene period, a large number of the same plants and animals inhabited the almost continuous circumpolar land; and that these plants and animals, both in the Old and New Worlds, began slowly to migrate southwards as the climate became less warm, long before the commencement of the Glacial period. We now see, as I believe, their descendants, mostly in a modified condition, in the central parts of Europe and the United States. On this view we can understand the relationship with very little identity, between the productions of North America and Europe, – a relationship which is highly remarkable, considering the distance of the two areas, and their separation by the whole Atlantic Ocean. We can further understand the singular fact remarked on by several observers that the productions of Europe and America during the later tertiary stages were more closely related to each other than they are at the present time; for during these warmer periods the northern parts of the Old and New Worlds will have been almost continuously united by land, serving as a bridge, since rendered impassable by cold, for the intermigration of their inhabitants.

During the slowly decreasing warmth of the Pliocene period, as soon as the species in common, which inhabited the New and Old Worlds, migrated south of the Polar Circle, they will have been completely cut off from each other.

This separation, as far as the more temperate productions are concerned, must have taken place long ages ago. As the plants and animals migrated southward, they will have become mingled in the one great region with the native American productions, and would have had to compete with them; and in the other great region, with those of the Old World. Consequently we have here everything favourable for much modification, – for far more modification than with the Alpine productions, left isolated, within a much more recent period, on the several mountain ranges and on the arctic lands of Europe and North America. Hence, it has come, that when we compare the now living productions of the temperate regions of the New and Old Worlds, we find very few identical species (though Asa Gray has lately shown that more plants are identical than was formerly supposed), but we find in every great class many forms, which some naturalists rank as geographical races, and others as distinct species; and a host of closely allied or representative forms which are ranked by all naturalists as specifically distinct.

As on the land, so in the waters of the sea, a slow southern migration of a marine fauna, which, during the Pliocene or even a somewhat earlier period, was nearly uniform along the continuous shores of the Polar Circle, will account, on the theory of modification, for many closely allied forms now living in marine areas completely sundered. Thus, I think, we can understand the presence of some closely allied, still existing and extinct tertiary forms, on the eastern and western shores of temperate North America; and the still more striking fact of many closely allied crustaceans (as described in Dana's admirable work), some fish and other marine animals, inhabiting the Mediterranean and the seas of Japan, – these two areas being now completely separated by the breadth of a whole continent and by wide spaces of ocean.

These cases of close relationship in species either now or formerly inhabiting the seas on the eastern and western shores of North America, the Mediterranean and Japan, and the temperate lands of North America and Europe, are inexplicable on the theory of creation. We cannot maintain that such species have been created alike, in correspondence with the nearly similar physical conditions of the areas; for if we compare, for instance, certain parts of South America with parts of South Africa or Australia, we see countries closely similar in all their physical conditions, with their inhabitants utterly dissimilar.

ALTERNATE GLACIAL PERIODS IN THE NORTH AND SOUTH

But we must return to our more immediate subject. I am convinced that Forbes's view may be largely extended. In Europe we meet with the plainest evidence of the Glacial period, from the western shores of Britain to the Ural range, and southward to the Pyrenees. We may infer from the frozen mammals and nature of the mountain vegetation, that Siberia was similarly affected. In the Lebanon, according to Dr. Hooker, perpetual snow formerly covered the central axis, and fed glaciers which rolled 4,000 feet down the valleys. The same observer has recently found great moraines at a low level on the Atlas range in North Africa. Along the Himalaya, at points 900 miles apart, glaciers have left the marks of their former low descent; and in Sikkim, Dr. Hooker saw maize growing on ancient and gigantic moraines. Southward of the Asiatic continent, on the opposite side of the equator, we know, from the excellent researches of Dr. J. Haast and Dr. Hector, that in New Zealand immense glaciers formerly descended to a low level; and the same plants, found by Dr. Hooker on widely separated mountains in this island tell the same story of a former cold period. From facts communicated to me by the Rev. W.B. Clarke, it appears also that there are traces of former glacial action on the mountains of the south-eastern corner of Australia.

Looking to America: in the northern half, ice-borne fragments of rock have been observed on the eastern side of the continent, as far south as latitude 36°–37°, and on the shores of the Pacific, where the climate is now so different, as far south as latitude 46°. Erratic boulders have, also, been noticed on the Rocky Mountains. In the Cordillera of South America, nearly under the equator, glaciers once extended far below their present level. In central Chile I examined a vast mound of detritus with great boulders, crossing the Portillo valley, which, there can hardly be a doubt, once formed a huge moraine; and Mr. D. Forbes informs me that he found in various parts of the Cordillera, from latitude 13° to 30°S., at about the height of 12,000 feet, deeply-furrowed rocks, resembling those with which he was familiar in Norway, and likewise great masses of detritus, including grooved pebbles. Along this whole space of the Cordillera true glaciers do not now exist even at much more considerable heights. Further south, on both sides of the continent, from latitude 41° to the southernmost extremity, we have the clearest evidence of former glacial action, in numerous immense boulders transported far from their parent source.

From these several facts, namely, from the glacial action having extended all round the northern and southern hemispheres – from the period having been in a geological sense recent in both hemispheres – from its having lasted in both during a great length of time, as may be inferred from the amount of work effected – and lastly, from glaciers having recently descended to a low level along the whole line of the Cordillera, it at one time appeared to me that we could not avoid the conclusion that the temperature of the whole world had been simultaneously lowered during the Glacial period. But now, Mr. Croll, in a series of admirable memoirs, has attempted to show that a glacial condition of climate is the result of various physical causes, brought into operation by an increase in the eccentricity of the earth's orbit. All these causes tend towards the same end; but the most powerful appears to be the indirect influence of the eccentricity of the orbit upon

oceanic currents. According to Mr. Croll, cold periods regularly recur every ten or fifteen thousand years; and these at long intervals are extremely severe, owing to certain contingencies, of which the most important, as Sir C. Lyell has shown, is the relative position of the land and water. Mr. Croll believes that the last great glacial period occurred about 240,000 years ago, and endured, with slight alterations of climate, for about 160,000 years. With respect to more ancient glacial periods, several geologists are convinced, from direct evidence, that such occurred during the Miocene and Eocene formations, not to mention still more ancient formations. But the most important result for us, arrived at by Mr. Croll, is that whenever the northern hemisphere passes through a cold period the temperature of the southern hemisphere is actually raised, with the winters rendered much milder, chiefly through changes in the direction of the ocean currents. So conversely it will be with the northern hemisphere, whilst the southern passes through a glacial period. This conclusion throws so much light on geographical distribution that I am strongly inclined to trust in it; but I will first give the facts which demand an explanation.

In South America, Dr. Hooker has shown that besides many closely allied species, between forty and fifty of the flowering plants of Tierra del Fuego, forming no inconsiderable part of its scanty flora, are common to North America and Europe, enormously remote as these areas in opposite hemispheres are from each other. On the lofty mountains of equatorial America a host of peculiar species belonging to European genera occur. On the Organ Mountains of Brazil some few temperate European, some Antarctic and some Andean genera were found by Gardner which do not exist in the low intervening hot countries. On the Silla of Caraccas, the illustrious Humboldt long ago found species belonging to genera characteristic of the Cordillera.

In Africa, several forms characteristic of Europe, and some few representatives of the flora of the Cape of Good Hope, occur on the mountains of Abyssinia. At the Cape of Good Hope a very few European species, believed not to have been introduced by man, and on the mountains several representative European forms are found which have not been discovered in the intertropical parts of Africa. Dr. Hooker has also lately shown that several of the plants living on the upper parts of the lofty island of Fernando Po, and on the neighbouring Cameroon Mountains, in the Gulf of Guinea, are closely related to those on the mountains of Abyssinia, and likewise to those of temperate Europe. It now also appears, as I hear from Dr. Hooker, that some of these same temperate plants have been discovered by the Rev. R.T. Lowe on the mountains of the Cape Verde Islands. This extension of the same temperate forms, almost under the equator, across the whole continent of Africa and to the mountains of the Cape Verde archipelago, is one of the most astonishing facts ever recorded in the distribution of plants.

On the Himalaya, and on the isolated mountain ranges of the peninsula of India, on the heights of Ceylon, and on the volcanic cones of Java, many plants occur either identically the same or representing each other, and at the same time representing plants of Europe not found in the intervening hot lowlands. A list of the genera of plants collected on the loftier peaks of Java, raises a picture of a collection made on a hillock in Europe. Still more striking is the fact that peculiar Australian forms are represented by certain plants growing on the summits of the mountains of Borneo. Some of these Australian forms, as I hear from Dr. Hooker, extend along the heights of the peninsula of Malacca, and are thinly scattered on the one hand over India, and on the other hand as far north as Japan.

On the southern mountains of Australia, Dr. F. Müller has discovered several European species; other species, not introduced by man, occur on the lowlands; and a long list can be given, as I am informed by Dr. Hooker, of European genera, found in Australia, but not in the intermediate torrid regions. In the admirable 'Introduction to the Flora of New Zealand,' by Dr. Hooker, analogous and striking facts are given in regard to the plants of that large island. Hence, we see that certain plants growing on the more lofty mountains of the tropics in all parts of the world, and on the temperate plains of the north and south, are either the same species or varieties of the same species. It should, however, be observed that these plants are not strictly arctic forms; for, as Mr. H.C. Watson has remarked, 'in receding from polar toward equatorial latitudes, the Alpine or mountain flora really become less and less Arctic.' Besides these identical and closely allied forms, many species inhabiting the same widely sundered areas, belong to genera not now found in the intermediate tropical lowlands.

These brief remarks apply to plants alone; but some few analogous facts could be given in regard to terrestrial animals. In marine productions, similar cases likewise occur; as an example, I may quote a statement by the highest authority, Professor Dana, that 'it is certainly a wonderful fact that New Zealand should have a closer resemblance in its crustacea to Great Britain, its antipode, than to any other part of the world.' Sir J. Richardson, also, speaks of the reappearance on the shores of New Zealand, Tasmania, etc., of northern forms of fish. Dr. Hooker informs me that twenty-five species of Algae are common to New Zealand and to Europe, but have not been found in the intermediate tropical seas.

From the foregoing facts, namely, the presence of temperate forms on the highlands across the whole of equatorial Africa, and along the Peninsula of India, to Ceylon and the Malay Archipelago, and in a less well-marked manner across the wide expanse of tropical South America, it appears almost certain that at some former period, no doubt during the most severe part of a Glacial period, the lowlands of these great continents were everywhere tenanted under the equator by a considerable number of temperate forms. At this period the equatorial climate at the level of the sea was probably about the same with that now experienced at the height of from five to six thousand feet under the same latitude, or perhaps even rather cooler. During this, the coldest period, the lowlands under the equator must have been clothed with a mingled tropical and temperate vegetation, like that described by Hooker as growing luxuriantly at the height of from four to five thousand feet on the lower slopes of the Himalaya, but with perhaps a still greater preponderance of temperate forms. So again in the mountainous island of Fernando Po, in the Gulf of Guinea, Mr. Mann found temperate European forms beginning to appear at the height of about five thousand feet. On the mountains of Panama, at the height of only two thousand feet, Dr. Seemann found the vegetation like that of Mexico, 'with forms of the torrid zone harmoniously blended with those of the temperate.'

Now let us see whether Mr. Croll's conclusion that when the northern hemisphere suffered from the extreme cold of the great Glacial period, the southern hemisphere was actually warmer, throws any clear light on the present apparently inexplicable distribution of various organisms in the temperate parts of both hemispheres, and on the mountains of the tropics. The Glacial period, as measured by years, must have been very long; and when we remember over what vast spaces some naturalised plants and animals have spread within a few centuries, this period will have been ample for any amount of migration. As the cold became more and more intense, we know that Arctic forms invaded the temperate regions; and from the facts just given, there can hardly be a doubt that some of the more vigorous, dominant and widest-spreading temperate forms invaded the equatorial lowlands. The inhabitants of these hot lowlands would at the same time have migrated to the tropical and subtropical regions of the south, for the southern hemisphere was at this period warmer. On the

decline of the Glacial period, as both hemispheres gradually recovered their former temperature, the northern temperate forms living on the lowlands under the equator, would have been driven to their former homes or have been destroyed, being replaced by the equatorial forms returning from the south. Some, however, of the northern temperate forms would almost certainly have ascended any adjoining high land, where, if sufficiently lofty, they would have long survived like the arctic forms on the mountains of Europe. They might have survived, even if the climate was not perfectly fitted for them, for the change of temperature must have been very slow, and plants undoubtedly possess a certain capacity for acclimatisation, as shown by their transmitting to their offspring different constitutional powers of resisting heat and cold.

In the regular course of events the southern hemisphere would in its turn be subjected to a severe Glacial period, with the northern hemisphere rendered warmer; and then the southern temperate forms would invade the equatorial lowlands. The northern forms which had before been left on the mountains would now descend and mingle with the southern forms. These latter, when the warmth returned, would return to their former homes, leaving some few species on the mountains, and carrying southward with them some of the northern temperate forms which had descended from their mountain fastnesses. Thus, we should have some few species identically the same in the northern and southern temperate zones and on the mountains of the intermediate tropical regions. But the species left during a long time on these mountains, or in opposite hemispheres, would have to compete with many new forms and would be exposed to somewhat different physical conditions; hence, they would be eminently liable to modification, and would generally now exist as varieties or as representative species; and this is the case. We must, also, bear in mind the occurrence in both hemispheres of former Glacial periods; for these will account, in accordance with the same principles, for the many quite distinct species inhabiting the same widely separated areas, and belonging to genera not now found in the intermediate torrid zones.

It is a remarkable fact, strongly insisted on by Hooker in regard to America, and by Alph. de Candolle in regard to Australia, that many more identical or slightly modified species have migrated from the north to the south, than in a reversed direction. We see, however, a few southern forms on the mountains of Borneo and Abyssinia. I suspect that this preponderant migration from the north to the south is due to the greater extent of land in the north, and to the northern forms having existed in their own homes in greater numbers, and having consequently been advanced through natural selection and competition to a higher stage of perfection, or dominating power, than the southern forms. And thus, when the two sets became commingled in the equatorial regions, during the alternations of the Glacial periods, the northern forms were the more powerful and were able to hold their places on the mountains, and afterwards migrate southward with the southern forms; but not so the southern in regard to the northern forms. In the same manner, at the present day, we see that very many European productions cover the ground in La Plata, New Zealand, and to a lesser degree in Australia, and have beaten the natives; whereas extremely few southern forms have become naturalised in any part of the northern hemisphere, though hides, wool, and other objects likely to carry seeds have been largely imported into Europe during the last two or three centuries from La Plata and during the last forty or fifty years from Australia. The Neilgherrie Mountains in India, however, offer a partial exception; for here, as I hear from Dr. Hooker, Australian forms are rapidly sowing themselves and becoming naturalised. Before the last great Glacial period, no doubt the intertropical mountains were stocked with endemic Alpine

forms; but these have almost everywhere yielded to the more dominant forms generated in the larger areas and more efficient workshops of the north. In many islands the native productions are nearly equalled, or even outnumbered, by those which have become naturalised; and this is the first stage towards their extinction. Mountains are islands on the land; and their inhabitants have yielded to those produced within the larger areas of the north, just in the same way as the inhabitants of real islands have everywhere yielded and are still yielding to continental forms naturalised through man's agency.

The same principles apply to the distribution of terrestrial animals and of marine productions, in the northern and southern temperate zones, and on the intertropical mountains. When, during the height of the Glacial period, the ocean-currents were widely different to what they now are, some of the inhabitants of the temperate seas might have reached the equator; of these a few would perhaps at once be able to migrate southwards, by keeping to the cooler currents, whilst others might remain and survive in the colder depths until the southern hemisphere was in its turn subjected to a glacial climate and permitted their further progress; in nearly the same manner as, according to Forbes, isolated spaces inhabited by Arctic productions exist to the present day in the deeper parts of the northern temperate seas.

I am far from supposing that all the difficulties in regard to the distribution and affinities of the identical and allied species, which now live so widely separated in the north and south, and sometimes on the intermediate mountain ranges, are removed on the views above given. The exact lines of migration cannot be indicated. We cannot say why certain species and not others have migrated; why certain species have been modified and have given rise to new forms, whilst others have remained unaltered. We cannot hope to explain such facts, until we can say why one species and not another becomes naturalised by man's agency in a foreign land; why one species ranges twice or thrice as far, and is twice or thrice as common, as another species within their own homes.

Various special difficulties also remain to be solved; for instance, the occurrence, as shown by Dr. Hooker, of the same plants at points so enormously remote as Kerguelen Land, New Zealand, and Fuegia; but icebergs, as suggested by Lyell, may have been concerned in their dispersal. The existence at these and other distant points of the southern hemisphere, of species, which, though distinct, belong to genera exclusively confined to the south, is a more remarkable case. Some of these species are so distinct, that we cannot suppose that there has been time since the commencement of the last Glacial period for their migration and subsequent modification to the necessary degree. The facts seem to indicate that distinct species belonging to the same genera have migrated in radiating lines from a common centre; and I am inclined to look in the southern, as in the northern hemisphere, to a former and warmer period, before the commencement of the last Glacial period, when the Antarctic lands, now covered with ice, supported a highly peculiar and isolated flora. It may be suspected that before this flora was exterminated during the last Glacial epoch, a few forms had been already widely dispersed to various points of the southern hemisphere by occasional means of transport, and by the aid, as halting-places, of now sunken islands. Thus the southern shores of America, Australia, and New Zealand may have become slightly tinted by the same peculiar forms of life.

Sir C. Lyell in a striking passage has speculated, in language almost identical with mine, on the effects of great alternations of climate throughout the world on geographical distribution. And we have now seen that Mr. Croll's conclusion that successive Glacial periods in the one hemisphere coincide with warmer periods in the opposite hemisphere, together with the admission of the slow modification of species, explains a multitude of facts in the distribution of the same and of the allied forms of life in all parts of the globe. The living waters have flowed during one period from the north and during another from the south, and in both cases have reached the equator; but the stream of life has flowed with greater force from the north than in the opposite direction, and has consequently more freely inundated the south. As the tide leaves its drift in horizontal lines, rising higher on the shores where the tide rises highest, so have the living waters left their living drift on our mountain summits, in a line gently rising from the arctic lowlands to a great latitude under the equator. The various beings thus left stranded may be compared with savage races of man, driven up and surviving in the mountain fastnesses of almost every land, which serves as a record, full of interest to us, of the former inhabitants of the surrounding lowlands.

Chapter XIII

GEOGRAPHICAL DISTRIBUTION – *continued*

Distribution of fresh-water productions–On the inhabitants of oceanic islands–Absence of batrachians and of terrestrial mammals–On the relation of the inhabitants of islands to those of the nearest mainland–On colonisation from the nearest source with subsequent modification–Summary of the last and present chapters.

FRESH-WATER PRODUCTIONS

As lakes and river systems are separated from each other by barriers of land, it might have been thought that fresh-water productions would not have ranged widely within the same country, and as the sea is apparently a still more formidable barrier, that they would never have extended to distant countries. But the case is exactly the reverse. Not only have many fresh-water species, belonging to different classes, an enormous range, but allied species prevail in a remarkable manner throughout the world. When first collecting in the fresh waters of Brazil, I well remember feeling much surprise at the similarity of the fresh-water insects, shells, etc., and at the dissimilarity of the surrounding terrestrial beings, compared with those of Britain.

But the wide ranging power of fresh-water productions can, I think, in most cases be explained by their having become fitted, in a manner highly useful to them, for short and frequent migrations from pond to pond, or from stream to stream, within their own countries; and liability to wide dispersal would follow from this capacity as an almost necessary consequence. We can here consider only a few cases; of these, some of the most difficult to explain are presented by fish. It was formerly believed that the same fresh-water species never existed on two continents distant from each other. But Dr. Günther has lately shown that the Galaxias attenuatus inhabits Tasmania, New Zealand, the Falkland Islands and the mainland of South America. This is a wonderful case, and probably indicates dispersal from an Antarctic centre during a former warm period. This case, however, is rendered in some degree less surprising by the species of this genus having the power of crossing by some unknown means considerable spaces of open ocean: thus there is one species common to New Zealand and to the Auckland Islands, though separated by a distance of about 230 miles. On the same continent fresh-water fish often range widely, and as if capriciously; for in two adjoining river-systems some of the species may be the same, and some wholly different.

It is probable that they are occasionally transported by what may be called accidental means. Thus fishes still alive are not very rarely dropped at distant points by whirlwinds; and it is known that the ova retain their vitality for a considerable time after removal from the water. Their dispersal may, however, be mainly attributed to changes in the level of the land within the recent period, causing rivers to flow into each other. Instances, also, could be given of this having occurred during floods, without any change of level. The wide difference of the fish on

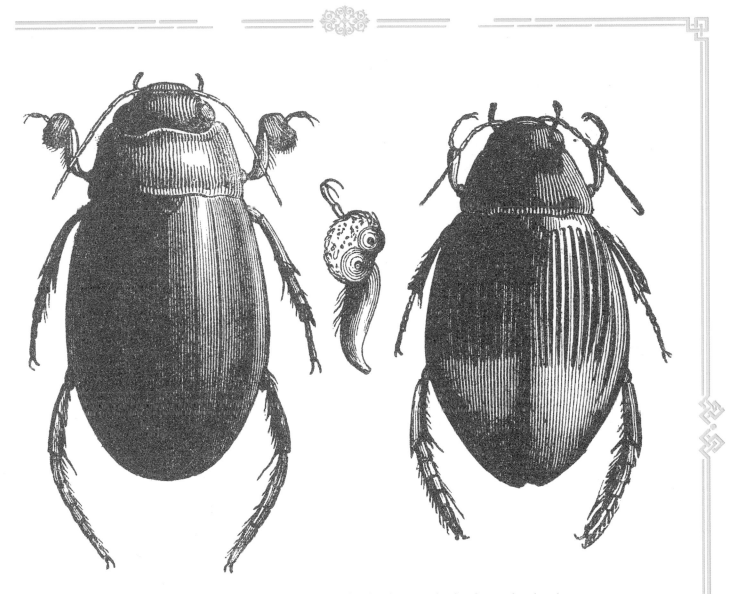

The Great Water Beetle: *left* male *right* female, also showing the forefoot pad and suckers

the opposite sides of most mountain-ranges, which are continuous and consequently must, from an early period, have completely prevented the inosculation of the river system on the two sides, leads to the same conclusion. Some fresh-water fish belong to very ancient forms, and in such cases there will have been ample time for great geographical changes, and consequently time and means for much migration. Moreover Dr. Günther has recently been led by several considerations to infer that with fishes the same forms have a long endurance. Salt water fish can with care be slowly accustomed to live in fresh water; and, according to Valenciennes, there is

hardly a single group of which all the members are confined to fresh water, so that a marine species belonging to a fresh-water group might travel far along the shores of the sea, and could, it is probable, become adapted without much difficulty to the fresh waters of a distant land.

Some species of fresh-water shells have very wide ranges, and allied species which, on our theory, are descended from a common parent, and must have proceeded from a single source, prevail throughout the world. Their distribution at first perplexed me much, as their ova are not likely to be transported by birds, and the ova, as well as the adults, are immediately killed by sea-water. I could not even understand how some naturalised species have spread rapidly throughout the same country. But two facts, which I have observed – and many others no doubt will be discovered – throw some light on this subject. When ducks suddenly emerge from a pond covered with duckweed, I have twice seen these little plants adhering to their backs; and it has happened to me, in removing a little duckweed from one aquarium to another, that I have unintentionally stocked the one with fresh-water shells from the other. But another agency is perhaps more effectual: I suspended the feet of a duck in an aquarium, where many ova of fresh-water shells were hatching; and I found that numbers of the extremely minute and just-hatched shells crawled on the feet, and clung to them so firmly that when taken out of the water they could not be jarred off, though at a somewhat more advanced age they would voluntarily drop off. These just-hatched molluscs, though aquatic in their nature, survived on the duck's feet, in damp air, from twelve to twenty hours; and in this length of time a duck or heron might fly at least six or seven hundred miles, and if blown across the sea to an oceanic island, or to any other distant point, would be sure to alight on a pool or rivulet. Sir Charles Lyell informs me that a Dytiscus has been caught with an Ancylus (a fresh-water shell like a limpet) firmly adhering to it; and a water beetle of the same family, a Colymbetes, once flew on board the 'Beagle,' when forty-five miles distant from the nearest land: how much farther it might have been blown by a favouring gale no one can tell.

With respect to plants, it has long been known what enormous ranges many fresh-water, and even marsh species, have, both over continents and to the most remote oceanic islands. This is strikingly illustrated, according to Alph. de Candolle, in those large groups of terrestrial plants, which have very few aquatic members; for the latter seem immediately to acquire, as if in consequence, a wide range. I think favourable

means of dispersal explain this fact. I have before mentioned that earth occasionally adheres in some quantity to the feet and beaks of birds. Wading birds, which frequent the muddy edges of ponds, if suddenly flushed, would be the most likely to have muddy feet. Birds of this order wander more than those of any other; and are occasionally found on the most remote and barren islands of the open ocean; they would not be likely to alight on the surface of the sea, so that any dirt on their feet would not be washed off; and when gaining the land, they would be sure to fly to their natural fresh-water haunts. I do not believe that botanists are aware how charged the mud of ponds is with seeds: I have tried several little experiments, but will here give only the most striking case: I took in February three tablespoonfuls of mud from three different points, beneath water, on the edge of a little pond; this mud when dry weighed only six and three-quarters ounces; I kept it covered up in my study for six months, pulling up and counting each plant as it grew; the plants were of many kinds, and were altogether 537 in number; and yet the viscid mud was all contained in a breakfast cup! Considering these facts, I think it would be an inexplicable circumstance if water-birds did not transport the seeds of fresh-water plants to unstocked ponds and streams, situated at very distant points. The same agency may have come into play with the eggs of some of the smaller fresh-water animals.

Other and unknown agencies probably have also played a part. I have stated that fresh-water fish eat some kinds of seeds, though they reject many other kinds after having swallowed them; even small fish swallow seeds of moderate size, as of the yellow water-lily and Potamogeton. Herons and other birds, century after century, have gone on daily devouring fish; they then take flight and go to other waters, or are blown across the sea; and we have seen that seeds retain their power of germination, when rejected many hours afterwards in pellets or in the excrement. When I saw the great size of the seeds of that fine water-lily, the Nelumbium, and remembered Alph. de Candolle's remarks on the distribution of this plant, I thought that the means of its dispersal must remain inexplicable; but Audubon states that he found the seeds of the great southern water-lily (probably according to Dr. Hooker, the Nelumbium luteum) in a heron's stomach. Now this bird must often have flown with its stomach thus well stocked to distant ponds, and, then getting a hearty meal of fish, analogy makes me believe that it would have rejected the seeds in the pellet in a fit state for germination.

In considering these several means of distribution, it should be remembered that when a pond or stream is first formed, for instance on a rising islet, it will be unoccupied; and a single seed or egg will have a good chance of succeeding. Although there will always be a struggle for life between the inhabitants of the same pond, however few in kind, yet as the number even in a well-stocked pond is small in comparison with the number of species inhabiting an equal area of land, the competition between them will probably be less severe than between terrestrial species; consequently an intruder from the waters of a foreign country would have a better chance of seizing on a new place, than in the case of terrestrial colonists. We should also remember that many fresh-water productions are low in the scale of nature, and we have reason to believe that such beings become modified more slowly than the high; and this will give time for the migration of aquatic species. We should not forget the probability of many fresh-water forms having formerly ranged continuously over immense areas, and then having become extinct at intermediate points. But the wide distribution of fresh-water plants, and of the lower animals, whether retaining the same identical form, or in some degree modified, apparently depends in main part on the wide dispersal of their seeds and eggs by animals, more especially by fresh-water birds, which have great powers of flight, and naturally travel from one piece of water to another.

ON THE INHABITANTS OF OCEANIC ISLANDS

We now come to the last of the three classes of facts, which I have selected as presenting the greatest amount of difficulty with respect to distribution, on the view that not only all the individuals of the same species have migrated from some one area, but that allied species, although now inhabiting the most distant points, have proceeded from a single area, the birthplace of their early progenitors. I have already given my reasons for disbelieving in continental extensions within the period of existing species on so enormous a scale that all the many islands of the several oceans were thus stocked with their present terrestrial inhabitants. This view removes many difficulties, but it does not accord with all the facts in regard to the productions of islands. In the following remarks I shall not confine myself to the mere question of dispersal, but shall consider some other cases bearing on the truth of the two theories of independent creation and of descent with modification.

The species of all kinds which inhabit oceanic islands are few in number compared with those on equal continental areas: Alph. de Candolle admits this for plants, and Wollaston for insects. New Zealand, for instance, with its lofty mountains and diversified stations, extending over 780 miles of latitude, together with the outlying islands of Auckland, Campbell and Chatham, contain altogether only 960 kinds of flowering plants; if we compare this moderate number with the species which swarm over equal areas in southwestern Australia or at the Cape of Good Hope, we must admit that some cause, independently of different physical conditions, has given rise to so great a difference in number. Even the uniform county of Cambridge has 847 plants, and the little island of Anglesea 764, but a few ferns and a few introduced plants are included in these numbers, and the comparison in some other respects is not quite fair. We have evidence that the barren island of Ascension aboriginally possessed less than half-a-dozen flowering plants; yet many species have now become naturalised on it, as they have in New Zealand and on every other oceanic island which can be named. In St. Helena there is reason to believe that the naturalised plants and animals have nearly or quite exterminated many native productions. He who admits the doctrine of the creation of each separate species, will have to admit that a sufficient number of the best adapted plants and animals were not created for oceanic islands; for man has unintentionally stocked them far more fully and perfectly than did nature.

Although in oceanic islands the species are few in number, the proportion of endemic kinds (i.e. those found nowhere else in the world) is often extremely large. If we compare, for instance, the number of endemic land shells in Madeira, or of endemic birds in the Galapagos Archipelago, with the number found on any continent, and then compare the area of the island with that of the continent, we shall see that this is true. This fact might have been theoretically expected, for, as already explained, species occasionally arriving, after long intervals of time in the new and isolated district, and having to compete with new associates, would be eminently liable to modification, and would often produce groups of modified descendants. But it by no means follows that, because in an island nearly all the species of one class are peculiar, those of another class, or of another section of the same class, are peculiar; and this difference seems to depend partly on the species which are not modified having immigrated in a body, so that their mutual relations have not been much disturbed; and partly on the frequent arrival of unmodified immigrants from the mother-country, with which the insular forms have intercrossed. It should be borne in mind that the offspring of such crosses would certainly gain in vigour; so that even an occasional cross would produce more effect than might have been anticipated. I will give a few illustrations of the foregoing remarks: in the Galapagos Islands there

are twenty-six land birds; of these twenty-one (or perhaps twenty-three) are peculiar; whereas of the eleven marine birds only two are peculiar; and it is obvious that marine birds could arrive at these islands much more easily and frequently than land-birds. Bermuda, on the other hand, which lies at about the same distance from North America as the Galapagos Islands do from South America, and which has a very peculiar soil, does not possess a single endemic land bird; and we know from Mr. J.M. Jones's admirable account of Bermuda, that very many North American birds occasionally or even frequently visit this island. Almost every year, as I am informed by Mr. E.V. Harcourt, many European and African birds are blown to Madeira; this island is inhabited by ninety-nine kinds, of which one alone is peculiar, though very closely related to a European form; and three or four other species are confined to this island and to the Canaries. So that the islands of Bermuda and Madeira have been stocked from the neighbouring continents with birds, which for long ages have there struggled together, and have become mutually co-adapted. Hence, when settled in their new homes, each kind will have been kept by the others to its proper place and habits, and will consequently have been but little liable to modification. Any tendency to modification will also have been checked by intercrossing with the unmodified immigrants, often arriving from the mother-country. Madeira again is inhabited by a wonderful number of peculiar land shells, whereas not one species of sea-shell is peculiar to its shores: now, though we do not know how sea shells are dispersed, yet we can see that their eggs or larvae, perhaps attached to seaweed or floating timber, or to the feet of wading birds, might be transported across three or four hundred miles of open sea far more easily than land shells. The different orders of insects inhabiting Madeira present nearly parallel cases.

Oceanic islands are sometimes deficient in animals of certain whole classes, and their places are occupied by other classes; thus in the

Galapagos Islands reptiles, and in New Zealand gigantic wingless birds, take, or recently took, the place of mammals. Although New Zealand is here spoken of as an oceanic island, it is in some degree doubtful whether it should be so ranked; it is of large size, and is not separated from Australia by a profoundly deep sea; from its geological character and the direction of its mountain ranges, the Rev. W.B. Clarke has lately maintained that this island, as well as New Caledonia, should be considered as appurtenances of Australia. Turning to plants, Dr. Hooker has shown that in the Galapagos Islands the proportional numbers of the different orders are very different from what they are elsewhere. All such differences in number, and the absence of certain whole groups of animals and plants, are generally accounted for by supposed differences in the physical conditions of the islands; but this explanation is not a little doubtful. Facility of immigration seems to have been fully as important as the nature of the conditions.

Many remarkable little facts could be given with respect to the inhabitants of oceanic islands. For instance, in certain islands not tenanted by a single mammal, some of the endemic plants have beautifully hooked seeds; yet few relations are more manifest than that hooks serve for the transportal of seeds in the wool or fur of quadrupeds. But a hooked seed might be carried to an island by other means; and the plant then becoming modified would form an endemic species, still retaining its hooks, which would form a useless appendage, like the shrivelled wings under the soldered wing-covers of many insular beetles. Again, islands often possess trees or bushes belonging to orders which elsewhere include only herbaceous species; now trees, as Alph. de Candolle has shown, generally have, whatever the cause may be, confined ranges. Hence trees would be little likely to reach distant oceanic islands; and an herbaceous plant, which had no chance of successfully competing with the many fully developed trees growing on a continent, might, when established on an island,

gain an advantage over other herbaceous plants by growing taller and taller and overtopping them. In this case, natural selection would tend to add to the stature of the plant, to whatever order it belonged, and thus first convert it into a bush and then into a tree.

ABSENCE OF BATRACHIANS AND TERRESTRIAL MAMMALS ON OCEANIC ISLANDS

With respect to the absence of whole orders of animals on oceanic islands, Bory St. Vincent long ago remarked that Batrachians (frogs, toads, newts) are never found on any of the many islands with which the great oceans are studded. I have taken pains to verify this assertion, and have found it true, with the exception of New Zealand, New Caledonia, the Andaman Islands, and perhaps the Solomon Islands and the Seychelles. But I have already remarked that it is doubtful whether New Zealand and New Caledonia ought to be classed as oceanic islands; and this is still more doubtful with respect to the Andaman and Solomon groups and the Seychelles. This general absence of frogs, toads and newts on so many true oceanic islands cannot be accounted for by their physical conditions; indeed it seems that islands are peculiarly fitted for these animals; for frogs have been introduced into Madeira, the Azores, and Mauritius, and have multiplied so as to become a nuisance. But as these animals and their spawn are immediately killed (with the exception, as far as known, of one Indian species) by sea-water, there would be great difficulty in their transportal across the sea, and therefore we can see why they do not exist on strictly oceanic islands. But why, on the theory of creation, they should not have been created there, it would be very difficult to explain.

Mammals offer another and similar case. I have carefully searched the oldest voyages, and have not found a single instance, free from doubt, of a terrestrial mammal (excluding domesticated animals kept by the natives) inhabiting an island situated above 300 miles from a continent or great

The batrachians (tailless amphibians): frogs and toads observed and drawn during the course of Darwin's voyage continental island; and many islands situated at a much less distance are equally barren. The Falkland Islands, which are inhabited by a wolf-like fox, come nearest to an exception; but this group cannot be considered as oceanic, as it lies on a bank in connection with the mainland at a distance of about 280 miles; moreover, icebergs formerly brought boulders to its western shores, and they may have formerly transported foxes, as now frequently happens in the arctic regions. Yet it cannot be said that small islands will not support at least small mammals, for they occur in many parts of the world on very small islands, when lying close to a continent; and hardly an island can be named on which our smaller quadrupeds have not become naturalised and greatly multiplied. It cannot be said, on the ordinary view of creation, that there has not been time for the creation of mammals, many volcanic islands are sufficiently ancient, as shown by the stupendous degradation which they have suffered, and by their tertiary strata: there has also been time for the production of endemic species belonging to other classes; and on continents it is known that new species of mammals appear and disappear at a quicker rate than other and lower animals. Although terrestrial mammals do not occur on oceanic islands, aerial mammals do occur on almost every island. New Zealand possesses two bats found nowhere else in the world: Norfolk Island, the Viti Archipelago, the Bonin Islands, the Caroline and Marianne Archipelagoes, and Mauritius, all possess their peculiar bats. Why, it may be asked, has the supposed creative force produced bats and no other mammals on remote islands? On my view this question can easily be answered; for no terrestrial mammal can be transported across a wide space of sea, but bats can fly across. Bats have been seen wandering by day far over the Atlantic Ocean; and two North American species, either regularly or occasionally, visit Bermuda, at the distance of 600 miles from the mainland. I hear from Mr. Tomes, who has specially studied

this family, that many species have enormous ranges, and are found on continents and on far distant islands. Hence, we have only to suppose that such wandering species have been modified in their new homes in relation to their new position, and we can understand the presence of endemic bats on oceanic islands, with the absence of all other terrestrial mammals.

Another interesting relation exists, namely, between the depth of the sea separating islands from each other, or from the nearest continent, and the degree of affinity of their mammalian inhabitants. Mr. Windsor Earl has made some striking observations on this head, since greatly extended by Mr. Wallace's admirable researches, in regard to the great Malay Archipelago, which is traversed near Celebes by a space of deep ocean, and this separates two widely distinct mammalian faunas. On either side, the islands stand on a moderately shallow submarine bank, and these islands are inhabited by the same or by closely allied quadrupeds. I have not as yet had time to follow up this subject in all quarters of the world; but as far as I have gone, the relation holds good. For instance, Britain is separated by a shallow channel from Europe, and the mammals are the same on both sides; and so it is with all the islands near the shores of Australia. The West Indian Islands, on the other hand, stand on a deeply submerged bank, nearly one thousand fathoms in depth, and here we find American forms, but the species and even the genera are quite distinct. As the amount of modification which animals of all kinds undergo partly depends on the lapse of time, and as the islands which are separated from each other, or from the mainland, by shallow channels, are more likely to have been continuously united within a recent period than the islands separated by deeper channels, we can understand how it is that a relation exists between the depth of the sea separating two mammalian faunas, and the degree of their affinity, a relation which is quite inexplicable on the theory of independent acts of creation.

The foregoing statements in regard to the inhabitants of oceanic islands, – namely, the fewness of the species, with a large proportion consisting of endemic forms – the members of certain groups, but not those of other groups in the same class, having been modified – the absence of certain whole orders, as of batrachians and of terrestrial mammals, notwithstanding the presence of aerial bats, – the singular proportions of certain orders of plants, – herbaceous forms having been developed into trees, etc., – seem to me to accord better with the belief in the efficiency of occasional means of transport, carried on during a long course of time, than with the belief in the former connection of all oceanic islands with the nearest continent; for on this latter view it is probable that the various classes would have immigrated more uniformly, and from the species having entered in a body, their mutual relations would not have been much disturbed, and consequently, they would either have not been modified, or all the species in a more equable manner.

I do not deny that there are many and serious difficulties in understanding how many of the inhabitants of the more remote islands, whether still retaining the same specific form or subsequently modified, have reached their present homes. But the probability of other islands having once existed as halting-places, of which not a wreck now remains, must not be overlooked. I will specify one difficult case. Almost all oceanic islands, even the most isolated and smallest, are inhabited by land shells, generally by endemic species, but sometimes by species found elsewhere – striking instances of which have been given by Dr. A.A. Gould in relation to the Pacific. Now it is notorious that land shells are easily killed by sea-water; their eggs, at least such as I have tried, sink in it and are killed. Yet there must be some unknown, but occasionally efficient means for their transportal. Would the just-hatched young sometimes adhere to the feet of birds roosting on the ground and thus get transported? It occurred

to me that land shells, when hybernating and having a membranous diaphragm over the mouth of the shell, might be floated in chinks of drifted timber across moderately wide arms of the sea. And I find that several species in this state withstand uninjured an immersion in sea-water during seven days. One shell, the Helix pomatia, after having been thus treated, and again hybernating, was put into sea-water for twenty days and perfectly recovered. During this length of time the shell might have been carried by a marine current of average swiftness to a distance of 660 geographical miles. As this Helix has a thick calcareous operculum, I removed it, and when it had formed a new membranous one, I again immersed it for fourteen days in sea-water, and again it recovered and crawled away. Baron Aucapitaine has since tried similar experiments. He placed 100 land shells, belonging to ten species, in a box pierced with holes, and immersed it for a fortnight in the sea. Out of the hundred shells twenty-seven recovered. The presence of an operculum seems to have been of importance, as out of twelve specimens of Cyclostoma elegans, which is thus furnished, eleven revived. It is remarkable, seeing how well the Helix pomatia resisted with me the salt water, that not one of fifty-four specimens belonging to four other species of Helix tried by Aucapitaine recovered. It is, however, not at all probable that land shells have often been thus transported; the feet of birds offer a more probable method.

ON THE RELATIONS OF THE INHABITANTS OF ISLANDS TO THOSE OF THE NEAREST MAINLAND

The most striking and important fact for us is the affinity of the species which inhabit islands to those of the nearest mainland, without being actually the same. Numerous instances could be given. The Galapagos Archipelago, situated under the equator, lies at a distance of between 500 and 600 miles from the shores of South America. Here almost every product of the land and of the water bears the unmistakable stamp of the American continent. There are twenty-six land birds. Of these twenty-one, or perhaps twenty-three, are ranked as distinct species, and would commonly be assumed to have been here created; yet the close affinity of most of these birds to American species is manifest in every character in their habits, gestures, and tones of voice. So it is with the other animals, and with a large proportion of the plants, as shown by Dr. Hooker in his admirable Flora of this archipelago. The naturalist, looking at the inhabitants of these volcanic islands in the Pacific, distant several hundred miles from the continent, feels that he is standing on American land. Why should this be so? Why should the species which are supposed to have been created in the Galapagos Archipelago, and nowhere else, bear so plainly the stamp of affinity to those created in America? There is nothing in the conditions of life, in the geological nature of the islands, in their height or climate, or in the proportions in which the several classes are associated together, which closely resembles the conditions of the South American coast. In fact, there is a considerable dissimilarity in all these respects. On the other hand, there is a considerable degree of resemblance in the volcanic nature of the soil, in the climate, height, and size of the islands, between the Galapagos and Cape Verde Archipelagos: but what an entire and absolute difference in their inhabitants! The inhabitants of the Cape Verde Islands are related to those of Africa, like those of the Galapagos to America. Facts, such as these, admit of no sort of explanation on the ordinary view of independent creation; whereas, on the view here maintained, it is obvious that the Galapagos Islands would be likely to receive colonists from America, whether by occasional means of transport or (though I do not believe in this doctrine) by formerly continuous land, and the Cape Verde Islands from Africa; such colonists would be liable to modification – the principle of inheritance still betraying their original birthplace.

Many analogous facts could be given: indeed it is an almost universal rule that the endemic productions of islands are related to those of the nearest continent, or of the nearest large island. The exceptions are few, and most of them can be explained. Thus, although Kerguelen Land stands nearer to Africa than to America, the plants are related, and that very closely, as we know from Dr. Hooker's account, to those of America: but on the view that this island has been mainly stocked by seeds brought with earth and stones on icebergs, drifted by the prevailing currents, this anomaly disappears. New Zealand in its endemic plants is much more closely related to Australia, the nearest mainland, than to any other region: and this is what might have been expected; but it is also plainly related to South America, which, although the next nearest continent, is so enormously remote, that the fact becomes an anomaly. But this difficulty partially disappears on the view that New Zealand, South America, and the other southern lands, have been stocked in part from a nearly intermediate though distant point, namely, from the antarctic islands, when they were clothed with vegetation, during a warmer tertiary period, before the commencement of the last Glacial period. The affinity, which, though feeble, I am assured by Dr. Hooker is real, between the flora of the south-western corner of Australia and of the Cape of Good Hope, is a far more remarkable case; but this affinity is confined to the plants, and will, no doubt, some day be explained.

The same law which has determined the relationship between the inhabitants of islands and the nearest mainland, is sometimes displayed on a small scale, but in a most interesting manner, within the limits of the same archipelago. Thus each separate island of the Galapagos Archipelago is tenanted, and the fact is a marvellous one, by many distinct species; but these species are related to each other in a very much closer manner than to the inhabitants of the American continent, or of any other quarter of the world. This is what might have been expected, for islands situated so near to each other would almost necessarily receive immigrants from the same original source, and from each other. But how is it that many of the immigrants have been differently modified, though only in a small degree, in islands situated within sight of each other, having the same geological nature, the same height, climate, etc.? This long appeared to me a great difficulty: but it arises in chief part from the deeply-seated error of considering the physical conditions of a country as the most important; whereas it cannot be disputed that the nature of the other species with which each has to compete, is at least as important, and generally a far more important element of success. Now if we look to the species which inhabit the Galapagos Archipelago, and are likewise found in other parts of the world, we find that they differ considerably in the several islands. This difference might indeed have been expected if the islands have been stocked by occasional means of transport – a seed, for instance, of one plant having been brought to one island, and that of another plant to another island, though all proceeding from the same general source. Hence, when in former times an immigrant first settled on one of the islands, or when it subsequently spread from one to another, it would undoubtedly be exposed to different conditions in the different islands, for it would have to compete with a different set of organisms; a plant, for instance, would find the ground best-fitted for it occupied by somewhat different species in the different islands, and would be exposed to the attacks of somewhat different enemies. If, then, it varied, natural selection would probably favour different varieties in the different islands. Some species, however, might spread and yet retain the same character throughout the group, just as we see some species spreading widely throughout a continent and remaining the same.

The really surprising fact in this case of the Galapagos Archipelago, and in a lesser degree in some analogous cases, is that each new species after being formed in any one island, did not spread quickly to the other islands. But the islands, though in sight of each other, are separated by deep arms of the sea, in most cases wider than the British Channel, and there is no reason to suppose that they have at any former period been continuously united. The currents of the sea are rapid and deep between the islands, and gales of wind are extraordinarily rare; so that the islands are far more effectually separated from each other than they appear on a map. Nevertheless, some of the species, both of those found in other parts of the world and of those confined to the archipelago, are common to the several islands; and we may infer from the present manner of distribution that they have spread from one island to the others. But we often take, I think, an erroneous view of the probability of closely allied species invading each other's territory, when put into free intercommunication. Undoubtedly, if one species has any advantage over another, it will in a very brief time wholly or in part supplant it; but if both are equally well fitted for their own places, both will probably hold their separate places for almost any length of time. Being familiar with the fact that many species, naturalised through man's agency, have spread with astonishing rapidity over wide areas, we are apt to infer that most species would thus spread; but we should remember that the species which become naturalised in new countries are not generally closely allied to the aboriginal inhabitants, but are very distinct forms, belonging in a large proportion of cases, as shown by Alph. de Candolle, to distinct genera. In the Galapagos Archipelago, many even of the birds, though so well adapted for flying from island to island, differ on the different islands; thus there are three closely allied species of mocking-thrush, each confined to its own island. Now let us suppose the mocking-thrush of Chatham Island to be blown to Charles Island, which has its own mocking-thrush; why should it succeed in establishing itself there? We may safely infer that Charles Island is well

The Mockingbird of the Galapagos Islands

stocked with its own species, for annually more eggs are laid and young birds hatched than can possibly be reared; and we may infer that the mocking-thrush peculiar to Charles Island is at least as well fitted for its home as is the species peculiar to Chatham Island. Sir C. Lyell and Mr. Wollaston have communicated to me a remarkable fact bearing on this subject; namely, that Madeira and the adjoining islet of Porto Santo possess many distinct but representative species of land shells, some of which live in crevices of stone; and although large quantities of stone are annually transported from Porto Santo to Madeira, yet this latter island has not become colonised by the Porto Santo species: nevertheless, both islands have been colonised by some European land shells, which no doubt had some advantage over the indigenous species. From these considerations I think we need not greatly marvel at the endemic species which inhabit the several islands of the Galapagos Archipelago not having all spread from island to island. On the same continent, also, pre-occupation has probably played an important part in checking the commingling of the species which inhabit different districts with nearly the same physical conditions. Thus, the south-east and south-west corners of Australia have nearly the same physical conditions, and are united by continuous land, yet they are inhabited by a vast number of distinct mammals, birds, and plants; so it is, according to Mr. Bates, with the butterflies and other animals inhabiting the great, open, and continuous valley of the Amazons.

The same principle which governs the general character of the inhabitants of oceanic islands, namely, the relation to the source whence colonists could have been most easily derived, together with their subsequent modification, is of the widest application throughout nature. We see this on every mountain-summit, in every lake and marsh. For Alpine species, excepting in as far as the same species have become widely spread during the Glacial epoch, are related to those of the surrounding lowlands; thus we have in South America, Alpine humming-birds, Alpine rodents, Alpine plants, etc., all strictly belonging to American forms; and it is obvious that a mountain, as it became slowly upheaved, would be colonised from the surrounding lowlands. So it is with the inhabitants of lakes and marshes, excepting in so far as great facility of transport has allowed the same forms to prevail throughout large portions of the world. We see the same principle in the character of most of the blind animals inhabiting the caves of America and of Europe. Other analogous facts could be given. It will, I believe, be found universally true, that wherever in two regions, let them be ever so distant, many closely allied or representative species occur, there will likewise be found some identical species; and wherever many closely-allied species occur, there will be found many forms which some naturalists rank as distinct species, and others as mere varieties; these doubtful forms showing us the steps in the process of modification.

The relation between the power and extent of migration in certain species, either at the present or at some former period, and the existence at remote points of the world of closely allied species, is shown in another and more general way. Mr. Gould remarked to me long ago, that in those genera of birds which range over the world, many of the species have very wide ranges. I can hardly doubt that this rule is generally true, though difficult of proof. Amongst mammals, we see it strikingly displayed in Bats, and in a lesser degree in the Felidae and Canidae. We see the same rule in the distribution of butterflies and beetles. So it is with most of the inhabitants of fresh water, for many of the genera in the most distinct classes range over the world, and many of the species have enormous ranges. It is not meant that all, but that some of the species have very wide ranges in the genera which range very widely. Nor is it meant that the species in such genera have, on an average, a very wide range; for this will largely depend on how far the process of modification has gone; for instance, two varieties of the same species inhabit America and Europe, and thus the species has an immense range; but, if variation were to be carried a little further, the two varieties would be ranked as distinct species, and their range would be greatly reduced. Still less is it meant, that species which have the capacity of crossing barriers and ranging widely, as in the case of certain powerfully-winged birds, will necessarily range widely; for we should never forget that to range widely implies not only the power of crossing barriers, but the more important power of being victorious in distant lands in the struggle for life with foreign associates. But according to the view that all the species of a genus, though distributed to the most remote points of the world, are descended from a single progenitor, we ought to find, and I believe as a general rule we do find, that some at least of the species range very widely.

We should bear in mind that many genera in all classes are of ancient origin, and the species in this case will have had ample time for dispersal and subsequent modification. There is also reason to believe, from geological evidence, that within each great class the lower organisms change at a slower rate than the higher; consequently they will have had a better chance of ranging widely and of still retaining the same specific character. This fact, together with that of the seeds and eggs of most lowly organised forms being very minute and better fitted for distant transportal, probably accounts for a law which has long been observed, and which has lately been discussed by Alph. de Candolle in regard to plants, namely, that the lower any group of organisms stands the more widely it ranges.

The relations just discussed, – namely, lower organisms ranging more widely than the higher, – some of the species of widely-ranging genera themselves ranging widely, – such facts, as alpine, lacustrine, and marsh productions being generally related to those which live on the surrounding low lands and dry lands,   the striking relationship between the inhabitants of islands and those of the nearest mainland – the still closer relationship of the distinct inhabitants of the islands of the same archipelago – are inexplicable on the ordinary view of the independent creation of each species, but are explicable if we admit colonisation from the nearest or readiest source, together with the subsequent adaptation of the colonists to their new homes.

SUMMARY OF THE LAST AND PRESENT CHAPTERS

In these chapters I have endeavoured to show that if we make due allowance for our ignorance of the full effects of changes of climate and of the level of the land, which have certainly occurred within the recent period, and of other changes which have probably occurred, – if we remember how ignorant we are with respect to the many curious means of occasional transport, – if we bear in mind, and this is a very important consideration, how often a species may have ranged continuously over a wide area, and then have become extinct in the intermediate tracts, – the difficulty is not insuperable in believing that all the individuals of the same species, wherever found, are descended from common parents. And we are led to this conclusion, which has been arrived at by many naturalists under the designation of single centres of creation, by various general considerations, more especially from the importance of barriers of all kinds, and from the analogical distribution of subgenera, genera, and families.

With respect to distinct species belonging to the same genus, which on our theory have spread from one parent-source; if we make the same allowances as before for our ignorance, and remember that some forms of life have changed very slowly, enormous periods of time having been thus granted for their migration, the difficulties are far from insuperable; though in this case, as in that of the individuals of the same species, they are often great.

As exemplifying the effects of climatical changes on distribution, I have attempted to show

how important a part the last Glacial period has played, which affected even the equatorial regions, and which, during the alternations of the cold in the north and the south, allowed the productions of opposite hemispheres to mingle, and left some of them stranded on the mountain-summits in all parts of the world. As showing how diversified are the means of occasional transport, I have discussed at some little length the means of dispersal of fresh-water productions.

If the difficulties be not insuperable in admitting that in the long course of time all the individuals of the same species, and likewise of the several species belonging to the same genus, have proceeded from some one source; then all the grand leading facts of geographical distribution are explicable on the theory of migration, together with subsequent modification and the multiplication of new forms. We can thus understand the high importance of barriers, whether of land or water, in not only separating but in apparently forming the several zoological and botanical provinces. We can thus understand the concentration of related species within the same areas; and how it is that under different latitudes, for instance, in South America, the inhabitants of the plains and mountains, of the forests, marshes, and deserts, are linked together in so mysterious a manner, and are likewise linked to the extinct beings which formerly inhabited the same continent. Bearing in mind that the mutual relation of organism to organism is of the highest importance, we can see why two areas, having nearly the same physical conditions, should often be inhabited by very different forms of life; for according to the length of time which has elapsed since the colonists entered one of the regions, or both; according to the nature of the communication which allowed certain forms and not others to enter, either in greater or lesser numbers; according or not as those which entered happened to come into more or less direct competition with each other and with the aborigines; and according as the immigrants were

capable of varying more or less rapidly, there would ensue in the two or more regions, independently of their physical conditions, infinitely diversified conditions of life; there would be an almost endless amount of organic action and reaction, and we should find some groups of beings greatly, and some only slightly modified; some developed in great force, some existing in scanty numbers – and this we do find in the several great geographical provinces of the world.

On these same principles we can understand, as I have endeavoured to show, why oceanic islands should have few inhabitants, but that of these, a large proportion should be endemic or peculiar; and why, in relation to the means of migration, one group of beings should have all its species peculiar, and another group, even within the same class, should have all its species the same with those in an adjoining quarter of the world. We can see why whole groups of organisms, as batrachians and terrestrial mammals, should be absent from oceanic islands, whilst the most isolated islands should possess their own peculiar species of aerial mammals or bats. We can see why, in islands, there should be some relation between the presence of mammals, in a more or less modified condition, and the depth of the sea between such islands and the mainland. We can clearly see why all the inhabitants of an archipelago, though specifically distinct on the several islets, should be closely related to each other, and should likewise be related, but less closely, to those of the nearest continent, or other source whence immigrants might have been derived. We can see why, if there exist very closely allied or representative species in two areas, however distant from each other, some identical species will almost always there be found.

As the late Edward Forbes often insisted, there is a striking parallelism in the laws of life throughout time and space; the laws governing the succession of forms in past times being nearly the same with those governing at the present time the differences in different areas. We see this in many facts. The endurance of each species and group of species is continuous in time; for the apparent exceptions to the rule are so few that they may fairly be attributed to our not having as yet discovered in an intermediate deposit certain forms which are absent in it, but which occur above and below: so in space, it certainly is the general rule that the area inhabited by a single species, or by a group of species, is continuous, and the exceptions, which are not rare, may, as I have attempted to show, be accounted for by former migrations under different circumstances, or through occasional means of transport, or by the species having become extinct in the intermediate tracts. Both in time and space species and groups of species have their points of maximum development. Groups of species, living during the same period of time, or living within the same area, are often characterised by trifling features in common, as of sculpture or colour. In looking to the long succession of past ages, as in looking to distant provinces throughout the world, we find that species in certain classes differ little from each other, whilst those in another class, or only in a different section of the same order, differ greatly from each other. In both time and space the lowly organised members of each class generally change less than the highly organised; but there are in both cases marked exceptions to the rule. According to our theory, these several relations throughout time and space are intelligible; for whether we look to the allied forms of life which have changed during successive ages, or to those which have changed after having migrated into distant quarters, in both cases they are connected by the same bond of ordinary generation; in both cases the laws of variation have been the same, and modifications have been accumulated by the same means of natural selection.

Chapter XIV

Mutual Affinities of Organic Beings: Morphology – Embryology – Rudimentary Organs

Classification, groups subordinate to groups–Natural system–Rules and difficulties in classification, explained on the theory of descent with modification–Classification of varieties–Descent always used in classification–Analogical or adaptive characters–Affinities, general, complex and radiating–Extinction separates and defines groups–Morphology, between members of the same class, between parts of the same individual–Embryology, laws of, explained by variations not supervening at an early age, and being inherited at a corresponding age–Rudimentary organs; their origin explained–Summary.

CLASSIFICATION

From the most remote period in the history of the world organic beings have been found to resemble each other in descending degrees, so that they can be classed in groups under groups. This classification is not arbitrary like the grouping of the stars in constellations. The existence of groups would have been of simple significance, if one group had been exclusively fitted to inhabit the land, and another the water; one to feed on flesh, another on vegetable matter, and so on; but the case is widely different, for it is notorious how commonly members of even the same sub-group have different habits. In the second and fourth chapters, on Variation and on Natural Selection, I have attempted to show that within each country it is the widely ranging, the much diffused and common, that is the dominant species, belonging to the larger genera in each class, which vary most. The varieties, or incipient species, thus produced, ultimately become converted into new and distinct species; and these, on the principle of inheritance, tend to produce other new and dominant species. Consequently the groups which are now large, and which generally include many dominant species, tend to go on increasing in size. I further attempted to show that from the varying descendants of each species trying to occupy as many and as different places as possible in the economy of nature, they constantly tend to diverge in character. This latter conclusion is supported by observing the great diversity of forms, which, in any small area, come into the closest competition, and by certain facts in naturalisation.

I attempted also to show that there is a steady tendency in the forms which are increasing in number and diverging in character, to supplant and exterminate the preceding, less divergent and less improved forms. I request the reader to turn to the diagram illustrating the action, as formerly explained, of these several principles; and he will see that the inevitable result is, that the modified descendants proceeding from one progenitor become broken up into groups subordinate to groups. In the diagram each letter on the uppermost line may represent a genus including several species; and the whole of the genera along this upper line form together one class, for all are descended from one ancient parent, and, consequently, have inherited something in common. But the three genera on the left hand have, on this same principle, much in common, and form a sub-family, distinct from that containing the next two genera on the right hand, which diverged from a common parent at the fifth stage of descent. These five genera have also much in common, though less than when grouped in subfamilies; and they form a family distinct from that containing the three genera still further to the right hand, which diverged at an earlier period. And all these genera, descended from (A), form an order distinct from the genera descended from (I). So that we here have many species descended from a single progenitor grouped into genera; and the genera into subfamilies, families and orders, all under one great class. The grand fact of the natural subordination of organic beings in groups under groups, which, from its familiarity, does not always sufficiently strike us, is in my judgment thus explained. No doubt organic beings, like all other objects, can be classed in many ways, either artificially by single characters, or more naturally by a number of characters. We know, for instance, that minerals and the elemental substances can be thus arranged. In this case there is of course no relation to genealogical succession, and no cause can at present be assigned for their falling into groups. But with organic beings the case is different, and the view above given accords with their natural arrangement in group under group; and no other explanation has ever been attempted.

Naturalists, as we have seen, try to arrange the species, genera and families in each class, on what is called the Natural System. But what is meant by this system? Some authors look at it merely as a scheme for arranging together those living objects which are most alike, and for separating those which are most unlike; or as an artificial method of enunciating, as briefly as possible, general propositions – that is, by one sentence to give the characters common, for instance, to all mammals, by another those common to all carnivora, by another those common to the dog-genus, and then, by adding a single sentence, a full description is given of each kind of dog. The ingenuity and utility of this system are indisputable. But many naturalists think that something more is meant by the Natural System; they believe that it reveals the plan of the Creator; but unless it be specified

whether order in time or space, or both, or what else is meant by the plan of the Creator, it seems to me that nothing is thus added to our knowledge. Expressions such as that famous one by Linnaeus, which we often meet with in a more or less concealed form, namely, that the characters do not make the genus, but that the genus gives the characters, seem to imply that some deeper bond is included in our classifications than mere resemblance. I believe that this is the case, and that community of descent – the one known cause of close similarity in organic beings – is the bond, which, though observed by various degrees of modification, is partially revealed to us by our classifications.

Let us now consider the rules followed in classification, and the difficulties which are encountered on the view that classification either gives some unknown plan of creation, or is simply a scheme for enunciating general propositions and of placing together the forms most like each other. It might have been thought (and was in ancient times thought) that those parts of the structure which determined the habits of life, and the general place of each being in the economy of nature, would be of very high importance in classification. Nothing can be more false. No one regards the external similarity of a mouse to a shrew, of a dugong to a whale, of a whale to a fish, as of any importance. These resemblances, though so intimately connected with the whole life of the being, are ranked as merely 'adaptive or analogical characters;' but to the consideration of these resemblances we shall recur. It may even be given as a general rule, that the less any part of the organisation is concerned with special habits, the more important it becomes for classification. As an instance: Owen, in speaking of the dugong, says, 'The generative organs, being those which are most remotely related to the habits and food of an animal, I have always regarded as affording very clear indications of its true affinities. We are least likely in the modifications of these organs to mistake a merely adaptive for an essential

Top: The American Marsh Shrew, a large shrew measuring 6 inches in length, excluding the tail
Bottom: Two types of South American mouse, published in *The Zoology of the Voyage of H.M.S. Beagle*

character.' With plants how remarkable it is that the organs of vegetation, on which their nutrition and life depend, are of little signification; whereas the organs of reproduction, with their product the seed and embryo, are of paramount importance! So again, in formerly discussing certain morphological characters which are not functionally important, we have seen that they are often of the highest service in classification. This depends on their constancy throughout many allied groups; and their constancy chiefly depends on any slight deviations not having been preserved and accumulated by natural selection, which acts only on serviceable characters.

That the mere physiological importance of an organ does not determine its classificatory value, is almost proved by the fact, that in allied groups, in which the same organ, as we have every reason to suppose, has nearly the same physiological value, its classificatory value is widely different. No naturalist can have worked at any group without being struck with this fact; and it has been fully acknowledged in the writings of almost every author. It will suffice to quote the highest authority, Robert Brown, who, in speaking of certain organs in the Proteaceae, says their generic importance, 'like that of all their parts, not only in this, but, as I apprehend in every natural family, is very unequal, and in some cases seems to be entirely lost.' Again, in another work he says, the genera of the Connaraceae 'differ in having one or more ovaria, in the existence or absence of albumen, in the imbricate or valvular aestivation. Any one of these characters singly is frequently of more than generic importance, though here even, when all taken together, they appear insufficient to separate Cnestis from Connarus.' To give an example amongst insects: in one great division of the Hymenoptera, the antennae, as Westwood has remarked, are most constant in structure; in another division they differ much, and the differences are of quite subordinate value in classification; yet no one will say that the antennae in these two divisions of the same order are of

unequal physiological importance. Any number of instances could be given of the varying importance for classification of the same important organ within the same group of beings.

Again, no one will say that rudimentary or atrophied organs are of high physiological or vital importance; yet, undoubtedly, organs in this condition are often of much value in classification. No one will dispute that the rudimentary teeth in the upper jaws of young ruminants, and certain rudimentary bones of the leg, are highly serviceable in exhibiting the close affinity between Ruminants and Pachyderms. Robert Brown has strongly insisted on the fact that the position of the rudimentary florets is of the highest importance in the classification of the Grasses.

Numerous instances could be given of characters derived from parts which must be considered of very trifling physiological importance, but which are universally admitted as highly serviceable in the definition of whole groups. For instance, whether or not there is an open passage from the nostrils to the mouth, the only character, according to Owen, which absolutely distinguishes fishes and reptiles – the inflection of the angle of the lower jaw in Marsupials – the manner in which the wings of insects are folded – mere colour in certain Algae – mere pubescence on parts of the flower in grasses – the nature of the dermal covering, as hair or feathers, in the Vertebrata. If the Ornithorhynchus had been covered with feathers instead of hair, this external and trifling character would have been considered by naturalists as an important aid in determining the degree of affinity of this strange creature to birds.

The importance, for classification, of trifling characters, mainly depends on their being correlated with many other characters of more or less importance. The value indeed of an aggregate of characters is very evident in natural history. Hence, as has often been remarked, a species may depart from its allies in several characters, both of high physiological importance, and of almost universal prevalence, and yet leave us in no doubt where it should be ranked. Hence, also, it has been found that a classification founded on any single character, however important that may be, has always failed; for no part of the organisation is invariably constant. The importance of an aggregate of characters, even when none are important, alone explains the aphorism enunciated by Linnaeus, namely, that the characters do not give the genus, but the genus gives the character; for this seems founded on the appreciation of many trifling points of resemblance, too slight to be defined. Certain plants, belonging to the Malpighiaceae, bear perfect and degraded flowers; in the latter, as A. de Jussieu has remarked, 'The greater number of the characters proper to the species, to the genus, to the family, to the class, disappear, and thus laugh at our classification.' When Aspicarpa produced in France, during several years, only these degraded flowers, departing so wonderfully in a number of the most important points of structure from the proper type of the order, yet M. Richard sagaciously saw, as Jussieu observes, that this genus should still be retained amongst the Malpighiaceae. This case well illustrates the spirit of our classifications.

Practically, when naturalists are at work, they do not trouble themselves about the physiological value of the characters which they use in defining a group or in allocating any particular species. If they find a character nearly uniform, and common to a great number of forms, and not common to others, they use it as one of high value; if common to some lesser number, they use it as of subordinate value. This principle has been broadly confessed by some naturalists to be the true one; and by none more clearly than by that excellent botanist, Aug. St. Hilaire. If several trifling characters are always found in combination, though no apparent bond of connexion can be discovered between them, especial value is set on them. As in most groups of animals, important organs, such as those for propelling the blood, or for aerating it, or those for propagating the race,

are found nearly uniform, they are considered as highly serviceable in classification; but in some groups all these, the most important vital organs, are found to offer characters of quite subordinate value. Thus, as Fritz Müller has lately remarked, in the same group of crustaceans, Cypridina is furnished with a heart, whilst in two closely allied genera, namely Cypris and Cytherea, there is no such organ; one species of Cypridina has well-developed branchiae, whilst another species is destitute of them.

We can see why characters derived from the embryo should be of equal importance with those derived from the adult, for a natural classification of course includes all ages. But it is by no means obvious, on the ordinary view, why the structure of the embryo should be more important for this purpose than that of the adult, which alone plays its full part in the economy of nature. Yet it has been strongly urged by those great naturalists, Milne Edwards and Agassiz, that embryological characters are the most important of all; and this doctrine has very generally been admitted as true. Nevertheless, their importance has sometimes been exaggerated, owing to the adaptive characters of larvae not having been excluded; in order to show this, Fritz Müller arranged, by the aid of such characters alone, the great class of crustaceans, and the arrangement did not prove a natural one. But there can be no doubt that embryonic, excluding larval characters, are of the highest value for classification, not only with animals but with plants. Thus the main divisions of flowering plants are founded on differences in the embryo – on the number and position of the cotyledons, and on the mode of development of the plumule and radicle. We shall immediately see why these characters possess so high a value in classification, namely, from the natural system being genealogical in its arrangement.

Our classifications are often plainly influenced by chains of affinities. Nothing can be easier than to define a number of characters common to all birds; but with crustaceans, any such definition has hitherto been found impossible. There are crustaceans at the opposite ends of the series, which have hardly a character in common; yet the species at both ends, from being plainly allied to others, and these to others, and so onwards, can be recognised as unequivocally belonging to this, and to no other class of the Articulata.

Geographical distribution has often been used, though perhaps not quite logically, in classification, more especially in very large groups of closely allied forms. Temminck insists on the utility or even necessity of this practice in certain groups of birds; and it has been followed by several entomologists and botanists.

Finally, with respect to the comparative value of the various groups of species, such as orders, suborders, families, subfamilies, and genera, they seem to be, at least at present, almost arbitrary. Several of the best botanists, such as Mr. Bentham and others, have strongly insisted on their arbitrary value. Instances could be given amongst plants and insects, of a group first ranked by practised naturalists as only a genus, and then raised to the rank of a sub-family or family; and this has been done, not because further research has detected important structural differences, at first overlooked, but because numerous allied species, with slightly different grades of difference, have been subsequently discovered.

All the foregoing rules and aids and difficulties in classification may be explained, if I do not greatly deceive myself, on the view that the natural system is founded on descent with modification, – that the characters which naturalists consider as showing true affinity between any two or more species, are those which have been inherited from a common parent, all true classification being genealogical, – that community of descent is the hidden bond which naturalists have been unconsciously seeking, and not some unknown plan of creation, or the enunciation of general propositions, and the mere putting together and separating objects more or less alike.

But I must explain my meaning more fully. I believe that the *arrangement* of the groups within each class, in due subordination and relation to each other, must be strictly genealogical in order to be natural; but that the *amount* of difference in the several branches or groups, though allied in the same degree in blood to their common progenitor, may differ greatly, being due to the different degrees of modification which they have undergone; and this is expressed by the forms being ranked under different genera, families, sections or orders. The reader will best understand what is meant, if he will take the trouble to refer to the diagram in the fourth chapter. We will suppose the letters A to L to represent allied genera existing during the Silurian epoch, and descended from some still earlier form. In three of these genera (A, F, and I) a species has transmitted modified descendants to the present day, represented by the fifteen genera (a^{11} to z^{11}) on the uppermost horizontal line. Now, all these modified descendants from a single species are related in blood or descent in the same degree. They may metaphorically be called cousins to the same millionth degree, yet they differ widely and in different degrees from each other. The forms descended from A, now broken up into two or three families, constitute a distinct order from those descended from I, also broken up into two families. Nor can the existing species descended from A be ranked in the same genus with the parent A, or those from I with parent I. But the existing genus F^{14} may be supposed to have been but slightly modified, and it will then rank with the parent genus F; just as some few still living organisms belong to Silurian genera. So that the comparative value of the differences between these organic beings, which are all related to each other in the same degree in blood, has come to be widely different. Nevertheless, their genealogical *arrangement* remains strictly true, not only at the present time, but at each successive period of descent. All the modified descendants from A will have inherited something in common from their common parent, as will all the descendants from I; so will it be with each subordinate branch of descendants at each successive stage. If, however, we suppose any descendant of A or of I to have become so much modified as to have lost all traces of its parentage in this case, its place in the natural system will be lost, as seems to have occurred with some few existing organisms. All the descendants of the genus F, along its whole line of descent, are supposed to have been but little modified, and they form a single genus. But this genus, though much isolated, will still occupy its proper intermediate position. The representation of the groups as here given in the diagram on a flat surface, is much too simple. The branches ought to have diverged in all directions. If the names of the groups had been simply written down in a linear series the representation would have been still less natural; and it is notoriously not possible to represent in a series, on a flat surface, the affinities which we discover in nature amongst the beings of the same group. Thus, the natural system is genealogical in its arrangement, like a pedigree. But the amount of modification which the different groups have undergone has to be expressed by ranking them under different so-called genera, subfamilies, families, sections, orders, and classes.

It may be worth while to illustrate this view of classification, by taking the case of languages. If we possessed a perfect pedigree of mankind, a genealogical arrangement of the races of man would afford the best classification of the various languages now spoken throughout the world; and if all extinct languages, and all intermediate and slowly changing dialects, were to be included, such an arrangement would be the only possible one. Yet it might be that some ancient languages had altered very little and had given rise to few new languages, whilst others had altered much owing to the spreading, isolation and state of civilisation of the several co-descended races, and had thus given rise to many new dialects and languages. The various degrees of difference between the

languages of the same stock would have to be expressed by groups subordinate to groups; but the proper or even the only possible arrangement would still be genealogical; and this would be strictly natural, as it would connect together all languages, extinct and recent, by the closest affinities, and would give the filiation and origin of each tongue.

In confirmation of this view, let us glance at the classification of varieties, which are known or believed to be descended from a single species. These are grouped under the species, with the subvarieties under the varieties; and in some cases, as with the domestic pigeon, with several other grades of difference. Nearly the same rules are followed as in classifying species. Authors have insisted on the necessity of arranging varieties on a natural instead of an artificial system; we are cautioned, for instance, not to class two varieties of the pineapple together, merely because their fruit, though the most important part, happens to be nearly identical; no one puts the Swedish and common turnip together, though the esculent and thickened stems are so similar. Whatever part is found to be most constant, is used in classing varieties: thus the great agriculturist Marshall says the horns are very useful for this purpose with cattle, because they are less variable than the shape or colour of the body, etc.; whereas with sheep the horns are much less serviceable, because less constant. In classing varieties, I apprehend that if we had a real pedigree, a genealogical classification would be universally preferred; and it has been attempted in some cases. For we might feel sure, whether there had been more or less modification, that the principle of inheritance would keep the forms together which were allied in the greatest number of points. In tumbler pigeons, though some of the subvarieties differ in the important character of the length of the beak, yet all are kept together from having the common habit of tumbling; but the short-faced breed has nearly or quite lost this habit; nevertheless, without any thought on the subject, these tumblers are kept in the same group, because allied in blood and alike in some other respects.

With species in a state of nature, every naturalist has in fact brought descent into his classification; for he includes in his lowest grade, that of species, the two sexes; and how enormously these sometimes differ in the most important characters is known to every naturalist: scarcely a single fact can be predicated in common of the adult males and hermaphrodites of certain cirripedes, and yet no one dreams of separating them. As soon as the three Orchidean forms, Monachanthus, Myanthus, and Catasetum, which had previously been ranked as three distinct genera, were known to be sometimes produced on the same plant, they were immediately considered as varieties; and now I have been able to show that they are the male, female, and hermaphrodite forms of the same species. The naturalist includes as one species the various larval stages of the same individual, however much they may differ from each other and from the adult; as well as the so-called alternate generations of Steenstrup, which can only in a technical sense be considered as the same individual. He includes monsters and varieties, not from their partial resemblance to the parent-form, but because they are descended from it.

As descent has universally been used in classing together the individuals of the same species, though the males and females and larvae are sometimes extremely different; and as it has been used in classing varieties which have undergone a certain, and sometimes a considerable amount of modification, may not this same element of descent have been unconsciously used in grouping species under genera, and genera under higher groups, all under the so-called natural system? I believe it has been unconsciously used; and thus only can I understand the several rules and guides which have been followed by our best systematists. As we have no written pedigrees, we are forced to trace community of descent by resemblances of any kind. Therefore, we choose

those characters which are the least likely to have been modified, in relation to the conditions of life to which each species has been recently exposed. Rudimentary structures on this view are as good as, or even sometimes better than other parts of the organisation. We care not how trifling a character may be – let it be the mere inflection of the angle of the jaw, the manner in which an insect's wing is folded, whether the skin be covered by hair or feathers – if it prevail throughout many and different species, especially those having very different habits of life, it assumes high value; for we can account for its presence in so many forms with such different habits, only by inheritance from a common parent. We may err in this respect in regard to single points of structure, but when several characters, let them be ever so trifling, concur throughout a large group of beings having different habits, we may feel almost sure, on the theory of descent, that these characters have been inherited from a common ancestor; and we know that such aggregated characters have especial value in classification.

We can understand why a species or a group of species may depart from its allies, in several of its most important characteristics, and yet be safely classed with them. This may be safely done, and is often done, as long as a sufficient number of characters, let them be ever so unimportant, betrays the hidden bond of community of descent. Let two forms have not a single character in common, yet, if these extreme forms are connected together by a chain of intermediate groups, we may at once infer their community of descent, and we put them all into the same class. As we find organs of high physiological importance – those which serve to preserve life under the most diverse conditions of existence – are generally the most constant, we attach especial value to them; but if these same organs, in another group or section of a group, are found to differ much, we at once value them less in our classification. We shall presently see why embryological characters are of such high classificatory importance. Geographical distribution may sometimes be brought usefully into play in classing large genera, because all the species of the same genus, inhabiting any distinct and isolated region, are in all probability descended from the same parents.

ANALOGICAL RESEMBLANCES

We can understand, on the above views, the very important distinction between real affinities and analogical or adaptive resemblances. Lamarck first called attention to this subject, and he has been ably followed by Macleay and others. The resemblance in the shape of the body and in the fin-like anterior limbs between dugongs and whales, and between these two orders of mammals and fishes, are analogical. So is the resemblance between a mouse and a shrew-mouse (Sorex), which belong to different orders; and the still closer resemblance, insisted on by Mr. Mivart, between the mouse and a small marsupial animal (Antechinus) of Australia. These latter resemblances may be accounted for, as it seems to me, by adaptation for similarly active movements through thickets and herbage, together with concealment from enemies.

Amongst insects there are innumerable instances; thus Linnaeus, misled by external appearances, actually classed an homopterous insect as a moth. We see something of the same kind even with our domestic varieties, as in the strikingly similar shape of the body in the improved breeds of the Chinese and common pig, which are descended from distinct species; and in the similarly thickened stems of the common and specifically distinct Swedish turnip. The resemblance between the greyhound and race-horse is hardly more fanciful than the analogies which have been drawn by some authors between widely different animals.

On the view of characters being of real importance for classification, only in so far as they reveal descent, we can clearly understand why

analogical or adaptive characters, although of the utmost importance to the welfare of the being, are almost valueless to the systematist. For animals, belonging to two most distinct lines of descent, may have become adapted to similar conditions, and thus have assumed a close external resemblance; but such resemblances will not reveal – will rather tend to conceal their blood-relationship. We can thus also understand the apparent paradox, that the very same characters are analogical when one group is compared with another, but give true affinities when the members of the same group are compared together: thus the shape of the body and fin-like limbs are only analogical when whales are compared with fishes, being adaptations in both classes for swimming through the water; but between the several members of the whale family, the shape of the body and the fin-like limbs offer characters exhibiting true affinity; for as these parts are so nearly similar throughout the whole family, we cannot doubt that they have been inherited from a common ancestor. So it is with fishes.

Numerous cases could be given of striking resemblances in quite distinct beings between single parts or organs, which have been adapted for the same functions. A good instance is afforded by the close resemblance of the jaws of the dog and Tasmanian wolf or Thylacinus – animals which are widely sundered in the natural system. But this resemblance is confined to general appearance, as in the prominence of the canines, and in the cutting shape of the molar teeth. For the teeth really differ much: thus the dog has on each side of the upper jaw four pre-molars and only two molars; whilst the Thylacinus has three pre-molars and four molars. The molars also differ much in the two animals in relative size and structure. The adult dentition is preceded by a widely different milk dentition. Any one may, of course, deny that the teeth in either case have been adapted for tearing flesh, through the natural selection of successive

Canine skulls: *1* jaws of a dog; *2* skull of a Tasmanian wolf (Thylacinus);
3 front teeth of a Tasmanian wolf

variations; but if this be admitted in the one case, it is unintelligible to me that it should be denied in the other. I am glad to find that so high an authority as Professor Flower has come to this same conclusion.

The extraordinary cases given in a former chapter, of widely different fishes possessing electric organs, – of widely different insects possessing luminous organs, – and of orchids and asclepiads having pollen-masses with viscid discs, come under this same head of analogical resemblances. But these cases are so wonderful that they were introduced as difficulties or objections to our theory. In all such cases some fundamental difference in the growth or development of the parts, and generally in their matured structure, can be detected. The end gained is the same, but the means, though appearing superficially to be the same, are essentially different. The principle formerly alluded to under the term of *analogical variation* has probably in these cases often come into play; that is, the members of the same class, although only distantly allied, have inherited so much in common in their constitution, that they are apt to vary under similar exciting causes in a similar manner; and this would obviously aid in the acquirement through natural selection of parts or organs, strikingly like each other, independently of their direct inheritance from a common progenitor.

As species belonging to distinct classes have often been adapted by successive slight modifications to live under nearly similar circumstances, – to inhabit, for instance, the three elements of land, air and water, – we can perhaps understand how it is that a numerical parallelism has sometimes been observed between the sub-groups of distinct classes. A naturalist, struck with a parallelism of this nature, by arbitrarily raising or sinking the value of the groups in several classes (and all our experience shows that their valuation is as yet arbitrary), could easily extend the parallelism over a wide range; and thus the septenary, quinary, quaternary and ternary classifications have probably arisen.

There is another and curious class of cases in which close external resemblance does not depend on adaptation to similar habits of life, but has been gained for the sake of protection. I allude to the wonderful manner in which certain butterflies imitate, as first described by Mr. Bates, other and quite distinct species. This excellent observer has shown that in some districts of South America, where, for instance, an Ithomia abounds in gaudy swarms, another butterfly, namely, a Leptalis, is often found mingled in the same flock; and the latter so closely resembles the Ithomia in every shade and stripe of colour, and even in the shape of its wings, that Mr. Bates, with his eyes sharpened by collecting during eleven years, was, though always on his guard, continually deceived. When the mockers and the mocked are caught and compared, they are found to be very different in essential structure, and to belong not only to distinct genera, but often to distinct families. Had this mimicry occurred in only one or two instances, it might have been passed over as a strange coincidence. But, if we proceed from a district where one Leptalis imitates an Ithomia, another mocking and mocked species, belonging to the same two genera, equally close in their resemblance, may be found. Altogether no less than ten genera are enumerated, which include species that imitate other butterflies. The mockers and mocked always inhabit the same region; we never find an imitator living remote from the form which it imitates. The mockers are almost invariably rare insects; the mocked in almost every case abounds in swarms. In the same district in which a species of Leptalis closely imitates an Ithomia, there are sometimes other Lepidoptera mimicking the same Ithomia: so that in the same place, species of three genera of butterflies and even a moth are found all closely resembling a butterfly belonging to a fourth genus. It deserves especial notice that many of the mimicking forms of the Leptalis, as well as of the mimicked forms, can be shown by a graduated series to be merely varieties of the same species; whilst others are undoubtedly distinct species. But why, it may be asked, are certain forms treated as the mimicked and others as the mimickers? Mr. Bates satisfactorily answers this question by showing that the form which is imitated keeps the usual dress of the group to which it belongs, whilst the counterfeiters have changed their dress and do not resemble their nearest allies.

We are next led to enquire what reason can be assigned for certain butterflies and moths so often assuming the dress of another and quite distinct form; why, to the perplexity of naturalists, has nature condescended to the tricks of the stage? Mr. Bates has, no doubt, hit on the true explanation. The mocked forms, which always abound in numbers, must habitually escape destruction to a large extent, otherwise they could not exist in such swarms; and a large amount of evidence has now been collected, showing that they are distasteful to birds and other insect-devouring animals. The mocking forms, on the other hand, that inhabit the same district, are comparatively rare, and belong to rare groups; hence, they must suffer habitually from some danger, for otherwise, from the number of eggs laid by all butterflies, they would in three or four generations swarm over the whole country. Now if a member of one of these persecuted and rare groups were to assume a dress so like that of a well-protected species that it continually deceived the practised eyes of an entomologist, it would often deceive predaceous birds and insects, and thus often escape destruction. Mr. Bates may almost be said to have actually witnessed the process by which the mimickers have come so closely to resemble the mimicked; for he found that some of the forms of Leptalis which mimic so many other butterflies, varied in an extreme degree. In one district several varieties occurred, and of these one alone resembled, to a certain extent, the common Ithomia of the same district. In another district there were two or three varieties, one of which was much commoner than the others, and this closely mocked another form of Ithomia. From facts of this nature, Mr. Bates concludes that the Leptalis first varies; and when a variety happens to resemble in some degree any common butterfly inhabiting the same district, this variety, from its resemblance to a flourishing and little persecuted kind, has a better chance of escaping destruction from predaceous birds and insects, and is consequently oftener preserved;

'the less perfect degrees of resemblance being generation after generation eliminated, and only the others left to propagate their kind.' So that here we have an excellent illustration of natural selection.

Messrs. Wallace and Trimen have likewise described several equally striking cases of imitation in the Lepidoptera of the Malay Archipelago and Africa, and with some other insects. Mr. Wallace has also detected one such case with birds, but we have none with the larger quadrupeds. The much greater frequency of imitation with insects than with other animals, is probably the consequence of their small size; insects cannot defend themselves, excepting indeed the kinds furnished with a sting, and I have never heard of an instance of such kinds mocking other insects, though they are mocked; insects cannot easily escape by flight from the larger animals which prey on them; therefore, speaking metaphorically, they are reduced, like most weak creatures, to trickery and dissimulation.

It should be observed that the process of imitation probably never commenced between forms widely dissimilar in colour. But, starting with species already somewhat like each other, the closest resemblance, if beneficial, could readily be gained by the above means, and if the imitated form was subsequently and gradually modified through any agency, the imitating form would be led along the same track, and thus be altered to almost any extent, so that it might ultimately assume an appearance or colouring wholly unlike that of the other members of the family to which it belonged. There is, however, some difficulty on this head, for it is necessary to suppose in some cases that ancient members belonging to several distinct groups, before they had diverged to their present extent, accidentally resembled a member of another and protected group in a sufficient degree to afford some slight protection, this having given the basis for the subsequent acquisition of the most perfect resemblance.

ON THE NATURE OF THE AFFINITIES CONNECTING ORGANIC BEINGS

As the modified descendants of dominant species, belonging to the larger genera, tend to inherit the advantages which made the groups to which they belong large and their parents dominant, they are almost sure to spread widely, and to seize on more and more places in the economy of nature. The larger and more dominant groups within each class thus tend to go on increasing in size, and they consequently supplant many smaller and feebler groups. Thus, we can account for the fact that all organisms, recent and extinct, are included under a few great orders and under still fewer classes. As showing how few the higher groups are in number, and how widely they are spread throughout the world, the fact is striking that the discovery of Australia has not added an insect belonging to a new class, and that in the vegetable kingdom, as I learn from Dr. Hooker, it has added only two or three families of small size.

In the chapter on Geological Succession I attempted to show, on the principle of each group having generally diverged much in character during the long-continued process of modification, how it is that the more ancient forms of life often present characters in some degree intermediate between existing groups. As some few of the old and intermediate forms having transmitted to the present day descendants but little modified, these constitute our so-called osculant or aberrant groups. The more aberrant any form is, the greater must be the number of connecting forms which have been exterminated and utterly lost. And we have evidence of aberrant groups having suffered severely from extinction, for they are almost always represented by extremely few species; and such species as do occur are generally very distinct from each other, which again implies extinction. The genera Ornithorhynchus and Lepidosiren, for example, would not have been less aberrant had each been represented by a dozen species, instead of as at present by a single one, or by two or three. We can, I think, account for this fact only by looking at aberrant groups as forms which have been conquered by more successful competitors, with a few members still preserved under unusually favourable conditions.

Mr. Waterhouse has remarked that when a member belonging to one group of animals exhibits an affinity to a quite distinct group, this affinity in most cases is general and not special: thus, according to Mr. Waterhouse, of all rodents, the bizcacha is most nearly related to marsupials; but in the points in which it approaches this order, its relations are general, that is, not to any one marsupial species more than to another. As these points of affinity are believed to be real and not merely adaptive, they must be due in accordance with our view to inheritance from a common progenitor. Therefore, we must suppose either that all rodents, including the bizcacha, branched off from some ancient marsupial, which will naturally have been more or less intermediate in character with respect to all existing marsupials; or that both rodents and marsupials branched off from a common progenitor, and that both groups have since undergone much modification in divergent directions. On either view we must suppose that the bizcacha has retained, by inheritance, more of the character of its ancient progenitor than have other rodents; and therefore it will not be specially related to any one existing marsupial, but indirectly to all or nearly all marsupials, from having partially retained the character of their common progenitor, or of some early member of the group. On the other hand, of all marsupials, as Mr. Waterhouse has remarked, the Phascolomys resembles most nearly, not any one species, but the general order of rodents. In this case, however, it may be strongly suspected that the resemblance is only analogical, owing to the Phascolomys having become adapted to habits like those of a rodent. The elder De Candolle has made nearly similar observations on the general nature of the affinities of distinct families of plants.

On the principle of the multiplication and gradual divergence in character of the species descended from a common progenitor, together with their retention by inheritance of some characters in common, we can understand the excessively complex and radiating affinities by which all the members of the same family or higher group are connected together. For the common progenitor of a whole family, now broken up by extinction into distinct groups and sub-groups, will have transmitted some of its characters, modified in various ways and degrees, to all the species; and they will consequently be related to each other by circuitous lines of affinity of various lengths (as may be seen in the diagram so often referred to), mounting up through many predecessors. As it is difficult to show the blood-relationship between the numerous kindred of any ancient and noble family, even by the aid of a genealogical tree, and almost impossible to do so without this aid, we can understand the extraordinary difficulty which naturalists have experienced in describing, without the aid of a diagram, the various affinities which they perceive between the many living and extinct members of the same great natural class.

Extinction, as we have seen in the fourth chapter, has played an important part in defining and widening the intervals between the several groups in each class. We may thus account for the distinctness of whole classes from each other – for instance, of birds from all other vertebrate animals – by the belief that many ancient forms of life have been utterly lost, through which the early progenitors of birds were formerly connected with the early progenitors of the other and at that time less differentiated vertebrate classes. There has been much less extinction of the forms of life which once connected fishes with batrachians. There has been still less within some whole classes, for instance the Crustacea, for here the most wonderfully diverse forms are still linked together by a long and only partially broken chain of affinities. Extinction has only defined the groups: it has by no means made them; for if every form which has ever lived on this earth were suddenly to reappear, though it would be quite impossible to give definitions by which each group could be distinguished, still a natural classification, or at least a natural arrangement, would be possible. We shall see this by turning to the diagram: the letters, A to L, may represent eleven Silurian genera, some of which have produced large groups of modified descendants, with every link in each branch and sub-branch still alive; and the links not greater than those between existing varieties. In this case it would be quite impossible to give definitions by which the several members of the several groups could be distinguished from their more immediate parents and descendants. Yet the arrangement in the diagram would still hold good and would be natural, for, on the principle of inheritance, all the forms descended, for instance from A, would have something in common. In a tree we can distinguish this or that branch, though at the actual fork the two unite and blend together. We could not, as I have said, define the several groups; but we could pick out types, or forms, representing most of the characters of each group, whether large or small, and thus give a general idea of the value of the differences between them. This is what we should be driven to, if we were ever to succeed in collecting all the forms in any one class which have lived throughout all time and space. Assuredly we shall never succeed in making so perfect a collection: nevertheless, in certain classes, we are tending toward this end; and Milne Edwards has lately insisted, in an able paper, on the high importance of looking to types, whether or not we can separate and define the groups to which such types belong.

Finally, we have seen that natural selection, which follows from the struggle for existence, and which almost inevitably leads to extinction and divergence of character in the descendants from any one parent-species, explains that great and

universal feature in the affinities of all organic beings, namely, their subordination in group under group. We use the element of descent in classing the individuals of both sexes and of all ages under one species, although they may have but few characters in common; we use descent in classing acknowledged varieties, however different they may be from their parents, and I believe that this element of descent is the hidden bond of connexion which naturalists have sought under the term of the Natural System. On this idea of the natural system being, in so far as it has been perfected, genealogical in its arrangement, with the grades of difference expressed by the terms genera, families, orders, etc., we can understand the rules which we are compelled to follow in our classification. We can understand why we value certain resemblances far more than others; why we use rudimentary and useless organs, or others of trifling physiological importance, why, in finding the relations between one group and another, we summarily reject analogical or adaptive characters, and yet use these same characters within the limits of the same group. We can clearly see how it is that all living and extinct forms can be grouped together within a few great classes; and how the several members of each class are connected together by the most complex and radiating lines of affinities. We shall never, probably, disentangle the inextricable web of the affinities between the members of any one class; but when we have a distinct object in view, and do not look to some unknown plan of creation, we may hope to make sure but slow progress.

Professor Haeckel in his 'Generelle Morphologie,' and in other works, has recently brought his great knowledge and abilities to bear on what he calls phylogeny, or the lines of descent of all organic beings. In drawing up the several series he trusts chiefly to embryological characters, but receives aid from homologous and rudimentary organs, as well as from the successive periods at which the various forms of life are believed to have first appeared in our geological formations. He has thus boldly made a great beginning, and shows us how classification will in the future be treated.

MORPHOLOGY

We have seen that the members of the same class, independently of their habits of life, resemble each other in the general plan of their organisation. This resemblance is often expressed by the term 'unity of type;' or by saying that the several parts and organs in the different species of the class are homologous. The whole subject is included under the general term of Morphology. This is one of the most interesting departments of natural history, and may almost be said to be its very soul. What can be more curious than that the hand of a man, formed for grasping, that of a mole for digging, the leg of the horse, the paddle of the porpoise, and the wing of the bat, should all be constructed on the same pattern, and should include similar bones, in the same relative positions? How curious it is, to give a subordinate though striking instance, that the hind feet of the kangaroo, which are so well fitted for bounding over the open plains, – those of the climbing, leaf-eating koala, equally well fitted for grasping the branches of trees, – those of the ground-dwelling, insect or root-eating, bandicoots, – and those of some other Australian marsupials, – should all be constructed on the same extraordinary type, namely with the bones of the second and third digits extremely slender and enveloped within the same skin, so that they appear like a single toe furnished with two claws. Notwithstanding this similarity of pattern, it is obvious that the hind feet of these several animals are used for as widely different purposes as it is possible to conceive. The case is rendered all the more striking by the American opossums, which follow nearly the same habits of life as some of their Australian relatives, having feet constructed on the ordinary plan. Professor Flower, from whom these statements are taken, remarks in conclusion: 'We may call this conformity to type,

Tropical Beetle: *left* male *right* female

without getting much nearer to an explanation of the phenomenon;' and he then adds 'but is it not powerfully suggestive of true relationship, of inheritance from a common ancestor?'

Geoffroy St. Hilaire has strongly insisted on the high importance of relative position or connexion in homologous parts; they may differ to almost any extent in form and size, and yet remain connected together in the same invariable order. We never find, for instance, the bones of the arm and forearm, or of the thigh and leg, transposed. Hence the same names can be given to the homologous bones in widely different animals. We see the same great law in the construction of the mouths of insects: what can be more different than the immensely long spiral proboscis of a sphinx-moth, the curious folded one of a bee or bug, and the great jaws of a beetle? Yet all these organs, serving for such widely different purposes, are formed by infinitely numerous modifications of an upper lip, mandibles, and two pairs of maxillae. The same law governs the construction of the mouths and limbs of crustaceans. So it is with the flowers of plants.

Nothing can be more hopeless than to attempt to explain this similarity of pattern in

The Brown Koala, including three views of its skull

Didelphis brachyura.

The Opossum, as it appeared in *The Zoology of the Voyage of H.M.S. Beagle* members of the same class, by utility or by the doctrine of final causes. The hopelessness of the attempt has been expressly admitted by Owen in his most interesting work on the 'Nature of Limbs.' On the ordinary view of the independent creation of each being, we can only say that so it is; that it has pleased the Creator to construct all the animals and plants in each great class on a uniform plan; but this is not a scientific explanation.

The explanation is to a large extent simple, on the theory of the selection of successive slight modifications, each being profitable in some way to the modified form, but often affecting by correlation other parts of the organisation. In changes of this nature, there will be little or no tendency to alter the original pattern, or to transpose the parts. The bones of a limb might be shortened and flattened to any extent, becoming at the same time enveloped in thick membrane, so as to serve as a fin; or a webbed hand might have all its bones, or certain bones, lengthened to any extent, with the membrane connecting them increased, so as to serve as a wing; yet all these modifications would not tend to alter the framework of the bones or the relative connexion of the parts. If we suppose that an early progenitor – the archetype, as it may be called – of all mammals, birds and reptiles, had its limbs constructed on the existing general pattern, for whatever purpose they served, we can at once perceive the plain signification of the homologous construction of the limbs throughout the class. So with the mouths of insects, we have only to suppose that their common progenitor had an upper lip, mandibles, and two pairs of maxillae, these parts being perhaps very simple in form; and then natural selection will account for the infinite diversity in structure and function of the mouths of insects. Nevertheless, it is conceivable that the general pattern of an organ might become so much obscured as to be finally lost, by the reduction and ultimately by the complete abortion of certain parts, by the fusion of other parts, and by the doubling or multiplication of others,

variations which we know to be within the limits of possibility. In the paddles of the gigantic extinct sea-lizards, and in the mouths of certain suctorial crustaceans, the general pattern seems thus to have become partially obscured.

There is another and equally curious branch of our subject; namely, serial homologies, or the comparison of the different parts or organs in the same individual, and not of the same parts or organs in different members of the same class. Most physiologists believe that the bones of the skull are homologous – that is, correspond in number and in relative connexion – with the elemental parts of a certain number of vertebrae. The anterior and posterior limbs in all the higher vertebrate classes are plainly homologous. So it is with the wonderfully complex jaws and legs of crustaceans. It is familiar to almost every one, that in a flower the relative position of the sepals, petals, stamens, and pistils, as well as their intimate structure, are intelligible on the view that they consist of metamorphosed leaves, arranged in a spire. In monstrous plants, we often get direct evidence of the possibility of one organ being transformed into another; and we can actually see, during the early or embryonic stages of development in flowers, as well as in crustaceans and many other animals, that organs, which when mature become extremely different are at first exactly alike.

How inexplicable are the cases of serial homologies on the ordinary view of creation! Why should the brain be enclosed in a box composed of such numerous and such extraordinarily shaped pieces of bone apparently representing vertebrae? As Owen has remarked, the benefit derived from the yielding of the separate pieces in the act of parturition by mammals, will by no means explain the same construction in the skulls of birds and reptiles. Why should similar bones have been created to form the wing and the leg of a bat, used as they are for such totally different purposes, namely flying and walking? Why should one crustacean, which has an extremely complex mouth formed of many parts, consequently always have fewer legs; or conversely, those with many legs have simpler mouths? Why should the sepals, petals, stamens, and pistils, in each flower, though fitted for such distinct purposes, be all constructed on the same pattern?

On the theory of natural selection, we can, to a certain extent, answer these questions. We need not here consider how the bodies of some animals first became divided into a series of segments, or how they became divided into right and left sides, with corresponding organs, for such questions are almost beyond investigation. It is, however, probable that some serial structures are the result of cells multiplying by division, entailing the multiplication of the parts developed from such cells. It must suffice for our purpose to bear in mind that an indefinite repetition of the same part or organ is the common characteristic, as Owen has remarked, of all low or little specialised forms; therefore the unknown progenitor of the Vertebrata probably possessed many vertebrae; the unknown progenitor of the Articulata, many segments; and the unknown progenitor of flowering plants, many leaves arranged in one or more spires. We have also formerly seen that parts many times repeated are eminently liable to vary, not only in number, but in form. Consequently such parts, being already present in considerable numbers, and being highly variable, would naturally afford the materials for adaptation to the most different purposes; yet they would generally retain, through the force of inheritance, plain traces of their original or fundamental resemblance. They would retain this resemblance all the more, as the variations, which afforded the basis for their subsequent modification through natural selection, would tend from the first to be similar; the parts being at an early stage of growth alike, and being subjected to nearly the same conditions. Such parts, whether more or less modified, unless their common origin became wholly obscured, would be serially homologous.

In the great class of molluscs, though the parts in distinct species can be shown to be homologous, only a few serial homologies; such as the valves of Chitons, can be indicated; that is, we are seldom enabled to say that one part is homologous with another part in the same individual. And we can understand this fact; for in molluscs, even in the lowest members of the class, we do not find nearly so much indefinite repetition of any one part as we find in the other great classes of the animal and vegetable kingdoms.

But morphology is a much more complex subject than it at first appears, as has lately been well shown in a remarkable paper by Mr. E. Ray Lankester, who has drawn an important distinction between certain classes of cases which have all been equally ranked by naturalists as homologous. He proposes to call the structures which resemble each other in distinct animals, owing to their descent from a common progenitor with subsequent modification, *homogenous*; and the resemblances which cannot thus be accounted for, he proposes to call *homoplastic*. For instance, he believes that the hearts of birds and mammals are as a whole homogenous, – that is, have been derived from a common progenitor; but that the four cavities of the heart in the two classes are homoplastic, – that is, have been independently developed. Mr. Lankester also adduces the close resemblance of the parts on the right and left sides of the body, and in the successive segments of the same individual animal; and here we have parts commonly called homologous which bear no relation to the descent of distinct species from a common progenitor. Homoplastic structures are the same with those which I have classed, though in a very imperfect manner, as analogous modifications or resemblances. Their formation may be attributed in part to distinct organisms, or to distinct parts of the same organism, having varied in an analogous manner; and in part to similar modifications, having been preserved for the same general purpose or function, of which many instances have been given.

Naturalists frequently speak of the skull as formed of metamorphosed vertebrae; the jaws of crabs as metamorphosed legs; the stamens and pistils in flowers as metamorphosed leaves; but it would in most cases be more correct, as Professor Huxley has remarked, to speak of both skull and vertebrae, jaws and legs, etc., as having been metamorphosed, not one from the other, as they now exist, but from some common and simpler element. Most naturalists, however, use such language only in a metaphorical sense: they are far from meaning that during a long course of descent, primordial organs of any kind – vertebrae in the one case and legs in the other – have actually been converted into skulls or jaws. Yet so strong is the appearance of this having occurred that naturalists can hardly avoid employing language having this plain signification. According to the views here maintained, such language may be used literally; and the wonderful fact of the jaws, for instance, of a crab retaining numerous characters, which they probably would have retained through inheritance, if they had really been metamorphosed from true though extremely simple legs, is in part explained.

DEVELOPMENT AND EMBRYOLOGY

This is one of the most important subjects in the whole round of natural history. The metamorphoses of insects, with which every one is familiar, are generally effected abruptly by a few stages; but the transformations are in reality numerous and gradual, though concealed. A certain ephemerous insect (Chlöeon) during its development, moults, as shown by Sir J. Lubbock, above twenty times, and each time undergoes a certain amount of change; and in this case we see the act of metamorphosis performed in a primary and gradual manner. Many insects, and especially certain crustaceans, show us what wonderful changes of structure can be effected during development. Such changes, however, reach their acme in the so called alternate generations of some of the lower animals. It is, for instance, an astonishing fact that a delicate branching coralline, studded with polypi and attached to a submarine rock, should produce, first by budding and then by transverse division, a host of huge floating jelly- fishes; and that these should produce eggs, from which are hatched swimming animalcules, which attach themselves to rocks and become developed into branching corallines; and so on in an endless cycle. The belief in the essential identity of the process of alternate generation and of ordinary metamorphosis has been greatly strengthened by Wagner's discovery of the larva or maggot of a fly, namely the Cecidomyia, producing asexually other larvae, and these others, which finally are developed into mature males and females, propagating their kind in the ordinary manner by eggs.

It may be worth notice that when Wagner's remarkable discovery was first announced, I was asked how was it possible to account for the larvae of this fly having acquired the power of asexual reproduction. As long as the case remained unique no answer could be given. But already Grimm has shown that another fly, a Chironomus, reproduces itself in nearly the same manner, and he believes that this occurs frequently in the order. It is the pupa, and not the larva, of the Chironomus which has this power; and Grimm further shows that this case, to a certain extent, 'unites that of the Cecidomyia with the parthenogenesis of the Coccidae;' the term parthenogenesis implying that the mature females of the Coccidae are capable of producing fertile eggs without the concourse of the male. Certain animals belonging to several classes are now known to have the power of ordinary reproduction at an unusually early age; and we have only to accelerate parthenogenetic reproduction by gradual steps to an earlier and earlier age – Chironomus showing us an almost exactly intermediate stage, viz., that of the pupa – and we can perhaps account for the marvellous case of the Cecidomyia.

It has already been stated that various parts in the same individual, which are exactly alike during

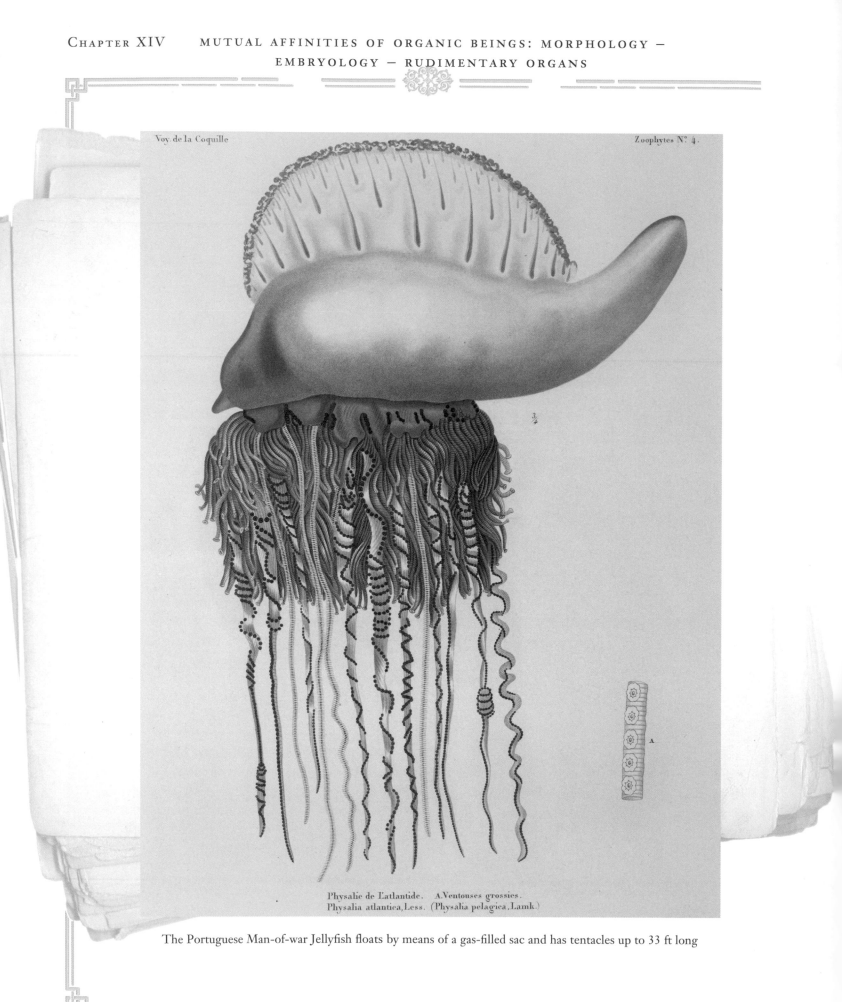

The Portuguese Man-of-war Jellyfish floats by means of a gas-filled sac and has tentacles up to 33 ft long

Birds of the Warbler family:
1 Barred Warbler *2* Lesser Whitethroat *3* Garden Warbler *4* Whitethroat *5* Blackcap

an early embryonic period, become widely different and serve for widely different purposes in the adult state. So again it has been shown that generally the embryos of the most distinct species belonging to the same class are closely similar, but become, when fully developed, widely dissimilar. A better proof of this latter fact cannot be given than the statement by Von Baer that 'the embryos of mammalia, of birds, lizards and snakes, probably also of chelonia, are in the earliest states exceedingly like one another, both as a whole and in the mode of development of their parts; so much so, in fact, that we can often distinguish the embryos only by their size. In my possession are two little embryos in spirit, whose names I have omitted to attach, and at present I am quite unable to say to what class they belong. They may be lizards or small birds, or very young mammalia, so complete is the similarity in the mode of formation of the head and trunk in these animals. The extremities, however, are still absent in these embryos. But even if they had existed in the earliest stage of their development we should learn nothing, for the feet of lizards and mammals, the wings and feet of birds, no less than the hands and feet of man, all arise from the same fundamental form.' The larvae of most crustaceans, at corresponding stages of development, closely resemble each other, however different the adults may become; and so it is with very many other animals. A trace of the law of embryonic resemblance occasionally lasts till a rather late age: thus birds of the same genus, and of allied genera, often resemble each other in their immature plumage; as we see in the spotted feathers in the young of the thrush group. In the cat tribe, most of the species when adult are striped or spotted in lines; and stripes or spots can be plainly distinguished in the whelp of the lion and the puma. We occasionally, though rarely, see something of the same kind in plants; thus the first leaves of the ulex or furze, and the first leaves of the phyllodineous acacias, are pinnate or divided like the ordinary leaves of the leguminosae.

Development of the West Indian Tree Frog:
A and *B* eggs with embryos *C* newly hatched froglet
D adult male

The points of structure, in which the embryos of widely different animals within the same class resemble each other, often have no direct relation to their conditions of existence. We cannot, for instance, suppose that in the embryos of the vertebrata the peculiar loop-like courses of the arteries near the branchial slits are related to similar conditions – in the young mammal which is nourished in the womb of its mother, in the egg of the bird which is hatched in a nest, and in the spawn of a frog under water. We have no more reason to believe in such a relation than we have to believe that the similar bones in the hand of a man, wing of a bat, and fin of a porpoise, are related to similar conditions of life. No one supposes that the stripes on the whelp of a lion, or the spots on the young blackbird, are of any use to these animals.

The case, however, is different when an animal, during any part of its embryonic career, is active, and has to provide for itself. The period of activity may come on earlier or later in life; but whenever it comes on, the adaptation of the larva to its conditions of life is just as perfect and as beautiful as in the adult animal. In how important a manner this has acted, has recently been well shown by Sir J. Lubbock in his remarks on the close similarity of the larvae of some insects belonging to very different orders, and on the dissimilarity of the larvae of other insects within the same order, according to their habits of life. Owing to such adaptations the similarity of the larvae of allied animals is sometimes greatly obscured; especially when there is a division of labour during the different stages of development, as when the same larva has during one stage to search for food, and during another stage has to search for a place of attachment. Cases can even be given of the larvae of allied species, or groups of species, differing more from each other than do the adults. In most cases, however, the larvae, though active, still obey, more or less closely, the law of common embryonic resemblance. Cirripedes afford a good instance of this: even the illustrious Cuvier did not perceive that a barnacle was a crustacean: but a glance at the larva shows this in an unmistakable manner. So again the two main divisions of cirripedes, the pedunculated and sessile, though differing widely in external appearance, have larvae in all their stages barely distinguishable.

The embryo in the course of development generally rises in organisation. I use this expression, though I am aware that it is hardly possible to define clearly what is meant by organisation being higher or lower. But no one probably will dispute that the butterfly is higher than the caterpillar. In some cases, however, the mature animal must be considered as lower in the scale than the larva, as with certain parasitic crustaceans. To refer once again to cirripedes. the larvae in the first stage have three pairs of locomotive organs, a simple single eye, and a prosciformed mouth, with which they feed largely, for they increase much in size. In the second stage, answering to the chrysalis stage of butterflies, they have six pairs of beautifully constructed natatory legs, a pair of magnificent compound eyes, and extremely complex antennae; but they have a closed and imperfect mouth, and cannot feed: their function at this stage is, to search out by their well-developed organs of sense, and to reach by their active powers of swimming, a proper place on which to become attached and to undergo their final metamorphosis. When this is completed they are fixed for life: their legs are now converted into prehensile organs; they again obtain a well-constructed mouth; but they have no antennae, and their two eyes are now reconverted into a minute, single, simple eye-spot. In this last and complete state, cirripedes may be considered as either more highly or more lowly organised than they were in the larval condition. But in some genera the larvae become developed into hermaphrodites having the ordinary structure, or into what I have called complemental males; and in the latter the development has assuredly been retrograde; for the male is a mere sack, which lives for a short time and is destitute of mouth, stomach, and every other organ of importance, excepting those for reproduction.

We are so much accustomed to see a difference in structure between the embryo and the adult, that we are tempted to look at this difference as in some necessary manner contingent on growth. But there is no reason why, for instance, the wing of a bat, or the fin of a porpoise, should not have been sketched out with all their parts in proper proportion, as soon as any part became visible. In some whole groups of animals and in certain members of other groups this is the case, and the embryo does not at any period differ widely from the adult: thus Owen has remarked in regard to cuttle-fish, 'there is no metamorphosis; the cephalopodic character is manifested long before the parts of the embryo are completed.' Land shells and fresh-water crustaceans are born having their proper forms, whilst the marine members of the same two great classes pass through considerable and often great changes during their development. Spiders, again, barely undergo any metamorphosis. The larvae of most insects pass through a worm-like stage, whether they are active and adapted to diversified habits, or are inactive from being placed in the midst of proper nutriment, or from being fed by their parents; but in some few cases, as in that of Aphis, if we look to the admirable drawings of the development of this insect, by Professor Huxley, we see hardly any trace of the vermiform stage.

Sometimes it is only the earlier developmental stages which fail. Thus, Fritz Müller has made the remarkable discovery that certain shrimp-like crustaceans (allied to Penoeus) first appear under the simple nauplius-form, and after passing through two or more zoea-stages, and then through the mysis-stage, finally acquire their mature structure: now in the whole great malacostracan order, to which these crustaceans belong, no other member is as yet known to be first developed under the nauplius-form, though many appear as zoeas; nevertheless Müller assigns reasons for his belief, that if there had been no suppression of development, all these crustaceans would have appeared as nauplii.

How, then, can we explain these several facts in embryology – namely, the very general, though not universal, difference in structure between the embryo and the adult; the various parts in the same individual embryo, which ultimately become very unlike, and serve for diverse purposes, being at an early period of growth alike; the common, but not invariable, resemblance between the embryos or larvae of the most distinct species in the same class; the embryo often retaining, whilst within the egg or womb, structures which are of no service to it, either at that or at a later period of life; on the other hand, larvae which have to provide for their own wants, being perfectly adapted to the surrounding conditions; and lastly, the fact of certain larvae standing higher in the scale of organisation than the mature animal into which they are developed? I believe that all these facts can be explained as follows.

It is commonly assumed, perhaps from monstrosities affecting the embryo at a very early period, that slight variations or individual differences necessarily appear at an equally early period. We have little evidence on this head, but what we have certainly points the other way; for it is notorious that breeders of cattle, horses and various fancy animals, cannot positively tell, until some time after birth, what will be the merits and demerits of their young animals. We see this plainly in our own children; we cannot tell whether a child will be tall or short, or what its precise features will be. The question is not, at what period of life any variation may have been caused, but at what period the effects are displayed. The cause may have acted, and I believe often has acted, on one or both parents before the act of generation. It deserves notice that it is of no importance to a very young animal, as long as it is nourished and protected by its parent, whether most of its characters are acquired a little earlier or later in life. It would not signify, for instance, to a bird which obtained its food by having a much-curved beak whether or not whilst young it possessed a beak of this shape, as long as it was fed by its parents.

I have stated in the first chapter, that at whatever age any variation first appears in the parent, it tends to reappear at a corresponding age in the offspring. Certain variations can only appear at corresponding ages; for instance, peculiarities in the caterpillar, cocoon, or imago states of the silk-moth; or, again, in the full-grown horns of cattle. But variations which, for all that we can see might have appeared either earlier or later in life, likewise tend to reappear at a corresponding age in the offspring and parent. I am far from meaning that this is invariably the case, and I could give several exceptional cases of

Birds. Pl. 19.

Bird with curved beak, identified as Upercerthia dumetaria. A ground-feeder, this bird uses its long beak to probe in the soil for food variations (taking the word in the largest sense) which have supervened at an earlier age in the child than in the parent.

These two principles, namely, that slight variations generally appear at a not very early period of life, and are inherited at a corresponding not early period, explain, as I believe, all the above specified leading facts in embryology. But first let us look to a few analogous cases in our domestic varieties. Some authors who have written on Dogs maintain that the greyhound and bull-dog, though so different, are really closely allied varieties, descended from the same wild stock, hence I was curious to see how far their puppies differed from each other. I was told by breeders that they differed just as much as their parents, and this, judging by the eye, seemed almost to be the case; but on actually measuring the old dogs and their six-days-old puppies, I found that the puppies had not acquired nearly their full amount of proportional difference. So, again, I was told that the foals of cart and race-horses – breeds which have been almost wholly formed by selection under domestication – differed as much as the full-grown animals; but having had careful measurements made of the dams and of three-days-old colts of race and heavy cart-horses, I find that this is by no means the case.

As we have conclusive evidence that the breeds of the Pigeon are descended from a single wild species, I compared the young pigeons within twelve hours after being hatched. I carefully measured the proportions (but will not here give the details) of the beak, width of mouth, length of nostril and of eyelid, size of feet and length of leg, in the wild parent species, in pouters, fantails, runts, barbs, dragons, carriers, and tumblers. Now, some of these birds, when mature, differ in so extraordinary a manner in the length and form of beak, and in other characters, that they would certainly have been ranked as distinct genera if found in a state of nature. But when the nestling birds of these several breeds were placed in a row, though most of them could just be distinguished,

the proportional differences in the above specified points were incomparably less than in the full-grown birds. Some characteristic points of difference – for instance, that of the width of mouth – could hardly be detected in the young. But there was one remarkable exception to this rule, for the young of the short-faced tumbler differed from the young of the wild rock-pigeon, and of the other breeds, in almost exactly the same proportions as in the adult stage.

These facts are explained by the above two principles. Fanciers select their dogs, horses, pigeons, etc., for breeding, when nearly grown up. They are indifferent whether the desired qualities are acquired earlier or later in life, if the full-grown animal possesses them. And the cases just given, more especially that of the pigeons, show that the characteristic differences which have been accumulated by man's selection, and which give value to his breeds, do not generally appear at a very early period of life, and are inherited at a corresponding not early period. But the case of the short-faced tumbler, which when twelve hours old possessed its proper characters, proves that this is not the universal rule; for here the characteristic differences must either have appeared at an earlier period than usual, or, if not so, the differences must have been inherited, not at a corresponding, but at an earlier age.

Now, let us apply these two principles to species in a state of nature. Let us take a group of birds, descended from some ancient form and modified through natural selection for different habits. Then, from the many slight successive variations having supervened in the several species at a not early age, and having been inherited at a corresponding age, the young will have been but little modified, and they will still resemble each other much more closely than do the adults, just as we have seen with the breeds of the pigeon. We may extend this view to widely distinct structures and to whole classes. The fore-limbs, for instance, which once served as legs to a remote progenitor, may have become, through a long course of modification, adapted in one descendant to act as hands, in another as paddles, in another as wings; but on the above two principles the fore-limbs will not have been much modified in the embryos of these several forms; although in each form the fore-limb will differ greatly in the adult state. Whatever influence long continued use or disuse may have had in modifying the limbs or other parts of any species, this will chiefly or solely have affected it when nearly mature, when it was compelled to use its full powers to gain its own living; and the effects thus produced will have been transmitted to the offspring at a corresponding nearly mature age. Thus the young will not be modified, or will be modified only in a slight degree, through the effects of the increased use or disuse of parts.

With some animals the successive variations may have supervened at a very early period of life, or the steps may have been inherited at an earlier age than that at which they first occurred. In either of these cases the young or embryo will closely resemble the mature parent-form, as we have seen with the short-faced tumbler. And this is the rule of development in certain whole groups, or in certain sub-groups alone, as with cuttle-fish, land-shells, fresh-water crustaceans, spiders, and some members of the great class of insects. With respect to the final cause of the young in such groups not passing through any metamorphosis, we can see that this would follow from the following contingencies: namely, from the young having to provide at a very early age for their own wants, and from their following the same habits of life with their parents; for in this case it would be indispensable for their existence that they should be modified in the same manner as their parents. Again, with respect to the singular fact that many terrestrial and fresh-water animals do not undergo any metamorphosis, whilst marine members of the same groups pass through various transformations, Fritz Müller has suggested that the process of slowly modifying and adapting an animal to live on the land or in fresh water, instead of in the sea,

would be greatly simplified by its not passing through any larval stage; for it is not probable that places well adapted for both the larval and mature stages, under such new and greatly changed habits of life, would commonly be found unoccupied or ill-occupied by other organisms. In this case the gradual acquirement at an earlier and earlier age of the adult structure would be favoured by natural selection; and all traces of former metamorphoses would finally be lost.

If, on the other hand, it profited the young of an animal to follow habits of life slightly different from those of the parent-form, and consequently to be constructed on a slightly different plan, or if it profited a larva already different from its parent to change still further, then, on the principle of inheritance at corresponding ages, the young or the larvae might be rendered by natural selection more and more different from their parents to any conceivable extent. Differences in the larva might, also, become correlated with successive stages of its development; so that the larva, in the first stage, might come to differ greatly from the larva in the second stage, as is the case with many animals. The adult might also become fitted for sites or habits, in which organs of locomotion or of the senses, etc., would be useless; and in this case the metamorphosis would be retrograde.

From the remarks just made we can see how by changes of structure in the young, in conformity with changed habits of life, together with inheritance at corresponding ages, animals might come to pass through stages of development, perfectly distinct from the primordial condition of their adult progenitors. Most of our best authorities are now convinced that the various larval and pupal stages of insects have thus been acquired through adaptation, and not through inheritance from some ancient form. The curious case of Sitaris – a beetle which passes through certain unusual stages of development – will illustrate how this might occur. The first larval form is described by M. Fabre, as an active,

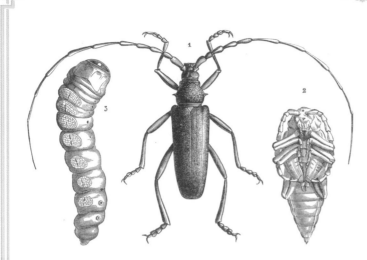

The Long-horned Oak Beetle:
1 adult *2* pupa *3* larva

minute insect, furnished with six legs, two long antennae, and four eyes. These larvae are hatched in the nests of bees; and when the male bees emerge from their burrows, in the spring, which they do before the females, the larvae spring on them, and afterwards crawl on to the females whilst paired with the males. As soon as the female bee deposits her eggs on the surface of the honey stored in the cells, the larvae of the Sitaris leap on the eggs and devour them. Afterwards they undergo a complete change; their eyes disappear; their legs and antennae become rudimentary, and they feed on honey; so that they now more closely resemble the ordinary larvae of insects; ultimately they undergo a further transformation, and finally emerge as the perfect beetle. Now, if an insect, undergoing transformations like those of the Sitaris, were to become the progenitor of a whole new class of insects, the course of development of the new class would be widely different from that of our existing insects; and the first larval stage certainly would not represent the former condition of any adult and ancient form.

On the other hand it is highly probable that with many animals the embryonic or larval stages show us, more or less completely, the condition of the progenitor of the whole group in its adult state. In the great class of the Crustacea, forms wonderfully distinct from each other, namely,

suctorial parasites, cirripedes, entomostraca, and even the malacostraca, appear at first as larvae under the nauplius-form; and as these larvae live and feed in the open sea, and are not adapted for any peculiar habits of life, and from other reasons assigned by Fritz Müller, it is probable that at some very remote period an independent adult animal, resembling the Nauplius, existed, and subsequently produced, along several divergent lines of descent, the above-named great Crustacean groups. So again, it is probable, from what we know of the embryos of mammals, birds, fishes, and reptiles, that these animals are the modified descendants of some ancient progenitor, which was furnished in its adult state with branchiae, a swim-bladder, four fin-like limbs, and a long tail, all fitted for an aquatic life.

As all the organic beings, extinct and recent, which have ever lived, can be arranged within a few great classes; and as all within each class have, according to our theory, been connected together by fine gradations, the best, and, if our collections were nearly perfect, the only possible arrangement, would be genealogical; descent being the hidden bond of connexion which naturalists have been seeking under the term of the Natural System. On this view we can understand how it is that, in the eyes of most naturalists, the structure of the embryo is even more important for classification than that of the adult. In two or more groups of animals, however much they may differ from each other in structure and habits in their adult condition, if they pass through closely similar embryonic stages, we may feel assured that they are all descended from one parent-form, and are therefore closely related. Thus, community in embryonic structure reveals community of descent; but dissimilarity in embryonic development does not prove discommunity of descent, for in one of two groups the developmental stages may have been suppressed, or may have been so greatly modified through adaptation to new habits of life as to be no longer recognisable. Even in groups, in which the adults have been modified to an extreme degree, community of origin is often revealed by the structure of the larvae; we have seen, for instance, that cirripedes, though externally so like shell-fish, are at once known by their larvae to belong to the great class of crustaceans. As the embryo often shows us more or less plainly the structure of the less modified and ancient progenitor of the group, we can see why ancient and extinct forms so often resemble in their adult state the embryos of existing species of the same class. Agassiz believes this to be a universal law of nature; and we may hope hereafter to see the law proved true. It can, however, be proved true only in those cases in which the ancient state of the progenitor of the group has not been wholly obliterated, either by successive variations having supervened at a very early period of growth, or by such variations having been inherited at an earlier age than that at which they first appeared. It should also be borne in mind, that the law may be true, but yet, owing to the geological record not extending far enough back in time, may remain for a long period, or for ever, incapable of demonstration. The law will not strictly hold good in those cases in which an ancient form became adapted in its larval state to some special line of life, and transmitted the same larval state to a whole group of descendants; for such larval state will not resemble any still more ancient form in its adult state.

Thus, as it seems to me, the leading facts in embryology, which are second to none in importance, are explained on the principle of variations in the many descendants from some one ancient progenitor, having appeared at a not very early period of life, and having been inherited at a corresponding period. Embryology rises greatly in interest, when we look at the embryo as a picture, more or less obscured, of the progenitor, either in its adult or larval state, of all the members of the same great class.

RUDIMENTARY, ATROPHIED, AND ABORTED ORGANS

Organs or parts in this strange condition, bearing the plain stamp of inutility, are extremely common, or even general, throughout nature. It would be impossible to name one of the higher animals in which some part or other is not in a rudimentary condition. In the mammalia, for instance, the males possess rudimentary mammae; in snakes one lobe of the lungs is rudimentary; in birds the 'bastard-wing' may safely be considered as a rudimentary digit, and in some species the whole wing is so far rudimentary that it cannot be used for flight. What can be more curious than the presence of teeth in foetal whales, which when grown up have not a tooth in their heads; or the teeth, which never cut through the gums, in the upper jaws of unborn calves?

Rudimentary organs plainly declare their origin and meaning in various ways. There are beetles belonging to closely allied species, or even to the same identical species, which have either full-sized and perfect wings, or mere rudiments of membrane, which not rarely lie under wing-covers firmly soldered together; and in these cases it is impossible to doubt, that the rudiments represent wings. Rudimentary organs sometimes retain their potentiality: this occasionally occurs with the mammae of male mammals, which have been known to become well developed and to secrete milk. So again in the udders of the genus Bos, there are normally four developed and two rudimentary teats; but the latter in our domestic cows sometimes become well developed and yield milk. In regard to plants, the petals are sometimes rudimentary, and sometimes well developed in the individuals of the same species. In certain plants having separated sexes Kölreuter found that by crossing a species, in which the male flowers included a rudiment of a pistil, with an hermaphrodite species, having of course a well-developed pistil, the rudiment in the hybrid offspring was much increased in size; and this clearly shows that the rudimentary and perfect pistils are essentially alike in nature. An animal may possess various parts in a perfect state, and yet they may in one sense be rudimentary, for they are useless: thus the tadpole of the common salamander or water-newt, as Mr. G.H. Lewes remarks, 'has gills, and passes its existence in the water; but the Salamandra atra, which lives high up amongst the mountains, brings forth its young full-formed. This animal never lives in the water. Yet if we open a gravid female, we find tadpoles inside her with exquisitely feathered gills; and when placed in water they swim about like the tadpoles of the water-newt. Obviously this aquatic organisation has no reference to the future life of the animal, nor has it any adaptation to its embryonic condition; it has solely reference to ancestral adaptations, it repeats a phase in the development of its progenitors.'

An organ, serving for two purposes, may become rudimentary or utterly aborted for one, even the more important purpose, and remain perfectly efficient for the other. Thus, in plants, the office of the pistil is to allow the pollen-tubes to reach the ovules within the ovarium. The pistil consists of a stigma supported on the style; but in some Compositae, the male florets, which of course cannot be fecundated, have a rudimentary pistil, for it is not crowned with a stigma; but the style remains well developed and is clothed in the usual manner with hairs, which serve to brush the pollen out of the surrounding and conjoined anthers. Again, an organ may become rudimentary for its proper purpose, and be used for a distinct one: in certain fishes the swim-bladder seems to be rudimentary for its proper function of giving buoyancy, but has become converted into a nascent breathing organ or lung. Many similar instances could be given.

Useful organs, however little they may be developed, unless we have reason to suppose that they were formerly more highly developed, ought not to be considered as rudimentary. They may be in a nascent condition, and in progress towards further development. Rudimentary organs, on the

other hand, are either quite useless, such as teeth which never cut through the gums, or almost useless, such as the wings of an ostrich, which serve merely as sails. As organs in this condition would formerly, when still less developed, have been of even less use than at present, they cannot formerly have been produced through variation and natural selection, which acts solely by the preservation of useful modifications. They have been partially retained by the power of inheritance, and relate to a former state of things. It is, however, often difficult to distinguish between rudimentary and nascent organs; for we can judge only by analogy whether a part is capable of further development, in which case alone it deserves to be called nascent. Organs in this condition will always be somewhat rare; for beings thus provided will commonly have been supplanted by their successors with the same organ in a more perfect state, and consequently will have become long ago extinct. The wing of the penguin is of high service, acting as a fin; it may, therefore, represent the nascent state of the wing: not that I believe this to be the case; it is more probably a reduced organ, modified for a new function: the wing of the Apteryx, on the other hand, is quite useless, and is truly rudimentary. Owen considers the simple filamentary limbs of the Lepidosiren as the 'beginnings of organs which attain full functional development in higher vertebrates;' but, according to the view lately advocated by Dr. Günther, they are probably remnants, consisting of the persistent axis of a fin, with the lateral rays or branches aborted. The mammary glands of the Ornithorhynchus may be considered, in comparison with the udders of a cow, as in a nascent condition. The ovigerous frena of certain cirripedes, which have ceased to give attachment to the ova and are feebly developed, are nascent branchiae.

Rudimentary organs in the individuals of the same species are very liable to vary in the degree of their development and in other respects. In closely allied species, also, the extent to which the same organ has been reduced occasionally differs much. This latter fact is well exemplified in the state of the wings of female moths belonging to the same family. Rudimentary organs may be utterly aborted; and this implies, that in certain animals or plants, parts are entirely absent which analogy would lead us to expect to find in them, and which are occasionally found in monstrous individuals. Thus in most of the Scrophulariaceae the fifth stamen is utterly aborted; yet we may conclude that a fifth stamen once existed, for a rudiment of it is found in many species of the family, and this rudiment occasionally becomes perfectly developed, as may sometimes be seen in the common snap-dragon. In tracing the homologies of any part in different members of the same class, nothing is more common, or, in order fully to understand the relations of the parts, more useful than the discovery of rudiments. This is well shown in the drawings given by Owen of the leg bones of the horse, ox, and rhinoceros.

It is an important fact that rudimentary organs, such as teeth in the upper jaws of whales and ruminants, can often be detected in the embryo, but afterwards wholly disappear. It is also, I believe, a universal rule, that a rudimentary part is of greater size in the embryo relatively to the adjoining parts, than in the adult; so that the organ at this early age is less rudimentary, or even cannot be said to be in any degree rudimentary. Hence rudimentary organs in the adult are often said to have retained their embryonic condition.

I have now given the leading facts with respect to rudimentary organs. In reflecting on them, every one must be struck with astonishment; for the same reasoning power which tells us that most parts and organs are exquisitely adapted for certain purposes, tells us with equal plainness that these rudimentary or atrophied organs are imperfect and useless. In works on natural history, rudimentary organs are generally said to have been created 'for the sake of symmetry,' or in order 'to complete the scheme of nature.' But this is not an explanation, merely a restatement of the fact. Nor is it consistent with itself: thus the boa-constrictor

has rudiments of hind limbs and of a pelvis, and if it be said that these bones have been retained 'to complete the scheme of nature,' why, as Professor Weismann asks, have they not been retained by other snakes, which do not possess even a vestige of these same bones? What would be thought of an astronomer who maintained that the satellites revolve in elliptic courses round their planets 'for the sake of symmetry,' because the planets thus revolve round the sun? An eminent physiologist accounts for the presence of rudimentary organs, by supposing that they serve to excrete matter in excess, or matter injurious to the system; but can we suppose that the minute papilla, which often represents the pistil in male flowers, and which is formed of mere cellular tissue, can thus act? Can we suppose that rudimentary teeth, which are subsequently absorbed, are beneficial to the rapidly growing embryonic calf by removing matter so precious as phosphate of lime? When a man's fingers have been amputated, imperfect nails have been known to appear on the stumps, and I could as soon believe that these vestiges of nails are developed in order to excrete horny matter, as that the rudimentary nails on the fin of the manatee have been developed for this same purpose.

On the view of descent with modification, the origin of rudimentary organs is comparatively simple; and we can understand to a large extent the laws governing their imperfect development. We have plenty of cases of rudimentary organs in our domestic productions, as the stump of a tail in tailless breeds, the vestige of an ear in earless breeds of sheep, the reappearance of minute dangling horns in hornless breeds of cattle, more especially, according to Youatt, in young animals, and the state of the whole flower in the cauliflower. We often see rudiments of various parts in monsters; but I doubt whether any of these cases throw light on the origin of rudimentary organs in a state of nature, further than by showing that rudiments can be produced; for the balance of evidence clearly indicates that species under nature do not undergo great and abrupt changes. But we learn from the study of our domestic productions that the disuse of parts leads to their reduced size; and that the result is inherited.

It appears probable that disuse has been the main agent in rendering organs rudimentary. It would at first lead by slow steps to the more and more complete reduction of a part, until at last it became rudimentary – as in the case of the eyes of animals inhabiting dark caverns, and of the wings of birds inhabiting oceanic islands, which have seldom been forced by beasts of prey to take flight, and have ultimately lost the power of flying. Again, an organ, useful under certain conditions, might become injurious under others, as with the wings of beetles living on small and exposed islands; and in this case natural selection will have aided in reducing the organ, until it was rendered harmless and rudimentary.

Any change in structure and function, which can be effected by small stages, is within the power of natural selection; so that an organ rendered, through changed habits of life, useless or injurious for one purpose, might be modified and used for another purpose. An organ might, also, be retained for one alone of its former functions. Organs, originally formed by the aid of natural selection, when rendered useless may well be variable, for their variations can no longer be checked by natural selection. All this agrees well with what we see under nature. Moreover, at whatever period of life either disuse or selection reduces an organ, and this will generally be when the being has come to maturity and has to exert its full powers of action, the principle of inheritance at corresponding ages will tend to reproduce the organ in its reduced state at the same mature age, but will seldom affect it in the embryo. Thus we can understand the greater size of rudimentary organs in the embryo relatively to the adjoining parts, and their lesser relative size in the adult. If, for instance, the digit of an adult animal was used less and less during many generations, owing to some change of habits, or if an organ or gland was less and less functionally exercised, we may infer that it would

become reduced in size in the adult descendants of this animal, but would retain nearly its original standard of development in the embryo.

There remains, however, this difficulty. After an organ has ceased being used, and has become in consequence much reduced, how can it be still further reduced in size until the merest vestige is left; and how can it be finally quite obliterated? It is scarcely possible that disuse can go on producing any further effect after the organ has once been rendered functionless. Some additional explanation is here requisite which I cannot give. If, for instance, it could be proved that every part of the organisation tends to vary in a greater degree towards diminution than toward augmentation of size, then we should be able to understand how an organ which has become useless would be rendered, independently of the effects of disuse, rudimentary and would at last be wholly suppressed; for the variations towards diminished size would no longer be checked by natural selection. The principle of the economy of growth, explained in a former chapter, by which the materials forming any part, if not useful to the possessor, are saved as far as is possible, will perhaps come into play in rendering a useless part rudimentary. But this principle will almost necessarily be confined to the earlier stages of the process of reduction; for we cannot suppose that a minute papilla, for instance, representing in a male flower the pistil of the female flower, and formed merely of cellular tissue, could be further reduced or absorbed for the sake of economising nutriment.

Finally, as rudimentary organs, by whatever steps they may have been degraded into their present useless condition, are the record of a former state of things, and have been retained solely through the power of inheritance – we can understand, on the genealogical view of classification, how it is that systematists, in placing organisms in their proper places in the natural system, have often found rudimentary parts as useful as, or even sometimes more useful than, parts of high physiological importance.

Rudimentary organs may be compared with the letters in a word, still retained in the spelling, but become useless in the pronunciation, but which serve as a clue for its derivation. On the view of descent with modification, we may conclude that the existence of organs in a rudimentary, imperfect, and useless condition, or quite aborted, far from presenting a strange difficulty, as they assuredly do on the old doctrine of creation, might even have been anticipated in accordance with the views here explained.

SUMMARY

In this chapter I have attempted to show that the arrangement of all organic beings throughout all time in groups under groups – that the nature of the relationships by which all living and extinct organisms are united by complex, radiating, and circuitous lines of affinities into a few grand classes, – the rules followed and the difficulties encountered by naturalists in their classifications, – the value set upon characters, if constant and prevalent, whether of high or of the most trifling importance, or, as with rudimentary organs of no importance, – the wide opposition in value between analogical or adaptive characters, and characters of true affinity; and other such rules, – all naturally follow if we admit the common parentage of allied forms, together with their modification through variation and natural selection, with the contingencies of extinction and divergence of character. In considering this view of classification, it should be borne in mind that the element of descent has been universally used in ranking together the sexes, ages, dimorphic forms, and acknowledged varieties of the same species, however much they may differ from each other in structure. If we extend the use of this element of descent, – the one certainly known cause of similarity in organic beings, – we shall understand what is meant by the Natural System: it is genealogical in its attempted arrangement, with the grades of acquired difference marked by the terms, varieties, species, genera, families, orders, and classes.

On this same view of descent with modification, most of the great facts in Morphology become intelligible, – whether we look to the same pattern displayed by the different species of the same class in their homologous organs, to whatever purpose applied, or to the serial and lateral homologies in each individual animal and plant.

On the principle of successive slight variations, not necessarily or generally supervening at a very early period of life, and being inherited at a corresponding period, we can understand the leading facts in Embryology; namely, the close resemblance in the individual embryo of the parts which are homologous, and which when matured become widely different in structure and function; and the resemblance of the homologous parts or organs in allied though distinct species, though fitted in the adult state for habits as different as is possible. Larvae are active embryos, which have become specially modified in a greater or less degree in relation to their habits of life, with their modifications inherited at a corresponding early age. On these same principles, and bearing in mind that when organs are reduced in size, either from disuse or through natural selection, it will generally be at that period of life when the being has to provide for its own wants, and bearing in mind how strong is the force of inheritance – the occurrence of rudimentary organs might even have been anticipated. The importance of embryological characters and of rudimentary organs in classification is intelligible, on the view that a natural arrangement must be genealogical.

Finally, the several classes of facts which have been considered in this chapter, seem to me to proclaim so plainly, that the innumerable species, genera and families, with which this world is peopled, are all descended, each within its own class or group, from common parents, and have all been modified in the course of descent, that I should without hesitation adopt this view, even if it were unsupported by other facts or arguments.

Chapter XV

RECAPITULATION AND CONCLUSION

Recapitulation of the objections to the theory of Natural Selection–Recapitulation of the general and special circumstances in its favour–Causes of the general belief in the immutability of species–How far the theory of Natural Selection may be extended–Effects of its adoption on the study of Natural History–Concluding remarks.

As this whole volume is one long argument, it may be convenient to the reader to have the leading facts and inferences briefly recapitulated.

That many and serious objections may be advanced against the theory of descent with modification through Variation and Natural Selection, I do not deny. I have endeavoured to give to them their full force. Nothing at first can appear more difficult to believe than that the more complex organs and instincts have been perfected, not by means superior to, though analogous with, human reason, but by the accumulation of innumerable slight variations, each good for the individual possessor. Nevertheless, this difficulty, though appearing to our imagination insuperably great, cannot be considered real if we admit the following propositions, namely, that all parts of the organisation and instincts offer, at least, individual differences – that there is a struggle for existence leading to the preservation of profitable deviations of structure or instinct – and, lastly, that gradations in the state of perfection of each organ may have existed, each good of its kind. The truth of these propositions cannot, I think, be disputed.

It is, no doubt, extremely difficult even to conjecture by what gradations many structures have been perfected, more especially amongst broken and failing groups of organic beings, which have suffered much extinction; but we see so many strange gradations in nature, that we ought to be extremely cautious in saying that any organ or instinct, or any whole structure, could not have arrived at its present state by many graduated steps. There are, it must be admitted, cases of special difficulty opposed to the theory of Natural Selection; and one of the most curious of these is the existence in the same community of two or three defined castes of workers or sterile female ants; but I have attempted to show how these difficulties can be mastered.

With respect to the almost universal sterility of species when first crossed, which forms so remarkable a contrast with the almost universal fertility of varieties when crossed, I must refer the reader to the recapitulation of the facts given at the end of the ninth chapter, which seem to me conclusively to show that this sterility is no more a special endowment than is the incapacity of two distinct kinds of trees to be grafted together; but that it is incidental on differences confined to the reproductive systems of the intercrossed species. We see the truth of this conclusion in the vast difference in the results of crossing the same two species reciprocally, – that is, when one species is first used as the father and then as the mother. Analogy from the consideration of dimorphic and trimorphic plants clearly leads to the same conclusion, for when the forms are illegitimately united, they yield few or no seed, and their offspring are more or less sterile; and these forms belong to the same undoubted species, and differ from each other in no respect except in their reproductive organs and functions.

Although the fertility of varieties when intercrossed, and of their mongrel offspring, has been asserted by so many authors to be universal, this cannot be considered as quite correct after the facts given on the high authority of Gärtner and Kölreuter. Most of the varieties which have been experimented on have been produced under domestication; and as domestication (I do not mean mere confinement) almost certainly tends to eliminate that sterility which, judging from analogy, would have affected the parent-species if intercrossed, we ought not to expect that domestication would likewise induce sterility in their modified descendants when crossed. This elimination of sterility apparently follows from the same cause which allows our domestic animals to breed freely under diversified circumstances; and this again apparently follows from their having been gradually accustomed to frequent changes in their conditions of life.

A double and parallel series of facts seems to throw much light on the sterility of species, when first crossed, and of their hybrid offspring. On the one side, there is good reason to believe that slight changes in the conditions of life give vigour and fertility to all organic beings. We know also that a cross between the distinct individuals of the same variety, and between distinct varieties, increases the number of their offspring, and certainly gives to them increased size and vigour. This is chiefly owing to the forms which are crossed having been exposed to somewhat different conditions of life; for I have ascertained by a labourious series of experiments that if all the individuals of the same variety be subjected during several generations to the same conditions, the good derived from crossing is often much diminished or wholly

disappears. This is one side of the case. On the other side, we know that species which have long been exposed to nearly uniform conditions, when they are subjected under confinement to new and greatly changed conditions, either perish, or if they survive, are rendered sterile, though retaining perfect health. This does not occur, or only in a very slight degree, with our domesticated productions, which have long been exposed to fluctuating conditions. Hence when we find that hybrids produced by a cross between two distinct species are few in number, owing to their perishing soon after conception or at a very early age, or if surviving that they are rendered more or less sterile, it seems highly probable that this result is due to their having been in fact subjected to a great change in their conditions of life, from being compounded of two distinct organisations. He who will explain in a definite manner why, for instance, an elephant or a fox will not breed under confinement in its native country, whilst the domestic pig or dog will breed freely under the most diversified conditions, will at the same time be able to give a definite answer to the question why two distinct species, when crossed, as well as their hybrid offspring, are generally rendered more or less sterile, whilst two domesticated varieties when crossed and their mongrel offspring are perfectly fertile.

Turning to geographical distribution, the difficulties encountered on the theory of descent with modification are serious enough. All the individuals of the same species, and all the species of the same genus, or even higher group, are descended from common parents; and therefore, in however distant and isolated parts of the world they may now be found, they must in the course of successive generations have travelled from some one point to all the others. We are often wholly unable even to conjecture how this could have been effected. Yet, as we have reason to believe that some species have retained the same specific form for very long periods of time, immensely long as measured by years, too much stress ought

SIAMESE BREED.

Sow, 3 Years old, imported from Singapore by Messrs Dudgale, Manchester:__The litter by a half-bred Chinese Male.

Published June 1840, by Longman, Orme, Brown, Green & Longmans, Paternoster Row, London.

The Siamese breed of domestic pig: a sow with litter by a half-bred Chinese boar not to be laid on the occasional wide diffusion of the same species; for during very long periods there will always have been a good chance for wide migration by many means. A broken or interrupted range may often be accounted for by the extinction of the species in the intermediate regions. It cannot be denied that we are as yet very ignorant as to the full extent of the various climatical and geographical changes which have affected the earth during modern periods; and such changes will often have facilitated migration. As an example, I have attempted to show how potent has been the influence of the Glacial period on the distribution of the same and of allied species throughout the world. We are as yet profoundly ignorant of the many occasional means of transport. With respect to distinct species of the same genus, inhabiting distant and isolated regions, as the process of modification has necessarily been slow, all the means of migration will have been possible during a very long period; and consequently the difficulty of the wide diffusion of the species of the same genus is in some degree lessened.

As according to the theory of natural selection an interminable number of intermediate forms must have existed, linking together all the species in each group by gradations as fine as our existing varieties, it may be asked, Why do we not see these linking forms all around us? Why are not all organic beings blended together in an inextricable chaos? With respect to existing forms, we should remember that we have no right to expect (excepting in rare cases) to discover *directly* connecting links between them, but only between each and some extinct and supplanted form. Even on a wide area, which has during a long period remained continuous, and of which the climatic and other conditions of life change insensibly in proceeding from a district occupied by one species into another district occupied by a closely allied species, we have no just right to expect often to find intermediate varieties in the intermediate zones. For we have reason to believe that only a few species of a genus ever undergo change; the other species becoming utterly extinct and leaving no modified progeny. Of the species which do change, only a few within the same country change at the same time; and all modifications are slowly effected. I have also shown that the intermediate varieties which probably at first existed in the intermediate zones, would be liable to be supplanted by the allied forms on either hand; for the latter, from existing in greater numbers, would generally be modified and improved at a quicker rate than the intermediate varieties, which existed in lesser numbers; so that the intermediate varieties would, in the long run, be supplanted and exterminated.

On this doctrine of the extermination of an infinitude of connecting links, between the living and extinct inhabitants of the world, and at each successive period between the extinct and still older species, why is not every geological formation charged with such links? Why does not every collection of fossil remains afford plain evidence of the gradation and mutation of the forms of life? Although geological research has undoubtedly revealed the former existence of many links, bringing numerous forms of life much closer together, it does not yield the infinitely many fine gradations between past and present species required on the theory, and this is the most obvious of the many objections which may be urged against it. Why, again, do whole groups of allied species appear, though this appearance is often false, to have come in suddenly on the successive geological stages? Although we now know that organic beings appeared on this globe, at a period incalculably remote, long before the lowest bed of the Cambrian system was deposited, why do we not find beneath this system great piles of strata stored with the remains of the progenitors of the Cambrian fossils? For on the theory, such strata must somewhere have been deposited at these ancient and utterly unknown epochs of the world's history.

I can answer these questions and objections only on the supposition that the geological record is far more imperfect than most geologists believe. The number of specimens in all our museums is absolutely as nothing compared with the countless generations of countless species which have certainly existed. The parent form of any two or more species would not be in all its characters directly intermediate between its modified offspring, any more than the rock-pigeon is directly intermediate in crop and tail between its descendants, the pouter and fantail pigeons. We should not be able to recognise a species as the parent of another and modified species, if we were to examine the two ever so closely, unless we possessed most of the intermediate links; and owing to the imperfection of the geological record, we have no just right to expect to find so many links. If two or three, or even more linking forms were discovered, they would simply be ranked by many naturalists as so many new species, more especially if found in different geological substages, let their differences be ever so slight. Numerous existing doubtful forms could be named which are probably varieties; but who will pretend that in future ages so many fossil links will be discovered, that naturalists will be able to decide whether or not these doubtful forms ought to be called varieties? Only a small portion of the world has been geologically explored. Only organic beings of certain classes can be preserved in a fossil condition, at least in any great number. Many species when once formed never undergo any further change but become extinct without leaving modified descendants; and the periods during which species have undergone modification, though long as measured by years, have probably been short in comparison with the periods during which they retained the same form. It is the dominant and widely ranging species which vary most frequently and vary most, and varieties are often at first local – both causes rendering the discovery of intermediate links in any one formation less likely. Local varieties will not spread into other and distant regions until they are considerably modified and improved; and when they have spread, and are discovered in a geological formation, they appear as if suddenly created there, and will be simply classed as new species. Most formations have been intermittent in their accumulation; and their duration has probably been shorter than the average duration of specific forms. Successive formations are in most cases separated from each other by blank intervals of time of great length, for fossiliferous formations thick enough to resist future degradation can, as a general rule, be accumulated only where much sediment is deposited on the subsiding bed of the sea. During the alternate periods of elevation and of stationary level the record will generally be blank. During these latter periods there will probably be more variability in the forms of life; during periods of subsidence, more extinction.

With respect to the absence of strata rich in fossils beneath the Cambrian formation, I can recur only to the hypothesis given in the tenth chapter; namely, that though our continents and oceans have endured for an enormous period in nearly their present relative positions, we have no reason to assume that this has always been the case; consequently formations much older than any now known may lie buried beneath the great oceans. With respect to the lapse of time not having been sufficient since our planet was consolidated for the assumed amount of organic change, and this objection, as urged by Sir William Thompson, is probably one of the gravest as yet advanced, I can only say, firstly, that we do not know at what rate species change, as measured by years, and secondly, that many philosophers are not as yet willing to admit that we know enough of the constitution of the universe and of the interior of our globe to speculate with safety on its past duration.

That the geological record is imperfect all will admit; but that it is imperfect to the degree required by our theory, few will be inclined to admit. If we look to long enough intervals of time, geology plainly declares that species have all changed; and they have changed in the manner required by the theory, for they have changed slowly and in a graduated manner. We clearly see this in the fossil remains from consecutive formations invariably being much more closely related to each other than are the fossils from widely separated formations.

Such is the sum of the several chief objections and difficulties which may justly be urged against the theory; and I have now briefly recapitulated the answers and explanations which, as far as I can see, may be given. I have felt these difficulties far too heavily during many years to doubt their weight. But it deserves especial notice that the more important objections relate to questions on which we are confessedly ignorant; nor do we know how ignorant we are. We do not know all the possible transitional gradations between the simplest and the most perfect organs; it cannot be pretended that we know all the varied means of Distribution during the long lapse of years, or that we know how imperfect is the Geological Record. Serious as these several objections are, in my judgment they are by no means sufficient to overthrow the theory of descent with subsequent modification.

Now let us turn to the other side of the argument. Under domestication we see much variability, caused, or at least excited, by changed conditions of life; but often in so obscure a manner, that we are tempted to consider the variations as spontaneous. Variability is governed by many complex laws, by correlated growth, compensation, the increased use and disuse of parts, and the definite action of the surrounding conditions. There is much difficulty in ascertaining how largely our domestic productions have been modified; but we may safely infer that the amount has been large, and that modifications can be inherited for long periods. As long as the conditions of life remain the same, we have reason to believe that a modification, which has already been inherited for

many generations, may continue to be inherited for an almost infinite number of generations. On the other hand, we have evidence that variability, when it has once come into play, does not cease under domestication for a very long period; nor do we know that it ever ceases, for new varieties are still occasionally produced by our oldest domesticated productions.

Variability is not actually caused by man; he only unintentionally exposes organic beings to new conditions of life and then nature acts on the organisation and causes it to vary. But man can and does select the variations given to him by nature, and thus accumulates them in any desired manner. He thus adapts animals and plants for his own benefit or pleasure. He may do this methodically, or he may do it unconsciously by preserving the individuals most useful or pleasing to him without any intention of altering the breed. It is certain that he can largely influence the character of a breed by selecting, in each successive generation, individual differences so slight as to be inappreciable except by an educated eye. This unconscious process of selection has been the great agency in the formation of the most distinct and useful domestic breeds. That many breeds produced by man have to a large extent the character of natural species, is shown by the inextricable doubts whether many of them are varieties or aboriginally distinct species.

There is no reason why the principles which have acted so efficiently under domestication should not have acted under nature. In the survival of favoured individuals and races, during the constantly recurrent Struggle for Existence, we see a powerful and ever-acting form of Selection. The struggle for existence inevitably follows from the high geometrical ratio of increase which is common to all organic beings. This high rate of increase is proved by calculation – by the rapid increase of many animals and plants during a succession of peculiar seasons, and when naturalised in new countries. More individuals are born than can possibly survive. A grain in the balance may determine which individuals shall live and which shall die – which variety or species shall increase in number, and which shall decrease, or finally become extinct. As the individuals of the same species come in all respects into the closest competition with each other, the struggle will generally be most severe between them; it will be almost equally severe between the varieties of the same species, and next in severity between the species of the same genus. On the other hand the struggle will often be severe between beings remote in the scale of nature. The slightest advantage in certain individuals, at any age or during any season, over those with which they come into competition, or better adaptation in however slight a degree to the surrounding physical conditions, will, in the long run, turn the balance.

With animals having separated sexes, there will be in most cases a struggle between the males for the possession of the females. The most vigorous males, or those which have most successfully struggled with their conditions of life, will generally leave most progeny. But success will often depend on the males having special weapons or means of defence or charms; and a slight advantage will lead to victory.

As geology plainly proclaims that each land has undergone great physical changes, we might have expected to find that organic beings have varied under nature, in the same way as they have varied under domestication. And if there has been any variability under nature, it would be an unaccountable fact if natural selection had not come into play. It has often been asserted, but the assertion is incapable of proof, that the amount of variation under nature is a strictly limited quantity. Man, though acting on external characters alone and often capriciously, can produce within a short period a great result by adding up mere individual differences in his domestic productions; and every one admits that species present individual differences. But, besides such differences, all naturalists admit that natural

varieties exist, which are considered sufficiently distinct to be worthy of record in systematic works. No one has drawn any clear distinction between individual differences and slight varieties; or between more plainly marked varieties and subspecies and species. On separate continents, and on different parts of the same continent, when divided by barriers of any kind, and on outlying islands, what a multitude of forms exist, which some experienced naturalists rank as varieties, others as geographical races or subspecies, and others as distinct, though closely allied species!

If, then, animals and plants do vary, let it be ever so slightly or slowly, why should not variations or individual differences, which are in any way beneficial, be preserved and accumulated through natural selection, or the survival of the fittest? If man can by patience select variations useful to him, why, under changing and complex conditions of life, should not variations useful to nature's living products often arise, and be preserved or selected? What limit can be put to this power, acting during long ages and rigidly scrutinising the whole constitution, structure, and habits of each creature, – favouring the good and rejecting the bad? I can see no limit to this power, in slowly and beautifully adapting each form to the most complex relations of life. The theory of natural selection, even if we look no further than this, seems to be in the highest degree probable. I have already recapitulated, as fairly as I could, the opposed difficulties and objections: now let us turn to the special facts and arguments in favour of the theory.

On the view that species are only strongly marked and permanent varieties, and that each species first existed as a variety, we can see why it is that no line of demarcation can be drawn between species, commonly supposed to have been produced by special acts of creation, and varieties which are acknowledged to have been produced by secondary laws. On this same view we can understand how it is that in a region where many species of a genus have been produced, and where they now flourish, these same species should present many varieties; for where the manufactory of species has been active, we might expect, as a general rule, to find it still in action; and this is the case if varieties be incipient species. Moreover, the species of the larger genera, which afford the greater number of varieties or incipient species, retain to a certain degree the character of varieties; for they differ from each other by a less amount of difference than do the species of smaller genera. The closely allied species also of a larger genera apparently have restricted ranges, and in their affinities they are clustered in little groups round other species – in both respects resembling varieties. These are strange relations on the view that each species was independently created, but are intelligible if each existed first as a variety.

As each species tends by its geometrical rate of reproduction to increase inordinately in number; and as the modified descendants of each species will be enabled to increase by as much as they become more diversified in habits and structure, so as to be able to seize on many and widely different places in the economy of nature, there will be a constant tendency in natural selection to preserve the most divergent offspring of any one species. Hence during a long-continued course of modification, the slight differences characteristic of varieties of the same species, tend to be augmented into the greater differences characteristic of the species of the same genus. New and improved varieties will inevitably supplant and exterminate the older, less improved and intermediate varieties; and thus species are rendered to a large extent defined and distinct objects. Dominant species belonging to the larger groups within each class tend to give birth to new and dominant forms; so that each large group tends to become still larger, and at the same time more divergent in character. But as all groups cannot thus go on increasing in size, for the world would not hold them, the more dominant groups beat the less dominant. This tendency in the large groups to go on increasing in size and diverging in

character, together with the inevitable contingency of much extinction, explains the arrangement of all the forms of life in groups subordinate to groups, all within a few great classes, which has prevailed throughout all time. This grand fact of the grouping of all organic beings under what is called the Natural System, is utterly inexplicable on the theory of creation.

As natural selection acts solely by accumulating slight, successive, favourable variations, it can produce no great or sudden modifications; it can act only by short and slow steps. Hence, the canon of 'Natura non facit saltum,' which every fresh addition to our knowledge tends to confirm, is on this theory intelligible. We can see why throughout nature the same general end is gained by an almost infinite diversity of means, for every peculiarity when once acquired is long inherited, and structures already modified in many different ways have to be adapted for the same general purpose. We can, in short, see why nature is prodigal in variety, though niggard in innovation. But why this should be a law of nature if each species has been independently created no man can explain.

Many other facts are, as it seems to me, explicable on this theory. How strange it is that a bird, under the form of a woodpecker, should prey on insects on the ground; that upland geese, which rarely or never swim, would possess webbed feet; that a thrush-like bird should dive and feed on sub-aquatic insects; and that a petrel should have the habits and structure fitting it for the life of an auk! and so in endless other cases. But on the view of each species constantly trying to increase in number, with natural selection always ready to adapt the slowly varying descendants of each to any unoccupied or ill-occupied place in nature, these facts cease to be strange, or might even have been anticipated.

We can to a certain extent understand how it is that there is so much beauty throughout nature; for this may be largely attributed to the agency of selection. That beauty, according to our sense of it, is not universal, must be admitted by every one who will look at some venomous snakes, at some fishes, and at certain hideous bats with a distorted resemblance to the human face. Sexual selection has given the most brilliant colours, elegant patterns, and other ornaments to the males, and sometimes to both sexes of many birds, butterflies and other animals. With birds it has often rendered the voice of the male musical to the female, as well as to our ears. Flowers and fruit have been rendered conspicuous by brilliant colours in contrast with the green foliage, in order that the flowers may be easily seen, visited and fertilised by insects, and the seeds disseminated by birds. How it comes that certain colours, sounds and forms should give pleasure to man and the lower animals, that is, how the sense of beauty in its simplest form was first acquired, we do not know any more than how certain odours and flavours were first rendered agreeable.

As natural selection acts by competition, it adapts and improves the inhabitants of each country only in relation to their co-inhabitants; so that we need feel no surprise at the species of any one country, although on the ordinary view supposed to have been created and specially adapted for that country, being beaten and supplanted by the naturalised productions from another land. Nor ought we to marvel if all the contrivances in nature be not, as far as we can judge, absolutely perfect; as in the case even of the human eye; or if some of them be abhorrent to our ideas of fitness. We need not marvel at the sting of the bee, when used against the enemy, causing the bee's own death; at drones being produced in such great numbers for one single act, and being then slaughtered by their sterile sisters; at the astonishing waste of pollen by our fir-trees; at the instinctive hatred of the queen bee for her own fertile daughters; at ichneumonidae feeding within the living bodies of caterpillars; and at other such cases. The wonder, indeed, is, on the theory of natural selection, that more cases of the want of absolute perfection have not been detected.

The complex and little known laws governing the production of varieties are the same, as far as we can judge, with the laws which have governed the production of distinct species. In both cases physical conditions seem to have produced some direct and definite effect, but how much we cannot say. Thus, when varieties enter any new station, they occasionally assume some of the characters proper to the species of that station. With both varieties and species, use and disuse seem to have produced a considerable effect; for it is impossible to resist this conclusion when we look, for instance, at the logger-headed duck, which has wings incapable of flight, in nearly the same condition as in the domestic duck; or when we look at the burrowing tucu-tucu, which is occasionally blind, and then at certain moles, which are habitually blind and have their eyes covered with skin; or when we look at the blind animals inhabiting the dark caves of America and Europe. With varieties and species, correlated variation seems to have played an important part, so that when one part has been modified other parts have been necessarily modified. With both varieties and species, reversions to long-lost characters occasionally occur. How inexplicable on the theory of creation is the occasional appearance of stripes on the shoulders and legs of the several species of the horse-genus and of their hybrids! How simply is this fact explained if we believe that these species are all descended from a striped progenitor, in the same manner as the several domestic breeds of the pigeon are descended from the blue and barred rock-pigeon!

On the ordinary view of each species having been independently created, why should specific characters, or those by which the species of the same genus differ from each other, be more variable than the generic characters in which they all agree? Why, for instance, should the colour of a flower be more likely to vary in any one species of a genus, if the other species possess differently coloured flowers, than if all possessed the same coloured flowers? If species are only well-marked varieties, of which the characters have become in a high degree permanent, we can understand this fact; for they have already varied since they branched off from a common progenitor in certain characters, by which they have come to be specifically distinct from each other; therefore these same characters would be more likely again to vary than the generic characters which have been inherited without change for an immense period. It is inexplicable on the theory of creation why a part developed in a very unusual manner in one species alone of a genus, and therefore, as we may naturally infer, of great importance to that species, should be eminently liable to variation; but, on our view, this part has undergone, since the several species branched off from a common progenitor, an unusual amount of variability and modification, and therefore we might expect the part generally to be still variable. But a part may be developed in the most unusual manner, like the wing of a bat, and yet not be more variable than any other structure, if the part be common to many subordinate forms, that is, if it has been inherited for a very long period; for in this case it will have been rendered constant by long-continued natural selection.

Glancing at instincts, marvellous as some are, they offer no greater difficulty than do corporeal structures on the theory of the natural selection of successive, slight, but profitable modifications. We can thus understand why nature moves by graduated steps in endowing different animals of the same class with their several instincts. I have attempted to show how much light the principle of gradation throws on the admirable architectural powers of the hive bee. Habit no doubt often comes into play in modifying instincts; but it certainly is not indispensable, as we see in the case of neuter insects, which leave no progeny to inherit the effects of long-continued habit. On the view of all the species of the same genus having descended from a common parent, and having inherited much in common, we can understand how it is that allied species, when placed under

widely different conditions of life, yet follow nearly the same instincts; why the thrushes of tropical and temperate South America, for instance, line their nests with mud like our British species. On the view of instincts having been slowly acquired through natural selection, we need not marvel at some instincts being not perfect and liable to mistakes, and at many instincts causing other animals to suffer.

If species be only well-marked and permanent varieties, we can at once see why their crossed offspring should follow the same complex laws in their degrees and kinds of resemblance to their parents – in being absorbed into each other by successive crosses, and in other such points – as do the crossed offspring of acknowledged varieties. This similarity would be a strange fact, if species had been independently created and varieties had been produced through secondary laws.

If we admit that the geological record is imperfect to an extreme degree, then the facts, which the record does give, strongly support the theory of descent with modification. New species have come on the stage slowly and at successive intervals; and the amount of change after equal intervals of time, is widely different in different groups. The extinction of species and of whole groups of species, which has played so conspicuous a part in the history of the organic world, almost inevitably follows from the principle of natural selection; for old forms are supplanted by new and improved forms. Neither single species nor groups of species reappear when the chain of ordinary generation is once broken. The gradual diffusion of dominant forms, with the slow modification of their descendants, causes the forms of life, after long intervals of time, to appear as if they had changed simultaneously throughout the world. The fact of the fossil remains of each formation being in some degree intermediate in character between the fossils in the formations above and below, is simply explained by their intermediate position in the chain of descent. The grand fact that all extinct beings can be classed with all recent beings, naturally follows from the living and the extinct being the offspring of common parents. As species have generally diverged in character during their long course of descent and modification, we can understand why it is that the more ancient forms, or early progenitors of each group, so often occupy a position in some degree intermediate between existing groups. Recent forms are generally looked upon as being, on the whole, higher in the scale of organisation than ancient forms; and they must be higher, in so far as the later and more improved forms have conquered the older and less improved forms in the struggle for life; they have also generally had their organs more specialised for different functions. This fact is perfectly compatible with numerous beings still retaining simple and but little improved structures, fitted for simple conditions of life; it is likewise compatible with some forms having retrograded in organisation, by having become at each stage of descent better fitted for new and degraded habits of life. Lastly, the wonderful law of the long endurance of allied forms on the same continent – of marsupials in Australia, of edentata in America, and other such cases – is intelligible, for within the same country the existing and the extinct will be closely allied by descent.

Looking to geographical distribution, if we admit that there has been during the long course of ages much migration from one part of the world to another, owing to former climatical and geographical changes and to the many occasional and unknown means of dispersal, then we can understand, on the theory of descent with modification, most of the great leading facts in Distribution. We can see why there should be so striking a parallelism in the distribution of organic beings throughout space, and in their geological succession throughout time; for in both cases the beings have been connected by the bond of ordinary generation, and the means of modification have been the same. We see the full meaning of the wonderful fact, which has struck

every traveller, namely, that on the same continent, under the most diverse conditions, under heat and cold, on mountain and lowland, on deserts and marshes, most of the inhabitants within each great class are plainly related; for they are the descendants of the same progenitors and early colonists. On this same principle of former migration, combined in most cases with modification, we can understand, by the aid of the Glacial period, the identity of some few plants, and the close alliance of many others, on the most distant mountains, and in the northern and southern temperate zones; and likewise the close alliance of some of the inhabitants of the sea in the northern and southern temperate latitudes, though separated by the whole intertropical ocean. Although two countries may present physical conditions as closely similar as the same species ever require, we need feel no surprise at their inhabitants being widely different, if they have been for a long period completely sundered from each other; for as the relation of organism to organism is the most important of all relations, and as the two countries will have received colonists at various periods and in different proportions, from some other country or from each other, the course of modification in the two areas will inevitably have been different.

On this view of migration, with subsequent modification, we see why oceanic islands are inhabited by only few species, but of these, why many are peculiar or endemic forms. We clearly see why species belonging to those groups of animals which cannot cross wide spaces of the ocean, as frogs and terrestrial mammals, do not inhabit oceanic islands; and why, on the other hand, new and peculiar species of bats, animals which can traverse the ocean, are often found on islands far distant from any continent. Such cases as the presence of peculiar species of bats on oceanic islands and the absence of all other terrestrial mammals, are facts utterly inexplicable on the theory of independent acts of creation.

Reptiles. Plate 19.

1. 1a. *Hylonia sylvatica.*
2. 2a. *Hyla agrestis.*
3. 3a. ———— *vanterii.*

Frogs observed and drawn during the voyage of H.M.S. *Beagle*

The existence of closely allied representative species in any two areas, implies, on the theory of descent with modification, that the same parent-forms formerly inhabited both areas; and we almost invariably find that wherever many closely allied species inhabit two areas, some identical species are still common to both. Wherever many closely allied yet distinct species occur, doubtful forms and varieties belonging to the same groups likewise occur. It is a rule of high generality that the inhabitants of each area are related to the inhabitants of the nearest source whence immigrants might have been derived. We see this in the striking relation of nearly all the plants and animals of the Galapagos Archipelago, of Juan Fernandez, and of the other American islands, to the plants and animals of the neighbouring American mainland; and of those of the Cape de Verde Archipelago, and of the other African islands to the African mainland. It must be admitted that these facts receive no explanation on the theory of creation.

The fact, as we have seen, that all past and present organic beings can be arranged within a few great classes, in groups subordinate to groups, and with the extinct groups often falling in between the recent groups, is intelligible on the theory of natural selection with its contingencies of extinction and divergence of character. On these same principles we see how it is that the mutual affinities of the forms within each class are so complex and circuitous. We see why certain characters are far more serviceable than others for classification; why adaptive characters, though of paramount importance to the beings, are of hardly any importance in classification; why characters derived from rudimentary parts, though of no service to the beings, are often of high classificatory value; and why embryological characters are often the most valuable of all. The real affinities of all organic beings, in contradistinction to their adaptive resemblances, are due to inheritance or community of descent. The Natural System is a genealogical arrangement, with the acquired grades of difference, marked by the terms, varieties, species, genera, families, etc.; and we have to discover the lines of descent by the most permanent characters, whatever they may be, and of however slight vital importance.

The similar framework of bones in the hand of a man, wing of a bat, fin of the porpoise, and leg of the horse, – the same number of vertebrae forming the neck of the giraffe and of the elephant, – and innumerable other such facts, at once explain themselves on the theory of descent with slow and slight successive modifications. The similarity of pattern in the wing and in the leg of a bat, though used for such different purpose, – in the jaws and legs of a crab, – in the petals, stamens, and pistils of a flower, is likewise, to a large extent, intelligible on the view of the gradual modification of parts or organs, which were aboriginally alike in an early progenitor in each of these classes. On the principle of successive variations not always supervening at an early age, and being inherited at a corresponding not early period of life, we clearly see why the embryos of mammals, birds, reptiles, and fishes should be so closely similar, and so unlike the adult forms. We may cease marvelling at the embryo of an air-breathing mammal or bird having branchial slits and arteries running in loops, like those of a fish which has to breathe the air dissolved in water by the aid of well-developed branchiae.

Disuse, aided sometimes by natural selection, will often have reduced organs when rendered useless under changed habits or conditions of life; and we can understand on this view the meaning of rudimentary organs. But disuse and selection will generally act on each creature, when it has come to maturity and has to play its full part in the struggle for existence, and will thus have little power on an organ during early life; hence the organ will not be reduced or rendered rudimentary at this early age. The calf, for instance, has inherited teeth, which never cut through the gums of the upper jaw, from an early progenitor having well-developed teeth; and we may believe, that the teeth in the mature animal were formerly reduced by disuse owing to the tongue and palate, or lips, having become excellently fitted through natural selection to browse without their aid; whereas in the calf, the teeth have been left unaffected, and on the principle of inheritance at corresponding ages have been inherited from a remote period to the present day. On the view of each organism with all its separate parts having been specially created, how utterly inexplicable is it that organs bearing the plain stamp of inutility, such as the teeth in the embryonic calf or the shrivelled wings under the soldered wing-covers of many beetles, should so frequently occur. Nature may be said to have taken pains to reveal her scheme of modification, by means of rudimentary organs, of embryological and homologous structures, but we are too blind to understand her meaning.

I have now recapitulated the facts and considerations which have thoroughly convinced me that species have been modified, during a long course of descent. This has been effected chiefly through the natural selection of numerous successive, slight, favourable variations; aided in an important manner by the inherited effects of the use and disuse of parts; and in an unimportant manner, that is, in relation to adaptive structures, whether past or present, by the direct action of external conditions, and by variations which seem to us in our ignorance to arise spontaneously. It appears that I formerly underrated the frequency and value of these latter forms of variation, as leading to permanent modifications of structure independently of natural selection. But as my conclusions have lately been much misrepresented, and it has been stated that I attribute the modification of species exclusively to natural selection, I may be permitted to remark that in the first edition of this work, and subsequently, I placed in a most conspicuous position – namely, at the close of the Introduction – the following words: 'I am convinced that natural selection has been the main but not the exclusive means of modification.' This has been of no avail. Great is the power of steady misrepresentation; but the history of science shows that fortunately this power does not long endure.

It can hardly be supposed that a false theory would explain, in so satisfactory a manner as does the theory of natural selection, the several large classes of facts above specified. It has recently been objected that this is an unsafe method of arguing; but it is a method used in judging of the common events of life, and has often been used by the greatest natural philosophers. The undulatory theory of light has thus been arrived at; and the belief in the revolution of the earth on its own axis was until lately supported by hardly any direct evidence. It is no valid objection that science as yet throws no light on the far higher problem of the essence or origin of life. Who can explain what is the essence of the attraction of gravity? No one now objects to following out the results consequent on this unknown element of attraction; notwithstanding that Leibnitz formerly accused Newton of introducing 'occult qualities and miracles into philosophy.'

I see no good reasons why the views given in this volume should shock the religious feelings of any one. It is satisfactory, as showing how transient such impressions are, to remember that the greatest discovery ever made by man, namely, the law of the attraction of gravity, was also attacked by Leibnitz, 'as subversive of natural, and inferentially of revealed, religion.' A celebrated author and divine has written to me that 'he has gradually learned to see that it is just as noble a conception of the Deity to believe that He created a few original forms capable of self-development into other and needful forms, as to believe that He required a fresh act of creation to supply the voids caused by the action of His laws.'

Why, it may be asked, until recently did nearly all the most eminent living naturalists and geologists disbelieve in the mutability of species? It cannot be asserted that organic beings in a state of nature are subject to no variation; it cannot be proved that the amount of variation in the course of long ages is a limited quantity; no clear distinction has been, or can be, drawn between species and well-marked varieties. It cannot be maintained that species when intercrossed are invariably sterile and varieties invariably fertile; or that sterility is a special endowment and sign of creation. The belief that species were immutable productions was almost unavoidable as long as the history of the world was thought to be of short duration; and now that we have acquired some idea of the lapse of time, we are too apt to assume, without proof, that the geological record is so perfect that it would have afforded us plain evidence of the mutation of species, if they had undergone mutation.

But the chief cause of our natural unwillingness to admit that one species has given birth to other and distinct species, is that we are always slow in admitting any great changes of which we do not see the steps. The difficulty is the same as that felt by so many geologists, when Lyell first insisted that long lines of inland cliffs had been formed, and great valleys excavated, by the agencies which we still see at work. The mind cannot possibly grasp the full meaning of the term of even a million years; it cannot add up and perceive the full effects of many slight variations, accumulated during an almost infinite number of generations.

Although I am fully convinced of the truth of the views given in this volume under the form of an abstract, I by no means expect to convince experienced naturalists whose minds are stocked with a multitude of facts all viewed, during a long course of years, from a point of view directly opposite to mine. It is so easy to hide our ignorance under such expressions as the 'plan of creation,' 'unity of design,' etc., and to think that we give an explanation when we only restate a fact. Any one whose disposition leads him to attach more weight to unexplained difficulties than to the explanation of a certain number of facts will

certainly reject the theory. A few naturalists, endowed with much flexibility of mind, and who have already begun to doubt the immutability of species, may be influenced by this volume; but I look with confidence to the future, to young and rising naturalists, who will be able to view both sides of the question with impartiality. Whoever is led to believe that species are mutable will do good service by conscientiously expressing his conviction; for thus only can the load of prejudice by which this subject is overwhelmed be removed.

Several eminent naturalists have of late published their belief that a multitude of reputed species in each genus are not real species; but that other species are real, that is, have been independently created. This seems to me a strange conclusion to arrive at. They admit that a multitude of forms, which till lately they themselves thought were special creations, and which are still thus looked at by the majority of naturalists, and which consequently have all the external characteristic features of true species – they admit that these have been produced by variation, but they refuse to extend the same view to other and slightly different forms. Nevertheless, they do not pretend that they can define, or even conjecture, which are the created forms of life, and which are those produced by secondary laws. They admit variation as a *vera causa* in one case, they arbitrarily reject it in another, without assigning any distinction in the two cases. The day will come when this will be given as a curious illustration of the blindness of preconceived opinion. These authors seem no more startled at a miraculous act of creation than at an ordinary birth. But do they really believe that at innumerable periods in the earth's history certain elemental atoms have been commanded suddenly to flash into living tissues? Do they believe that at each supposed act of creation one individual or many were produced? Were all the infinitely numerous kinds of animals and plants created as eggs or seed, or as full grown? and in the case of mammals, were they created bearing the false marks of nourishment from the mother's womb? Undoubtedly some of these same questions cannot be answered by those who believe in the appearance or creation of only a few forms of life or of some one form alone. It has been maintained by several authors that it is as easy to believe in the creation of a million beings as of one; but Maupertuis' philosophical axiom 'of least action' leads the mind more willingly to admit the smaller number; and certainly we ought not to believe that innumerable beings within each great class have been created with plain, but deceptive, marks of descent from a single parent.

As a record of a former state of things, I have retained in the foregoing paragraphs, and elsewhere, several sentences which imply that naturalists believe in the separate creation of each species; and I have been much censured for having thus expressed myself. But undoubtedly this was the general belief when the first edition of the present work appeared. I formerly spoke to very many naturalists on the subject of evolution, and never once met with any sympathetic agreement. It is probable that some did then believe in evolution, but they were either silent or expressed themselves so ambiguously that it was not easy to understand their meaning. Now, things are wholly changed, and almost every naturalist admits the great principle of evolution. There are, however, some who still think that species have suddenly given birth, through quite unexplained means, to new and totally different forms. But, as I have attempted to show, weighty evidence can be opposed to the admission of great and abrupt modifications. Under a scientific point of view, and as leading to further investigation, but little advantage is gained by believing that new forms are suddenly developed in an inexplicable manner from old and widely different forms, over the old belief in the creation of species from the dust of the earth.

It may be asked how far I extend the doctrine of the modification of species. The question is difficult to answer, because the more

distinct the forms are which we consider, by so much the arguments in favour of community of descent become fewer in number and less in force. But some arguments of the greatest weight extend very far. All the members of whole classes are connected together by a chain of affinities, and all can be classed on the same principle, in groups subordinate to groups. Fossil remains sometimes tend to fill up very wide intervals between existing orders.

Organs in a rudimentary condition plainly show that an early progenitor had the organ in a fully developed condition, and this in some cases implies an enormous amount of modification in the descendants. Throughout whole classes various structures are formed on the same pattern, and at a very early age the embryos closely resemble each other. Therefore I cannot doubt that the theory of descent with modification embraces all the members of the same great class or kingdom. I believe that animals are descended from at most only four or five progenitors, and plants from an equal or lesser number.

Analogy would lead me one step further, namely, to the belief that all animals and plants are descended from some one prototype. But analogy may be a deceitful guide. Nevertheless all living things have much in common, in their chemical composition, their cellular structure, their laws of growth, and their liability to injurious influences. We see this even in so trifling a fact as that the same poison often similarly affects plants and animals; or that the poison secreted by the gall-fly produces monstrous growths on the wild rose or oak-tree. With all organic beings, excepting perhaps some of the very lowest, sexual reproduction seems to be essentially similar. With all, as far as is at present known, the germinal vesicle is the same; so that all organisms start from a common origin. If we look even to the two main divisions – namely, to the animal and vegetable kingdoms – certain low forms are so far intermediate in character that naturalists have disputed to which kingdom they should be referred. As Professor Asa Gray has remarked, 'the spores and other reproductive bodies of many of the lower algae may claim to have first a characteristically animal, and then an unequivocally vegetable existence.' Therefore, on the principle of natural selection with divergence of character, it does not seem incredible that, from some such low and intermediate form, both animals and plants may have been developed; and, if we admit this, we must likewise admit that all the organic beings which have ever lived on this earth may be descended from some one primordial form. But this inference is chiefly grounded on analogy, and it is immaterial whether or not it be accepted. No doubt it is possible, as Mr. G.H. Lewes has urged, that at the first commencement of life many different forms were evolved; but if so, we may conclude that only a very few have left modified descendants. For, as I have recently remarked in regard to the members of each great kingdom, such as the Vertebrata, Articulata, etc., we have distinct evidence in their embryological, homologous, and rudimentary structures, that within each kingdom all the members are descended from a single progenitor.

When the views advanced by me in this volume, and by Mr. Wallace, or when analogous views on the origin of species are generally admitted, we can dimly foresee that there will be a considerable revolution in natural history. Systematists will be able to pursue their labours as at present; but they will not be incessantly haunted by the shadowy doubt whether this or that form be a true species. This, I feel sure and I speak after experience, will be no slight relief. The endless disputes whether or not some fifty species of British brambles are good species will cease. Systematists will have only to decide (not that this will be easy) whether any form be sufficiently constant and distinct from other forms, to be capable of definition; and if definable, whether the differences be sufficiently important to deserve a specific name. This latter point will become a far more essential consideration than it is at present;

for differences, however slight, between any two forms, if not blended by intermediate gradations, are looked at by most naturalists as sufficient to raise both forms to the rank of species.

Hereafter we shall be compelled to acknowledge that the only distinction between species and well-marked varieties is, that the latter are known, or believed to be connected at the present day by intermediate gradations, whereas species were formerly thus connected. Hence, without rejecting the consideration of the present existence of intermediate gradations between any two forms, we shall be led to weigh more carefully and to value higher the actual amount of difference between them. It is quite possible that forms now generally acknowledged to be merely varieties may hereafter be thought worthy of specific names; and in this case scientific and common language will come into accordance. In short, we shall have to treat species in the same manner as those naturalists treat genera, who admit that genera are merely artificial combinations made for convenience. This may not be a cheering prospect; but we shall at least be freed from the vain search for the undiscovered and undiscoverable essence of the term species.

The other and more general departments of natural history will rise greatly in interest. The terms used by naturalists, of affinity, relationship, community of type, paternity, morphology, adaptive characters, rudimentary and aborted organs, etc., will cease to be metaphorical and will have a plain signification. When we no longer look at an organic being as a savage looks at a ship, as something wholly beyond his comprehension; when we regard every production of nature as one which has had a long history; when we contemplate every complex structure and instinct as the summing up of many contrivances, each useful to the possessor, in the same way as any great mechanical invention is the summing up of the labour, the experience, the reason, and even the blunders of numerous workmen; when we thus view each organic being, how far more interesting

– I speak from experience – does the study of natural history become!

A grand and almost untrodden field of inquiry will be opened, on the causes and laws of variation, on correlation, on the effects of use and disuse, on the direct action of external conditions, and so forth. The study of domestic productions will rise immensely in value. A new variety raised by man will be a far more important and interesting subject for study than one more species added to the infinitude of already recorded species. Our classifications will come to be, as far as they can be so made, genealogies; and will then truly give what may be called the plan of creation. The rules for classifying will no doubt become simpler when we have a definite object in view. We possess no pedigree or armorial bearings; and we have to discover and trace the many diverging lines of descent in our natural genealogies, by characters of any kind which have long been inherited. Rudimentary organs will speak infallibly with respect to the nature of long-lost structures. Species and groups of species which are called aberrant, and which may fancifully be called living fossils, will aid us in forming a picture of the ancient forms of life. Embryology will often reveal to us the structure, in some degree obscured, of the prototypes of each great class.

When we can feel assured that all the individuals of the same species, and all the closely allied species of most genera, have, within a not very remote period descended from one parent, and have migrated from some one birth-place; and when we better know the many means of migration, then, by the light which geology now throws, and will continue to throw, on former changes of climate and of the level of the land, we shall surely be enabled to trace in an admirable manner the former migrations of the inhabitants of the whole world. Even at present, by comparing the differences between the inhabitants of the sea on the opposite sides of a continent, and the nature of the various inhabitants of that continent in relation to their apparent means of immigration, some light can be thrown on ancient geography.

The noble science of geology loses glory from the extreme imperfection of the record. The crust of the earth, with its embedded remains, must not be looked at as a well-filled museum, but as a poor collection made at hazard and at rare intervals. The accumulation of each great fossiliferous formation will be recognised as having depended on an unusual occurrence of favourable circumstances, and the blank intervals between the successive stages as having been of vast duration. But we shall be able to gauge with some security the duration of these intervals by a comparison of the preceding and succeeding organic forms. We must be cautious in attempting to correlate as strictly contemporaneous two formations, which do not include many identical species, by the general succession of the forms of life. As species are produced and exterminated by slowly acting and still existing causes, and not by miraculous acts of creation; and as the most important of all causes of organic change is one which is almost independent of altered and perhaps suddenly altered physical conditions, namely, the mutual relation of organism to organism – the improvement of one organism entailing the improvement or the extermination of others; it follows, that the amount of organic change in the fossils of consecutive formations probably serves as a fair measure of the relative, though not actual lapse of time. A number of species, however, keeping in a body might remain for a long period unchanged, whilst within the same period, several of these species, by migrating into new countries and coming into competition with foreign associates, might become modified; so that we must not overrate the accuracy of organic change as a measure of time.

In the future I see open fields for far more important researches. Psychology will be securely based on the foundation already well laid by Mr. Herbert Spencer, that of the necessary acquirement of each mental power and capacity by gradation. Much light will be thrown on the origin of man and his history.

Authors of the highest eminence seem to be fully satisfied with the view that each species has been independently created. To my mind it accords better with what we know of the laws impressed on matter by the Creator, that the production and extinction of the past and present inhabitants of the world should have been due to secondary causes, like those determining the birth and death of the individual. When I view all beings not as special creations, but as the lineal descendants of some few beings which lived long before the first bed of the Cambrian system was deposited, they seem to me to become ennobled. Judging from the past, we may safely infer that not one living species will transmit its unaltered likeness to a distinct futurity. And of the species now living very few will transmit progeny of any kind to a far distant futurity; for the manner in which all organic beings are grouped, shows that the greater number of species in each genus, and all the species in many genera, have left no descendants, but have become utterly extinct. We can so far take a prophetic glance into futurity as to foretell that it will be the common and widely spread species, belonging to the larger and dominant groups within each class, which will ultimately prevail and procreate new and dominant species. As all the living forms of life are the lineal descendants of those which lived long before the Cambrian epoch, we may feel certain that the ordinary succession by generation has never once been broken, and that no

cataclysm has desolated the whole world. Hence, we may look with some confidence to a secure future of great length. And as natural selection works solely by and for the good of each being, all corporeal and mental endowments will tend to progress towards perfection.

It is interesting to contemplate a tangled bank, clothed with many plants of many kinds, with birds singing on the bushes, with various insects flitting about, and with worms crawling through the damp earth, and to reflect that these elaborately constructed forms, so different from each other, and dependent upon each other in so complex a manner, have all been produced by laws acting around us. These laws, taken in the largest sense, being Growth with Reproduction; Inheritance which is almost implied by reproduction; Variability from the indirect and direct action of the conditions of life, and from use and disuse; a Ratio of Increase so high as to lead to a Struggle for Life, and as a consequence to Natural Selection, entailing Divergence of Character and the Extinction of less improved forms. Thus, from the war of nature, from famine and death, the most exalted object which we are capable of conceiving, namely, the production of the higher animals, directly follows. There is grandeur in this view of life, with its several powers, having been originally breathed by the Creator into a few forms or into one; and that, whilst this planet has gone cycling on according to the fixed law of gravity, from so simple a beginning endless forms most beautiful and most wonderful have been, and are being, evolved.

Birds. Pl. 35.

Songbirds of the genus Pipilo, from the *Zoology of the Voyage of H.M.S. Beagle*

GLOSSARY OF THE PRINCIPAL SCIENTIFIC TERMS USED IN THE PRESENT VOLUME

(I am indebted to the kindness of Mr. W.S. Dallas for this Glossary, which has been given because several readers have complained to me that some of the terms used were unintelligible to them. Mr. Dallas has endeavoured to give the explanations of the terms in as popular a form as possible.)

Aberrant – Forms or groups of animals or plants which deviate in important characters from their nearest allies so as not to be easily included in the same group with them, are said to be aberrant.

Aberration (in Optics) – In the refraction of light by a convex lens the rays passing through different parts of the lens are brought to a focus at slightly different distances – this is called spherical aberration; at the same time the coloured rays are separated by the prismatic action of the lens and likewise brought to a focus at different distances – this is chromatic aberration.

Abnormal – Contrary to the general rule.

Aborted – An organ is said to be aborted, when its development has been arrested at a very early stage.

Albinism – Albinos are animals in which the usual colouring matters characteristic of the species have not been produced in the skin and its appendages. Albinism is the state of being an albino.

Algae – A class of plants including the ordinary seaweeds and the filamentous fresh-water weeds.

Alternation of Generations – This term is applied to a peculiar mode of reproduction which prevails amongst many of the lower animals, in which the egg produces a living form quite different from its parent, but from which the parent-form is reproduced by a process of budding, or by the division of the substance of the first product of the egg.

Ammonites – A group of fossil, spiral, chambered shells, allied to the existing pearly nautilus, but having the partitions between the chambers waved in complicated patterns at their junction with the outer wall of the shell.

Analogy – That resemblance of structures which depends upon similarity of function, as in the wings of insects and birds. Such structures are said to be *analogous*, and to be *analogues* of each other.

Animalcule – A minute animal: generally applied to those visible only by the microscope.

Annelids – A class of worms in which the surface of the body exhibits a more or less distinct division into rings or segments, generally provided with appendages for locomotion and with gills. It includes the ordinary Marine Worms, the Earthworms, and the Leeches.

Antennae – Jointed organs appended to the head in Insects, Crustacea, and Centipedes, and not belonging to the mouth.

Anthers – The summits of the stamens of flowers, in which the pollen or fertilising dust is produced.

Aplacentalia, Aplacentata or Aplacental Mammals – See *Mammalia*.

Archetypal – Of or belonging to the Archetype, or ideal primitive form upon which all the beings of a group seem to be organised.

Articulata – A great division of the Animal Kingdom characterised generally by having the surface of the body divided into rings called segments, a greater or less number of which are furnished with jointed legs (*Examples*, Insects, Crustaceans, and Centipedes).

Asymmetrical – Having the two sides unlike.

Atrophied – Arrested in development at a very early stage.

Balanus – The genus including the common Acorn-shells which live in abundance on the rocks of the sea coast.

Batrachians – A class of animals allied to the Reptiles, but undergoing a peculiar metamorphosis, in which the young animal is generally aquatic and breathes by gills. (*Examples*, Frogs, Toads, and Newts.)

Boulders – Large transported blocks of stone generally embedded in clays or gravels.

Brachiopoda – A class of marine *Mollusca*, or soft-bodied animals, furnished with a bivalve shell, attached to submarine objects by a stalk which passes through an aperture in one of the valves, and furnished with fringed arms, by the action of which food is carried to the mouth.

Branchiae – Gills or organs for respiration in water.

Branchial – Pertaining to gills or *branchiae*.

Cambrian system – A series of very ancient Palaeozoic rocks, between the Laurentian and the Silurian. Until recently these were regarded as the oldest fossiliferous rocks.

Canidae – The Dog family, including the Dog, Wolf, Fox, Jackal, etc.

Carapace – The shell enveloping the anterior part of the body in Crustaceans generally; applied also to the hard shelly pieces of the Cirripedes.

Carboniferous – This term is applied to the great formation which includes, amongst other rocks, the coal-measures. It belongs to the oldest, or Palaeozoic, system of formations.

Caudal – Of or belonging to the tail.

Cephalopods – The highest class of the *Mollusca*, or soft-bodied animals, characterised by having the mouth surrounded by a greater or less number of fleshy arms or tentacles, which, in most living species, are furnished with sucking-cups. (*Examples*, Cuttle-fish, Nautilus.)

Cetacea – An order of *Mammalia*, including the Whales, Dolphins, etc., having the form of the body fish-like, the skin naked, and only the fore limbs developed.

Chelonia – An order of Reptiles including the Turtles, Tortoises, etc.

Cirripedes – An order of Crustaceans including the Barnacles and Acorn-shells. Their young resemble those of many other Crustaceans in form; but when mature they are always attached to other objects, either directly or by means of a stalk; and their bodies are enclosed by a calcareous shell composed of several pieces, two of which can open to give issue to a bunch of curled, jointed tentacles, which represent the limbs.

Coccus – The genus of Insects including the Cochineal. In these the male is a minute, winged fly, and the female generally a motionless, berry-like mass.

Cocoon – A case usually of silky material, in which insects are frequently enveloped during the second or resting-stage (pupa) of their existence. The term *cocoon-stage* is here used as equivalent to *pupa-stage*.

Coelospermous – A term applied to those fruits of the *Umbelliferae* which have the seed hollowed on the inner face.

Coleoptera – Beetles, an order of Insects having a biting mouth and the first pair of wings more or less horny, forming sheaths for the second pair, and usually meeting in a straight line down the middle of the back.

Column – A peculiar organ in the flowers of Orchids, in which the stamens, style and stigma (or the reproductive parts) are united.

Compositae or Compositous Plants – Plants in which the inflorescence consists of numerous small flowers (florets) brought together into a dense head, the base of which is enclosed by a common envelope. (*Examples*, the Daisy, Dandelion, etc.)

Confervae – The filamentous weeds of fresh water.

Conglomerate – A rock made up of fragments of rock or pebbles, cemented together by some other material.

Corolla – The second envelope of a flower usually composed of coloured, leaf-like organs (petals), which may be united by their edges either in the basal part or throughout.

Correlation – The normal coincidence of one phenomenon, character, etc., with another.

Corymb – A bunch of flowers in which those springing from the lower part of the flower stalks are supported on long stalks so as to be nearly on a level with the upper ones.

Cotyledons – The first or seed-leaves of plants.

Crustaceans – A class of articulated animals, having the skin of the body generally more or less hardened by the deposition of calcareous matter, breathing by means of gills. (*Examples*, Crab, Lobster, Shrimp, etc.)

Curculio – The old generic term for the Beetles known as Weevils, characterised by their four-jointed feet, and by the head being produced into a sort of beak, upon the sides of which the antennae are inserted.

Cutaneous – Of or belonging to the skin.

Degradation – The wearing down of land by the action of the sea or of meteoric agencies.

Denudation – The wearing away of the surface of the land by water.

Devonian System or Formation – A series of Palaeozoic rocks, including the Old Red Sandstone.

Dicotyledons, or Dicotyledonous Plants – A class of plants characterised by having two seed-leaves, by the formation of new wood between the bark and the old wood (exogenous growth) and by the reticulation of the veins of the leaves. The parts of the flowers are generally in multiples of five.

Differentation – The separation or discrimination of parts or organs which in simpler forms of life are more or less united.

Dimorphic – Having two distinct forms. Dimorphism is the condition of the appearance of the same species under two dissimilar forms.

Dioecious – Having the organs of the sexes upon distinct individuals.

Diorite – A peculiar form of Greenstone.

Dorsal – Of or belonging to the back.

Edentata – A peculiar order of Quadrupeds, characterised by the absence of at least the middle incisor (front) teeth in both jaws. (*Examples*, the Sloths and Armadillos.)

Elytra – The hardened fore-wings of Beetles, serving as sheaths for the membranous hind-wings, which constitute the true organs of flight.

Embryo – The young animal undergoing development within the egg or womb.

Embryology – The study of the development of the embryo.

Endemic – Peculiar to a given locality.

Entomostraca – A division of the class Crustacea, having all the segments of the body usually distinct, gills attached to the feet or organs of the mouth, and the feet fringed with fine hairs. They are generally of small size.

Eocene – The earliest of the three divisions of the Tertiary epoch of geologists. Rocks of this age contain a small proportion of shells identical with species now living.

Ephemerous Insects – Insects allied to the May fly.

Fauna – The totality of the animals naturally inhabiting a certain country or region, or which have lived during a given geological period.

Felidae – The Cat family.

Feral – Having become wild from a state of cultivation or domestication.

Flora – The totality of the plants growing naturally in a country, or during a given geological period.

Florets – Flowers imperfectly developed in some respects, and collected into a dense spike or head, as in the Grasses, the Dandelion, etc.

Foetal – Of or belonging to the foetus, or embryo in course of development.

Foraminifera – A class of animals of very low organisation and generally of small size, having a jelly-like body, from the surface of which delicate filaments can be given off and retracted for the prehension of external objects; and having a calcareous or sandy shell, usually divided into chambers and perforated with small apertures.

Fossiliferous – Containing fossils.

Fossorial – Having a faculty of digging. The fossorial Hymenoptera are a group of wasp-like insects, which burrow in sandy soil to make nests for their young.

Frenum (pl. *Frena*) – A small band or fold of skin.

Fungi (sing. *Fungus*) – A class of cellular plants, of which Mushrooms, Toadstools, and Moulds, are familiar examples.

Furcula – The forked bone formed by the union of the collar-bones in many birds, such as the Common Fowl.

Gallinaceous Birds – An order of Birds of which the Common Fowl, Turkey, and Pheasant, are well-known examples.

Gallus – The genus of Birds which includes the Common Fowl.

Ganglion – A swelling or knot from which nerves are given off as from a centre.

Ganoid Fishes – Fishes covered with peculiar enamelled bony scales. Most of them are extinct.

Germinal Vesicle – A minute vesicle in the eggs of animals, from which the development of the embryo proceeds.

Glacial Period – A period of great cold and of enormous extension of ice upon the surface of the earth. It is believed that glacial periods have occurred repeatedly during the geological history of the earth, but the term is generally applied to the close of the Tertiary epoch, when nearly the whole of Europe was subjected to an arctic climate.

Gland – An organ which secretes or separates some peculiar product from the blood or sap of animals or plants.

Glottis – The opening of the windpipe into the oesophagus or gullet.

Gneiss – A rock approaching granite in composition, but more or less laminated, and really produced by the alteration of a sedimentary deposit after its consolidation.

Grallatores – The so-called wading birds (Storks, Cranes, Snipes, etc.), which are generally furnished with long legs, bare of feathers above the heel, and have no membranes between the toes.

Granite – A rock consisting essentially of crystals of felspar and mica in a mass of quartz.

Habitat – The locality in which a plant or animal naturally lives.

Hemiptera – An order or sub-order of Insects, characterised by the possession of a jointed beak or rostrum, and by having the fore-wings horny in the basal portion and membranous at the extremity, where they cross each other. This group includes the various species of bugs.

Hermaphrodite – Possessing the organs of both sexes.

Homology – That relation between parts which results from their development from corresponding embryonic parts, either in different animals, as in the case of the arm of man, the fore-leg of a quadruped, and the wing of a bird; or in the same individual, as in the case of the fore and hind legs in quadrupeds, and the segments or rings and their appendages of which the body of a Worm, a Centipede, etc., is composed. The latter is called *serial homology*. The parts which stand in such a relation to each other are said to be *homologous*, and one such part or organ is called the *homologue* of the other. In different plants the parts of the flower are homologous, and in general these parts are regarded as homologous with leaves.

Homoptera – An order or sub-order of Insects having (like the Hemiptera) a jointed beak, but in which the fore-wings are either wholly membranous or wholly leathery, The Cicadae, Frog-hoppers, and Aphides, are well-known examples.

Hybrid – The offspring of the union of two distinct species.

Hymenoptera – An order of Insects possessing biting jaws and usually four membranous wings in which there are a few veins. Bees and Wasps are familiar examples of this group.

Hypertrophied – Excessively developed.

Ichneumonidae – A family of Hymenopterous insects, the members of which lay their eggs in the bodies or eggs of other insects.

Imago – The perfect (generally winged) reproductive state of an insect.

Indigenes – The aboriginal animal or vegetable inhabitants of a country or region.

Inflorescence – The mode of arrangement of the flowers of plants.

Infusoria – A class of microscopic animalcules, so called from their having originally been observed in infusions of vegetable matters. They consist of a gelatinous material enclosed in a delicate membrane, the whole or part of which is furnished with short vibrating hairs (called *cilia*), by means of which the animalcules swim through the water or convey the minute particles of their food to the orifice of the mouth.

Insectivorous – Feeding on Insects.

Invertebrata, or Invertebrate Animals – Those animals which do not possess a backbone or spinal column.

Lacunae – Spaces left amongst the tissues in some of the lower animals and serving in place of vessels for the circulation of the fluids of the body.

Lamellated – Furnished with lamellae or little plates.

Larva (pl. *Larvae*) – The first condition of an insect at its issuing from the egg, when it is usually in the form of a Grub, Caterpillar, or Maggot.

Larynx – The upper part of the windpipe opening into the gullet.

Laurentian – A group of greatly altered and very ancient rocks, which is greatly developed along the course of the St. Laurence, whence the name. It is in these that the earliest known traces of organic bodies have been found.

Leguminosae – An order of plants represented by the common Peas and Beans, having an irregular flower in which one petal stands up like a wing, and the stamens and pistil are enclosed in a sheath formed by two other petals. The fruit is a pod (or legume).

Lemuridae – A group of four-handed animals, distinct from the Monkeys and approaching the insectivorous quadrupeds in some of their characters and habits. Its members have the nostrils curved or twisted, and a claw instead of a nail upon the first finger of the hind hands.

Lepidoptera – An order of Insects, characterised by the possession of a spiral proboscis, and of four large more or less scaly wings. It includes the well-known Butterflies and Moths.

Littoral – Inhabiting the seashore.

Loess – A marly deposit of recent (Post-Tertiary) date, which occupies a great part of the valley of the Rhine.

Malacostraca – The higher division of the Crustacea, including the ordinary Crabs, Lobsters, Shrimps, etc., together with the Woodlice and Sand-hoppers.

Mammalia – The highest class of animals, including the ordinary hairy quadrupeds, the Whales and Man, and characterised by the production of living young which are nourished after birth by milk from the teats (*mammae, mammary glands*) of the mother. A striking difference in embryonic development has led to the division of this class into two great groups; in one of these, when the embryo has attained a certain stage, a vascular connection, called the *placenta*, is formed between the embryo and the mother; in the other this is wanting, and the young are produced in a very incomplete state. The former, including the greater part of the class, are called *placental mammals*; the latter, or *aplacental mammals*, include the Marsupials and Monotremes (Ornithorhynchus).

Mammiferous – Having mammae or teats (see *Mammalia*).

Mandibles In insects the first or uppermost pair of jaws, which are generally solid, horny, biting organs. In birds the term is applied to both jaws with their horny coverings. In quadrupeds the mandible is properly the lower jaw.

Marsupials – An order of Mammalia in which the young are born in a very incomplete state of development, and carried by the mother, whilst sucking, in a ventral pouch (*marsupium*), such as the Kangaroos, Opossums, etc. (see *Mammalia*).

Maxillae – in insects, the second or lower pair of jaws, which are composed of several joints and furnished with peculiar jointed appendages called *palpi*, or feelers.

Melanism – The opposite of albinism; an undue development of colouring material in the skin and its appendages.

Metamorphic Rocks – Sedimentary rocks which have undergone alteration, generally by the action of heat, subsequently to their deposition and consolidation.

Mollusca – One of the great divisions of the Animal Kingdom, including those animals which have a soft body, usually furnished with a shell, and in which the nervous ganglia, or centres, present no definite general arrangement. They are generally known under the denomination of *shellfish*; the Cuttle-fish, and the common Snails, Whelks, Oysters, Mussels, and Cockles, may serve as examples of them.

Monocotyledons, or Monocotyledonous Plants – Plants in which the seed sends up only a single seed-leaf (or *cotyledon*); characterised by the absence of consecutive layers of wood in the stem (endogenous growth), by the veins of the leaves being generally straight, and by the parts of the flowers being generally in multiples of three. (*Examples*, Grasses, Lilies, Orchids, Palms, etc.)

Moraines – The accumulations of fragments of rock brought down by glaciers.

Morphology – The law of form or structure independent of function.

Mysis-stage – A stage in the development of certain Crustaceans (Prawns), in which they closely resemble the adults of a genus (*Mysis*) belonging to a slightly lower group.

Nascent – Commencing development.

Natatory – Adapted for the purpose of swimming.

Nauplius-form – The earliest stage in the development of many *Crustacea*, especially belonging to the lower groups. In this stage the animal has a short body, with indistinct indications of a division into segments, and three pairs of fringed limbs. This form of the common fresh-water *Cyclops* was described as a distinct genus under the name of Nauplius.

Neuration – The arrangement of the veins or nervures in the wings of Insects.

Neuters – Imperfectly developed females of certain social insects (such as Ants and Bees), which perform all the labours of the community. Hence, they are also called Workers.

Nictitating Membrane – A semi-transparent membrane, which can be drawn across the eye in Birds and Reptiles, either to moderate the effects of a strong light or to sweep particles of dust, etc., from the surface of the eye.

Ocelli – The simple eyes or *stemmata* of Insects, usually situated on the crown of the head between the great compound eyes.

Oesophagus – The gullet.

Oolitic – A great series of secondary rocks, so called from the texture of some of its members, which appear to be made up of a mass of small *egg-like* calcareous bodies.

Operculum – A calcareous plate employed by many Mollusca to close the aperture of their shell. The *opercular valves* of Cirripedes are those which close the aperture of the shell.

Orbit – The bony cavity for the reception of the eye.

Organism – An organised being, whether plant or animal.

Orthospermous – A term applied to those fruits of the *Umbelliferae* which have the seed straight.

Osculant – Forms or groups apparently intermediate between and connecting other groups are said to be osculant.

Ova – Eggs.

Ovarium or Ovary (in plants) – The lower part of the pistil or female organ of the flower, containing the ovules or incipient seeds; by growth after the other organs of the flower have fallen, it usually becomes converted into the fruit.

Ovigerous – Egg-bearing.

Ovules (of plants) – The seeds in the earliest condition.

Pachyderms – A group of Mammalia, so called from their thick skins, and including the Elephant, Rhinoceros, Hippopotamus, etc.

Palaeozoic – The oldest system of fossiliferous rocks.

Palpi – Jointed appendages to some of the organs of the mouth in Insects and Crustacea.

Papilionaceae – An order of plants (see Leguminosae), The flowers of these plants are called *papilionaceous*, or butterfly-like, from the fancied resemblance of the expanded superior petals to the wings of a butterfly.

Parasite – An animal or plant living upon or in, and at the expense of, another organism.

Parthenogenesis – The production of living organisms from unimpregnated eggs or seeds.

Pedunculated – Supported upon a stem or stalk. The pedunculated oak has its acorns borne upon a footstool.

Peloria or Pelorism – The appearance of regularity of structure in the flowers of plants which normally bear irregular flowers.

Pelvis The bony arch to which the hind limbs of vertebrate animals are articulated.

Petals – The leaves of the corolla, or second circle of organs in a flower. They are usually of delicate texture and brightly coloured.

Phyllodineous – Having flattened, leaf-like twigs or leafstalks instead of true leaves.

Pigment – The colouring material produced generally in the superficial parts of animals. The cells secreting it are called *pigment cells*.

Pinnate – Bearing leaflets on each side of a central stalk.

Pistils – The female organs of a flower, which occupy a position in the centre of the other floral organs. The pistil is generally divisible into the ovary or germen, the style and the stigma.

Placentalia, Placentata or Placental Mammals – See *Mammalia*.

Plantigrades – Quadrupeds which walk upon the whole sole of the foot, like the Bears.

Plastic – Readily capable of change.

Pleistocene Period – The latest portion of the Tertiary epoch.

Plumule (in Plants) – The minute bud between the seed-leaves of newly-germinated plants.

Plutonic Rocks – Rocks supposed to have been produced by igneous action in the depths of the earth.

Pollen – The male element in flowering plants; usually a fine dust produced by the anthers, which, by contact with the stigma effects the fecundation of the seeds. This impregnation is brought about by means of tubes (*pollen-tubes*) which issue from the pollen-grains adhering to the stigma, and penetrate through the tissues until they reach the ovary.

Polyandrous (Flowers) – Flowers having many stamens.

Polygamous Plants – Plants in which some flowers are unisexual and others hermaphrodite. The unisexual (male and female) flowers, may be on the same or on different plants.

Polymorphic – Presenting many forms.

Polyzoary – The common structure formed by the cells of the *Polyzoa*, such as the well-known Sea-mats.

Prehensile – Capable of grasping.

Prepotent – Having a superiority of power.

Primaries – The feathers forming the tip of the wing of a bird, and inserted upon that part which represents the hand of man.

Processes – Projecting portions of bones, usually for the attachment of muscles, ligaments, etc.

Propolis – A resinous material collected by the hive bees from the opening buds of various trees.

Protean – Exceedingly variable.

Protozoa – The lowest great division of the Animal Kingdom. These animals are composed of a gelatinous material, and show scarcely any trace of distinct organs. The *Infusoria, Foraminifera*, and Sponges, with some other forms, belong to this division.

Pupa (pl. Pupae) – The second stage in the development of an Insect, from which it emerges in the perfect (winged) reproductive form. In most insects the *pupal* stage is passed in perfect repose. The *chrysalis* is the pupal state of butterflies.

Radicle – The minute root of an embryo plant.

Ramus – One half of the lower jaw in the *Mammalia*. The portion which rises to articulate with the skull is called the *ascending ramus*.

Range – The extent of country over which a plant or animal is naturally spread.

Range in time expresses the distribution of a species or group through the fossiliferous beds of the earth's crust.

Retina – The delicate inner coat of the eye, formed by nervous filaments spreading from the optic nerve, and serving for the perception of the impressions produced by light.

Retrogression – Backward development. When an animal, as it approaches maturity, becomes less perfectly organised than might be expected from its early stages and known relationships, it is said to undergo a *retrograde development* or *metamorphosis*.

Rhizopods – A class of lowly organised animals (*Protozoa*), having a gelatinous body, the surface of which can be protruded in the form of root-like processes or filaments, which serve for locomotion and the prehension of food. The most important order is that of the *Foraminifera*.

Rodents – The gnawing *Mammalia*, such as the Rats, Rabbits, and Squirrels. They are especially characterised by the possession of a single pair of chisel-like cutting teeth in each jaw, between which and the grinding teeth there is a great gap.

Rubus – The Bramble genus.

Rudimentary – Very imperfectly developed.

Ruminants – The group of Quadrupeds which ruminate or chew the cud, such as Oxen, Sheep, and Deer. They have divided hoofs, and are destitute of front teeth in the upper jaw.

Sacral – Belonging to the sacrum, or the bone composed usually of two or more united vertebrae to which the sides of the pelvis in vertebrate animals are attached.

Sarcode – The gelatinous material of which the bodies of the lowest animals (*Protozoa*) are composed.

Scutellae – The horny plates with which the feet of birds are generally more or less covered, especially in front.

Sedimentary Formations – Rocks deposited as sediments from water.

Segments – The transverse rings of which the body of an articulate animal or annelid is composed.

Sepals – The leaves or segments of the calyx, or outermost envelope of an ordinary flower. They are usually green, but sometimes brightly coloured.

Serratures – Teeth like those of a saw.

Sessile – Not supported on a stem or footstalk.

Silurian System – A very ancient system of fossiliferous rocks belonging to the earlier part of the Palaeozoic series.

Specialisation – The setting apart of a particular organ for the performance of a particular function.

Spinal Cord – The central portion of the nervous system in the *Vertebrata*, which descends from the brain through the arches of the vertebrae, and gives off nearly all the nerves to the various organs of the body.

Stamens – The male organs of flowering plants, standing in a circle within the petals. They usually consist of a filament and an anther, the anther being the essential part in which the pollen, or fecundating dust, is formed.

Sternum – The breast-bone.

Stigma – The apical portion of the pistil in flowering plants.

Stipules – Small leafy organs placed at the base of the footstalks of the leaves in many plants.

Style – The middle portion of the perfect pistil, which rises like a column from the ovary and supports the stigma at its summit.

Subcutaneous – Situated beneath the skin.

Suctorial – Adapted for sucking.

Sutures (in the skull) – The lines of junction of the bones of which the skull is composed.

Tarsus (pl. Tarsi) – The jointed feet of articulate animals, such as insects.

Teleostean Fishes – Fishes of the kind familiar to us in the present day, having the skeleton usually completely ossified and the scales horny.

Tentacula or Tentacles – Delicate fleshy organs of prehension or touch possessed by many of the lower animals.

Tertiary – The latest geological epoch, immediately preceding the establishment of the present order of things.

Trachea – The windpipe or passage for the admission of air to the lungs.

Tridactyle – Three-fingered, or composed of three movable parts attached to a common base.

Trilobites – A peculiar group of extinct crustaceans, somewhat resembling the woodlice in external form, and, like some of them, capable of rolling themselves up into a ball. Their remains are found only in the Palaeozoic rocks, and most abundantly in those of Silurian age.

Trimorphic – Presenting three distinct forms.

Umbelliferae – An order of plants in which the flowers, which contain five stamens and a pistil with two styles, are supported upon footstalks which spring from the top of the flower stem and spread out like the wires of an umbrella, so as to bring all the flowers in the same head (*umbel*) nearly to the same level. (*Examples*, Parsley and Carrot.)

Ungulata – Hoofed quadrupeds.

Unicellular – Consisting of a single cell.

Vascular – Containing blood-vessels.

Vermiform – Like a worm.

Vertebrata; or Vertebrate Animals – The highest division of the Animal Kingdom, so called from the presence in most cases of a backbone composed of numerous joints or *vertebrae*, which constitutes the centre of the skeleton and at the same time supports and protects the central parts of the nervous system.

Whorls – The circles or spiral lines in which the parts of plants are arranged upon the axis of growth.

Workers – See *Neuters*.

Zoea-stage – The earliest stage in the development of many of the higher *Crustacea*, so called from the name of *Zoea* applied to these young animals when they were supposed to constitute a peculiar genus.

Zooids – In many of the lower animals (such as the Corals, Medusae, etc.) reproduction takes place in two ways, namely, by means of eggs and by a process of budding with or without separation from the parent of the product of the latter, which is often very different from that of the egg. The individuality of the species is represented by the whole of the form produced between two sexual reproductions; and these forms, which are apparently individual animals, have been called *zooids*.

Charles Darwin as a young man

Picture credits
AKG Images 92, 165, 193; Art Archive 47 (top left, top right, bottom left, bottom right), 57, 118 (top right, bottom left), 125, 144, 211, 215, 217, 222, 258, 346; Clipart 15; Corbis 127, 236; Getty Images 95, 108, 271; Mary Evans 16, 45, 54, 118 (top left, bottom right), 138, 177, 228, 264, 269, 286, 325, 330; Illustrations from *The Natural History of Animals* J.R. Ainsworth Davis (1904) 28, 41, 53, 61, 67, 72, 80, 94, 99, 100, 102, 114, 129 (top, bottom), 132, 139, 140, 151, 152, 158, 167, 168, 173, 174, 176, 179, 188, 189, 199, 206, 250, 251, 257, 260, 283, 295, 319, 324, 331, 332, 338; Science Photo Library 230, 243; Topfoto 383; Illustrations from *The Zoology of the Voyage of H.M.S. Beagle* 7, 20, 25 (top, bottom), 32, 64, 106 (top, bottom), 111, 131, 135 (top, bottom), 148, 150 (top left, top right, bottom left, bottom right), 154, 162, 191, 276, 300, 305, 312 (top, bottom), 326, 335, 355, 363